Handbook on
Fish Processing and Preservation

"Handbook on Fish Processing and Preservation" will be helpful to enlighten the students, industrialists and entrepreneurs on different aspects of fish handling, processing, preservation, storage and marketing. The book covers fourteen Chapters such as 1. Physical characteristics of fish, 2. Chemical characteristics of fish, 3. Biological characteristics of fish, 4. Post mortem changes in fish, 5. Microbiology of fish and fis products, 6. Fish handling, 7. Chilling and Chill store, 8. Freezing and Cold store, 9. Fish canning, 10. Fish paste products, 11. Fish curing, 12. Fish products, 13. Fish by-products and 14. Fish additives. This will be helpful for the students in fisheries discipline, industrialists and also entrepreneurs involved in fish processing and preservation.

Dr. Rabinarayan Mishra, born in the year 1963, obtained his Master's degree and Ph.D. from the University of Agricultural Sciences, Bangalore and Karnataka University, Dharwad of Karnataka respectively. At present he is working as Professor in Fishery Engineering and Technology at College of Fisheries (Odisha University of Agriculture & Technology) located at Rangeilunda P.O., Berhampur, Dist. Ganjam, Odisha. He has over 33 years of experience in teaching, research and extension in the field of fisheries including 5 years of successful engagement in Krishi Vigyan Kendra, Ganjam and Puri during 2004 to 2009. To his credit, he has several research papers in National and International Journals and has presented at numerous workshops and conferences.

Handbook on Fish Processing and Preservation

Rabinarayan Mishra

Fishery Engineering & Technology
College of Fisheries
Odisha University of Agriculture & Technology
Rangailunda, Berhampur - 760007, Odisha

NARENDRA PUBLISHING HOUSE

First published 2022
by CRC Press
4 Park Square, Milton Park, Abingdon, Oxon, OX14 4RN

and by CRC Press
6000 Broken Sound Parkway NW, Suite 300, Boca Raton, FL 33487-2742

CRC Press is an imprint of Informa UK Limited

British Library Cataloguing-in-Publication Data
A catalogue record for this book is available from the British Library

Library of Congress Cataloging-in-Publication Data
A catalog record has been requested

ISBN: 978-1-032-20472-7 (hbk)
ISBN: 978-1-003-26371-5 (ebk)

DOI: 10.1201/9781003263715

DEDICATED

TO

MY PARENTS

Contents

List of Figures

List of Photographs

List of Tables

Preface

Fisheries and aquaculture remain an important source of food, nutrition, employment and income for millions, especially for the rural population. In fact, the sector provides livelihood to about 28 million fishers and fish farmers at the primary level and twice the number along the value chain. Fish being an affordable and rich source of animal protein, is one of the healthiest options to mitigate hunger and nutrient deficiency. The Fisheries sector of India has shown impressive growth with an average annual growth rate of 10.88% during the year 2014-15 to 2018-19 and stood at an all-time high of 137.58 lakh metric tons during the year 2018-19. The export of marine products stood at 13.93 lakh metric tons and valued at Rs.46,589 crores (USD 6.73 billion) during the year 2018-19 with an impressive average annual growth rate of about 10% in recent years.

However, post harvest losses of fish reach about 20 to 25% of the total catch. Post-harvest losses of fish occur during the numerous steps from catch to market. The catch is further reduced due to lack of appropriate knowledge on handling, unsatisfactory preservation methods, appropriate storage and exposure during marketing process. Thus, reducing these losses could increase protein availability, improve nutritional status, and eliminate some of the need to import food.

Fish processing refers to the processes associated with fish and fish products between the time fish are caught or harvested, and the time the final product is delivered to the customer. The operations of importance in fish processing are washing, degutting, salting, fermentation, drying, and smoking. These operations contribute to the development of flavour, texture, colour, and improved storage characteristics of the products. The aim of fish processing and preservation is to slow down or prevent this enzymatic, bacterial, and chemical deterioration, and to maintain the fish flesh in a condition as near as possible to that of fresh fish. Most often "quality" refers to the aesthetic appearance and freshness or degree of spoilage which the fish has undergone. It may also involve safety aspects such as being free from harmful bacteria, parasites or chemicals. When working with fish it is essential that proper handling and storage are used to reduce the risk of food-borne illness and ensure a quality product.

The present book "Handbook on Fish Processing and Preservation" throws light on various aspects of fish, its processing and preservation methods. It contains mainly 14 chapters such as Physical, chemical, biological characteristics of fish, post-mortem changes, microbiology of fish and fishery products , fish handling, chilling and chill store, freezing and cold store, fish canning, sausage, curing of fish, preparation of fish products, by-products and fish additives. This book will be helpful to the students of fisheries, industrialists and entrepreneurs as well.

Rabinarayan Mishra

Abbreviations

Abbreviated form	Full form
AAN	Alpha amino nitrogen
AFPP	Active fish protein powder
AOM	Active oxygen method
ADP	Adenosine diphosphate
AMP	Adenosine monophosphate
ATP	Adenosine triphosphate
AF_2	Furyl furamide
AFGP	Anti-freeze glycoproteins
BV	Biological value
BHA	Butyl hydroxy anisole
BHT	Butyl hydroxy toluene
CFD	Cabinet fish dryer
CAF	Calcium activated factor
CVD	Cardiovascular diseases
CFW	Chilled freshwater
CSW	Chilled seawater
CoA	Coenzyme A
DMA	Dimethylamine
ETC	Electron transport chain

EA	Enzyme activity
EMC	Equilibrium moisture content
ELC	Essential light chain
FPC	Fish protein concentrate
FAD	Flavine adenine dinucleotide
FMN	Flavine mononucleotide
FAA	Free amino acids
FMC	Free moisture content
GHP	Good hygienic practices
GDP	Gross domestic product
GRS	Gross registered space
GUS	Gross usable space
HMM	Heavy meromyosin
HDL	High density lipoprotein
HQSL	High quality storage life
HTST	High temperature short time
HUFA	Highly unsaturated fatty acids
Hx	Hypoxanthine
IQF	Individual quick frozen
Ino	Inosine
IMP	Inosine monophosphate
IP	Insoluble proteins
IF	Intrinsic factor
LR	Lethal rate

LMM	Light meromyosin
LDL	Low density lipoprotein
MIC	Maximum ice crystal formation
MA	Methylamine
MSG	Monosodium glutamate
MFP	Myofibrillar proteins
MyBP	Myosin binding protein
MHC	Myosin heavy chain
MLC	Myosin light chains
NOS	Net occupied space
NPU	Net protein utilization
NTD	Neural tube defect
NADPH	Nicotinamide ademine denucleotide phosphate
NAD	Nicotinamide adenine dinucleotide
NPN	Non-protein nitrogenous
NDGA	Nordihydroguaiaretic acid
PHD	Partially hydrolysed and deodorised
PAH	Polycyclic aromatic hydrocarbons
PUFA	Polyunsaturated fatty acids
PSL	Practical storage life
PN	Protein nitrogen
QD	Quarter dingley
RBC	Red blood cells
RSW	Refrigerated seawater

RLC	Regulatory light chain
SSP	Salt soluble proteins
SPP	Sarcoplasmic proteins
NaCl	Sodium chloride
SUF	Space utilisation factor
SSO	Specific spoilage organisms
SOP	Standard operating procedures
SP	Stroma proteins
TAP	Thermal arrest period
TDT	Thermal death time
TBA	Thiobarbituric acid
TFM	Total fatty matter
TN	Total nitrogen
TVB-N	Total volatile base-nitrogen
TMA	Trimethylamine
TMAO	Trimethylamine oxide
UHT	Ultra high temperature
WHC	Water holding capacity
WSP	Water soluble proteins

1 Physical Characteristics of Fish

- Fish is a highly perishable food item. Today, there is an increasing awareness on fish consumption as it is considered as health food.
- There are three major classes of fishes, A) the round mouth eels (Cyclostomata), B) the cartilaginous fishes (Chondriachthys) and C) the bony fish (Osteichthys). Out of these, cartilaginous fish and bony fishes are considered as true fish.
- In case of bony fish, the skeleton is composed of bones, i.e. majority of fishes caught (catla, rohu, sardine, mackerel, mullet etc.) and in case of cartilaginous fish, skeleton is composed of cartilage instead of bones e.g. shark, skates, rays and some deep sea fishes etc.
- Considerable variations are seen in the body shape of fishes.
    a. When length increased considerably – snake shape Ex. Eel.
    b. When thickness increased considerably – squat shape Ex. Lump fish
    c. When width increased considerably – flattened shape Ex. Rays
- The swimming ability of fishes reduced when deviations are seen from the elongated shape of fish. Generally fish swim by flexing the tail fin or caudal fin from side to side.
- Small fishes up to 30 cm long can swim up to 10 times their body length per second and speed diminishes among larger fishes.
- There are two types of swimming speed of fishes.
    a. **Cruising speed**: 3-4 times of body length per second. It is for a long period.

b. **Burst speed**: 10 times of body size per second. It is for short period.

» The outer layer of fish is a skin which often covered with scales. The thickness of the skin varies from 0.3 to 1.1 mm based on different species. The skin is found to be thicker in aggressive species. It gets thickened when fish is not feeding. In case of carps, fish skin can be eaten. In case of sharks, the skin is used as a by-product e.g. shark skin.

» The skin of the fish contains 16% protein, a little fat and minerals and around 80% water.

» The physical characteristics of fishes are important during their processing. The details are described below.

PHYSICAL CHARACTERISTICS

1. Shape
2. Size
3. Weight composition
4. Density
5. Angle of repose
6. Angle of slip and Coefficient of friction
7. Texture
8. Freezing point
9. Specific heat
10. Latent heat of fusion
11. Thermal conductivity
12. Thermal diffusivity
13. Enthalpy

1. Shape of the fish:

- Fishes are of four basic shapes such as (1) Torpedo-shaped (spindle shaped), (2) Arrow-shaped, (3) Flattened: (a) Bi-laterally flattened or (b) Dorso-ventrally flattened and (4) Snake-shaped (Figure 1.1).
- The factors influencing shape of fishes are extent of movement in water, quickness of movement, habitat and sluggishness.
- The shape of fish has great influence on its processing in different methods. It determines the specific surface area of fishes. Fishes with greater specific surface area have several advantages and disadvantage during their processing. The advantages are faster rate of cooling during chilling and freezing preservation, faster rate of heating during cooking, frying and canning as well as faster rate of moisture loss during drying and salting preservation. The disadvantages are faster rate of cold storage dehydration, faster rate of lipid autoxidation and faster rate of attack by microorganisms leading to faster rate of spoilage. The shape of fish also determines the style of its dressing. For example, the shape of a fish determines whether it has to be filleted, cut into steaks or processed whole.

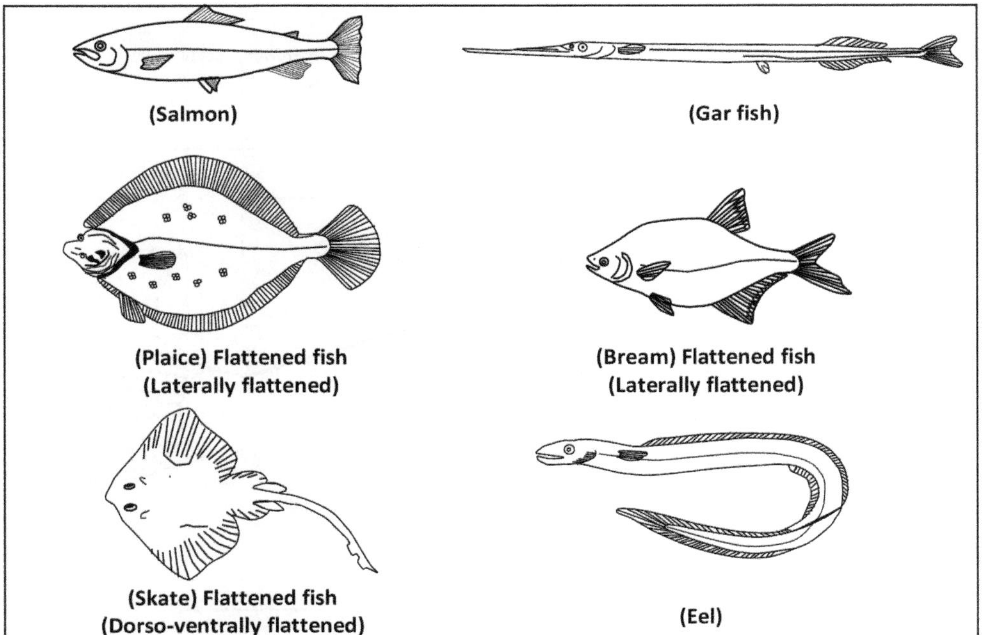

Figure 1.1: Shape of fishes

2. Size of the fish:

» The size of fish is expressed either in terms of length or weight.

» Measurement of length of fishes involves parameters such as (i) Overall length (absolute length), (ii)Length of fish, (iii)Length of head, (iv)Length of body (trunk), (v)Maximum cross-section, (vi)Maximum thickness etc. (Figure 1.2).

» The factors influencing size of fishes are species, genetic makeup, feeding conditions, environmental conditions, age and sex.

» The size of fishes has great importance in their processing. For example, the size of a fish has a great bearing on its price. It also determines the most suitable way of its processing and the container suitable for its packing.

» In designing fish processing machines, like filleting machine, gutting machine and steak cutting machine, it is important to provide information on the size of the fish to be used, size relation of the different parts of its body (*i.e.* between head, body, tailfin), its cross-section and thickness.

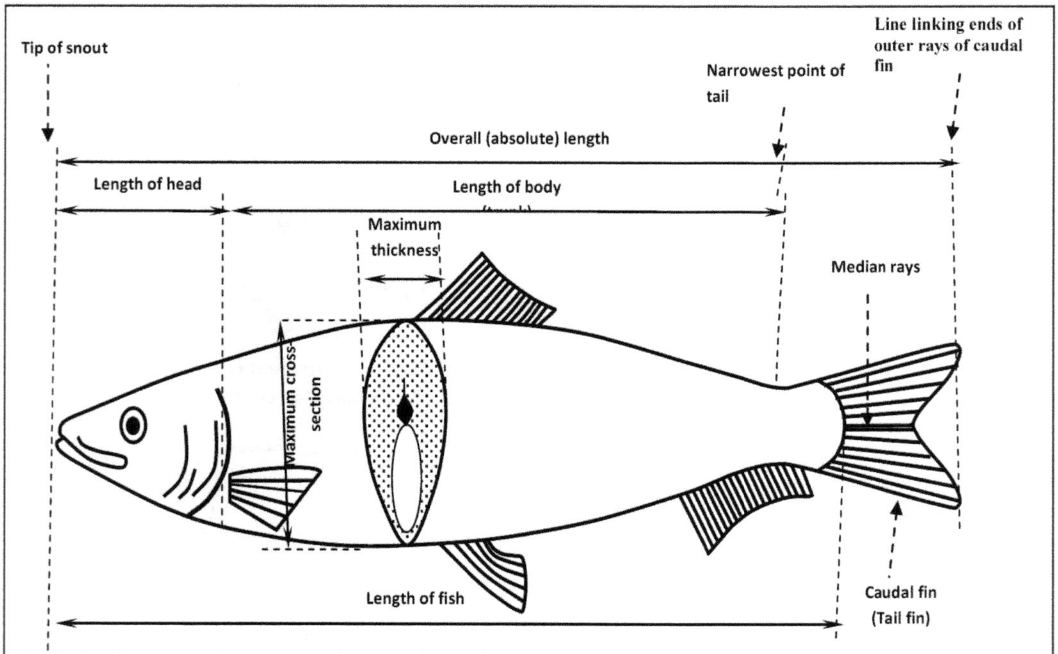

Figure 1.2: Parameters for measuring size of fish

3. Weight composition of fish:

- The weight composition of a fish is the proportion of the weight of different parts or organs of its body as percentage of its total body weight. It is calculated using the following equation.

$$X(\%) = \frac{Mx}{Mf} \times 100$$

Where,

X(%) = Proportion of weight (mass) of a certain part or organ, X, of the body of a fish as percentage of its total body weight.
Mx = Weight (Mass) of a certain part or organ, X, of the body of a fish (g)
Mf = Weight (Mass) of the fish (g)

- The weight composition of few fishes is given below.

Table 1.1: Weight composition of fish

Species	Weight composition (%)						
	Head	Flesh	Skin	Bone and cartilage	Fins	Scales	Viscera
Carp	18.0	47.0	4.0	8.5	3.0	5.0	13.0
Anchovy	20.5	54.0	3.0	4.0	2.0	1.0	14.0
Tuna	18.0	60.0	4.0	8.0	2.0	-	8.0

4. Density of fish:

- The density of fish is the mass of fish per unit volume. It is expressed by the following equation.

$$\rho_f = \frac{M_f}{V_f}$$

Where,

ρ_f = Density of fish (g/cm^3)
M_f = Mass of fish (g)
V_f = Volume of fish (cm^3)

- The density of water is comparatively lower than that of many other substances.
- At 20⁰C, it is 0.998 g/cm^3 and at 4⁰C it has its maximum value of 1.000 g/cm^3.
- At 0⁰C, water is transformed to ice with an increase in its volume. As a result, there is a decrease in its density by 9 % when water changes to ice.
- At 0⁰C the density of ice is 0.917 g/cm^3.
- Evisceration of fish leads to an increase in its density due to the removal of low-density air bladder and other internal organs. Thus, the density of eviscerated fish is higher and closer to that of fish meat as shown below.

Fish	Evisceration	Density (g/cm^3)
Wild carp	Whole fish	0.987
	Eviscerated fish	1.082
	Fish meat	1.065

- There is little change in the density of fish with change of temperature from ambient to about 0⁰C without freezing. However, freezing of fish leads to a decrease in its density. This is due to an increase in its volume as a result of the increase in the volume of water (moisture) present in its body upon freezing.

Fish	Temperature (⁰C)	Density (g/cm^3)
Wild carp	15	0.987
	0	0.980
	– 8	0.922

- The density of fishes is important in their processing in several ways. For example, it is the density of a fish that determines the possibility of its transportation in a flow of water along channels in a processing plant. It can also help to predict whether a fish is well-nourished or not. In fish processing, well-nourished fishes are preferred to poorly-nourished ones. The density of fish is also important in the thawing of fish in that, frozen fish floats in water while it sinks when thawing is completed. It is also used to find out the true volume of a bulk of fish (i.e. volume of fish excluding space between them) for a given mass.

5. Angle of repose:

» The angle of repose of fish is the angle (θ) formed between the base and the inclined surface of a cone-shaped heap of fish produced when a bulk of fish is freely piled on a horizontal platform (Figure 1.3).

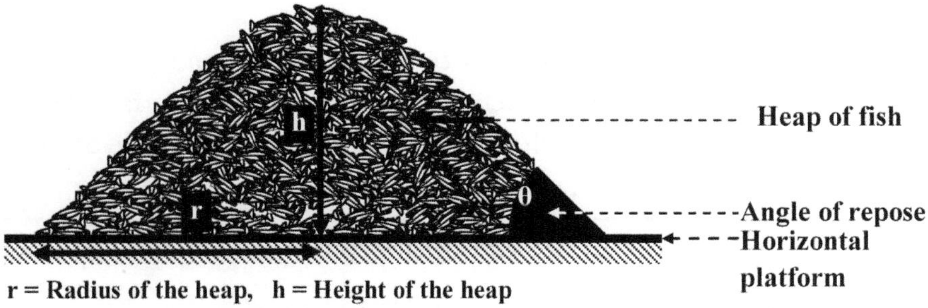

Figure 1.3: Angle of repose of fish

The factors influencing angle of repose of fishes are species, condition of fish (Dead fishes have greater angle of repose) and freezing (Freezing increases the angle of repose).

In fish processing, angle of repose is important to find out the size of platform required to pile a given quantity (mass or volume) of fish in a landing center or on a deck. The radius (r) of the circular platform required for piling a given volume (V) of fish having angle repose 'θ', can be given by the following equation.

$$r = \sqrt[3]{\frac{3V_F}{\pi \tan \theta}}$$

Where,

r = Radius (minimum) of the platform required to pile fish
V_F = Volume of fish including interstitial spaces
θ = Angle of repose of fish

6. Angle of slip and coefficient of friction:

» The angle of slip (β) of a fish is the angle formed between a horizontal plane and an inclined surface at the moment the fish lying on it just starts sliding down under gravity as its elevated end is raised gradually (Figure 1.4).

» Coefficient of friction (μ) is the tangent of the angle of slip (β). It is given by the following equation.

$$\mu = \tan \beta$$

Where,

μ = Coefficient of friction

β = Angle of slip

The factors influencing angle of slip and coefficient of friction of fishes are species, material and condition of the surface on which fish slips, wetness of the surface (Wet surfaces give smaller angle of slip), size of the fish (Larger fishes have smaller angle of slip) and its freshness (Live or very fresh fishes have smaller angle of slip).

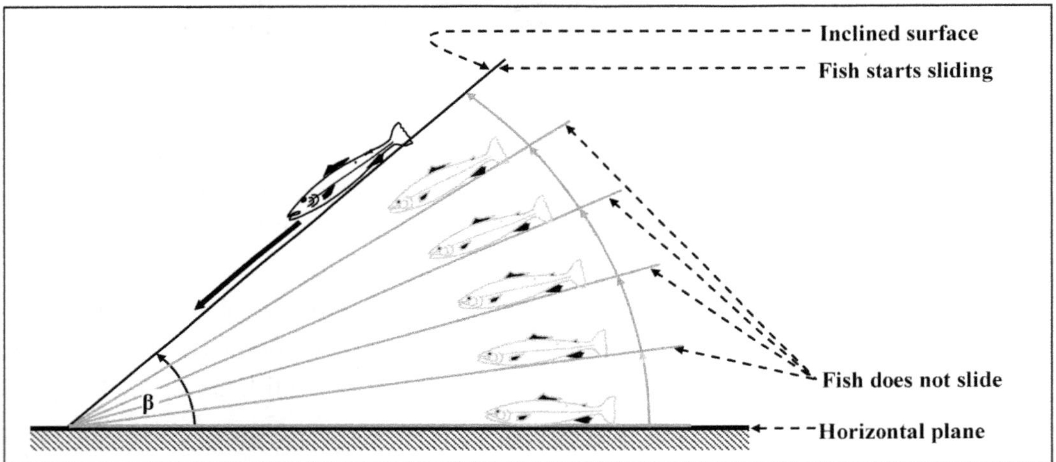

Figure 1.4: Angle of slip of fish

» With respect to importance in fish processing, angle of slip and coefficient of friction are useful in designing installations for moving and processing of fish. For example, if a rubber conveyor belt has to convey fish upward, then the angle of the belt with the horizontal must be less than the angle of slip between the fish and rubber, otherwise the fish slides down the belt and cannot move upwards (Figure 1.5).

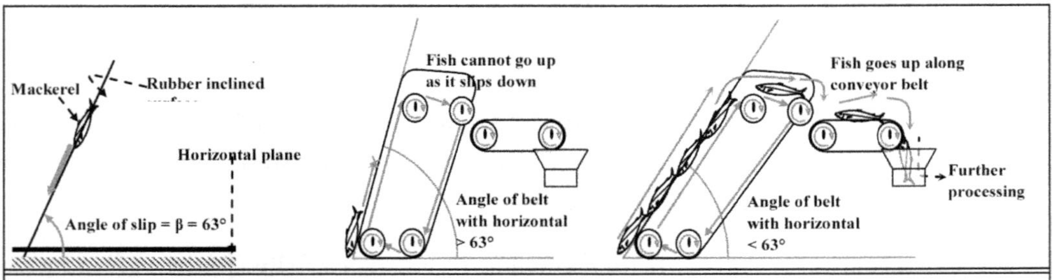

Figure 1.5: Application of angle of slip of fish in conveyor belt

7. Texture (consistency) of fish:

» Texture of fish refers to the firmness or softness of its whole body.

» The factors which influence the texture of a fish are its post mortem stage (Figure 1.6) (Fish in pre-rigor stage is firm and elastic, in on-rigor stage is hard and stiff and in post-rigor stage is soft and limp retaining finger indentation), extent of its handling (Rough handled fish quickly become soft), and its temperature (If the temperature of fish body is lowered by chilling the loss of firmness slows down and good texture is retained for a longer time).

» In the processing of fish, texture (tissue softness) is important in assessing the freshness of fish. A fish with firm elastic muscles, which do not retain finger indentation, is judged to be very fresh. Texture is also important in determining the cutting efforts required. A fresh fish with firm elastic muscles can be cut with a sharp knife with less effort giving smooth edge and surface of the cuts.

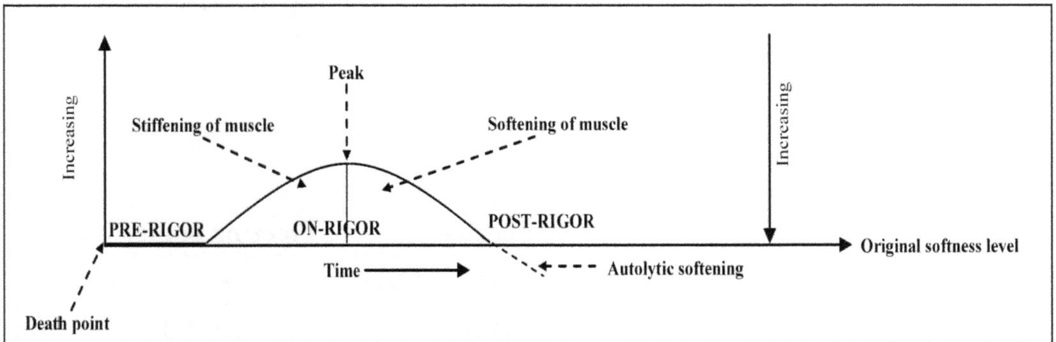

Figure 1.6: Influence of post mortem change on texture of fish

8. Freezing point of fish:

» Freezing point (initial freezing point) of fish is the temperature at which, an infinite small amount of ice crystals starts to form at its tissue water during freezing at normal atmospheric pressure.

» Pure substances like water have definite temperatures at which they can completely freeze. When heat is removed continuously from a given amount water at ambient temperature, at the outset, its temperature goes on decreasing till 0°C. Thereafter, decrease in temperature stops and change of phase starts. As a result, water changes from its liquid phase (liquid water) to solid phase (ice) by the process of freezing.

Freezing of water to ice continues at the constant temperature of 0^0C till all the water is converted to ice.

» In contrast, if heat is removed continuously from a dilute aqueous solution, it does not freeze 'completely' at any constant temperature; rather it only 'starts' to freeze at a definite temperature and thereafter continues to freeze over a range of decreasing temperature. Due to this, solutions do not have 'freezing points'; rather they have 'initial freezing points' where the freezing process starts.

» Fish can be treated similar to a dilute aqueous solution as it contains a large portion of water in which several solutes remain dissolved. Therefore, they do not have freezing points; rather they have initial freezing points. However, very often, the term 'freezing point' is loosely used instead of 'initial freezing point'.

» As fish can be treated as a dilute aqueous solution, there is a depression of freezing point and the freezing point of fish is always lower than that of water ($<0^0$C). However, for all practical purposes, the freezing point (initial freezing point) of fishes is assumed to be -1^0C. The initial freezing points of some fishes are given in the following table 1.2.

Table 1.2: Freezing point of some fish species

Species	Freezing point (^{0}C)	Species	Freezing point (^{0}C)
Carp	− 0.67	Shark and rays	− 1.8 to − 2.0
Eel	− 0.83	Sardines	− 1.3
Herring	− 1.3 to − 1.52	Tuna	− 1.3
Cod	− 1.5 to − 2.0	Crab	− 2.0

» The freezing point of a fish depends on the concentration of solutes in its body. Fishes with higher concentration of solutes have lower freezing points than those with lower concentrations. This is because of the depression of the freezing point by the solutes. Moreover, the depression of freezing point also depends on the nature of the solutes. Based on the type of solutes, fish tissue can be considered to be made of two types of solutions as mentioned below.

1. **True solution:** It is made of salts and sugars dissolved in the tissue moisture. These solutes contribute greatly towards the depression of freezing point.

2. **Colloidal solution:** It is made of tissue colloids (proteins and fats) dissolved in the tissue moisture. These colloids do not contribute much towards the depression of freezing point.

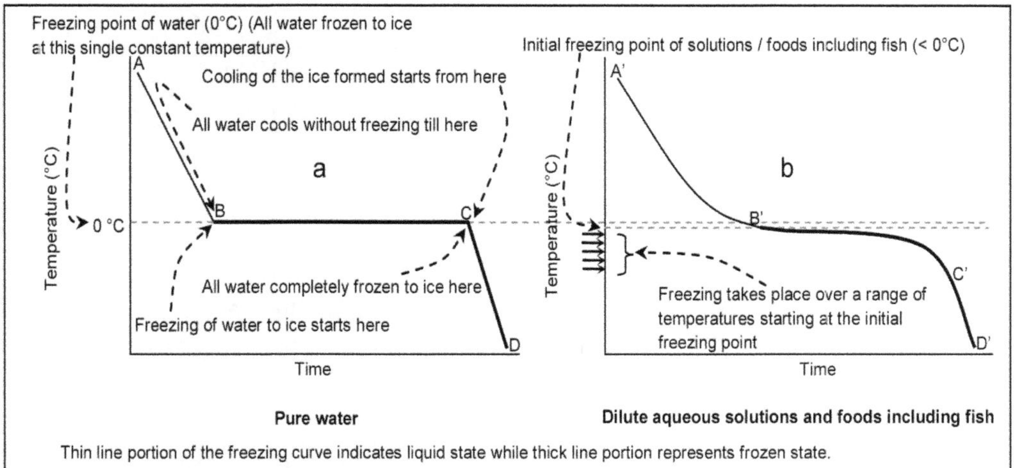

Figure 1.7: Freezing curve of pure water vs dilute aqueous solutions and foods including fish

» Generally marine fishes have lower freezing points than fresh water fishes as given below. This is due to the higher concentrations of salts (solutes) in marine fishes where the habitat is salty.

Habitat	Species	Freezing points (^{0}C)
Fresh water	Carp	− 0.67
Marine	Sardine	− 1.3
	Tuna	− 1.3

» Roughly speaking, the freezing points of surface-dwelling pelagic fishes are higher than those of the bottom-dwelling demersal fishes. This is because, pelagic fishes are usually fatty, i.e., their tissue contain more fat while demersal fishes are usually lean i.e., their tissue contain less fat. As fats, which make colloidal solution with the tissue moisture, do not contribute much towards the depression of freezing points, fatty fishes have higher freezing points as given in Table 1.3.

Table 1.3: Freezing point of fish species with respect to position in water column

Position in the water column	Species	Freezing points (^{0}C)
Pelagic	Sardine	− 1.3
	Tuna	− 1.3
Demersal	Cod	− 1.5 to − 2.0
	Shark and rays	− 1.8 to − 2.0
	Eel	− 1.95

9. Specific heat (c) of fish:

» Specific heat of fish is the amount of heat that must be supplied to or removed from unit mass of fish to change its temperature by unit degree. It is expressed by the following equation.

$$c_f = \frac{Q}{M_f \Delta t}$$

Where,

c_f = Specific heat of fish
Q = Quantity of heat required to be added (or removed)
M_f = Mass of fish
Δt = Change in temperature = $t_2 - t_1$
and, t_1 = Initial temperature of fish
t_2 = Final temperature of fish

Table 1.4: The specific heat of some fishes

Species	Specific heat (kcal/kg.^{0}C)	Species	Specific heat (kcal/kg.^{0}C)
Haddock, cod	0.82	Salmon	0.71
Halibut	0.80	Tuna	0.76
Herring (Kippered)	0.76	Mackerel	0.66
Herring (Smoked)	0.71	Shrimp	0.86
Menhaden	0.70	Lobster (American)	0.83

» The specific heat of fish varies from species to species depending on their chemical composition, because the specific heats of different constituents of fish which contribute to the chemical composition are different as given in Table 1.5.

Table 1.5: The specific heat of chemical components of some fishes

Components	Specific heat (kcal/kg.^{0}C)
Moisture	1.0
Fat	0.5
Protein & ash (*i.e.* dry non-fatty matter)	0.36
Dry matter (Fat + protein + ash)	0.20

» Lean fishes, which contain less fat and more water, have higher specific heat than fatty fishes, which contain more fat and less water.

» Specific heat of fishes also depends on season, as the chemical composition of fish varies with season. Thus, for a given species of fish, specific heat is low in the season when fat content of tissue is high.

» The specific heat of fish does not show any remarkable change in the temperature range from 30^{0}C to 0^{0}C. However, when cooled further, it gets frozen and its specific heat falls drastically to about half that of unfrozen fish. This is because, when fish is frozen, its tissue water changes to ice whose specific heat is much lower (half *i.e.* 0.5 kcal/kg.^{0}C) than that of water (1.0 kcal/kg.^{0}C). The specific heat of unfrozen and frozen fish is given below.

Fish	Temperature	Specific heat (kcal/kg.^{0}C)
Pike perch	Chilled fish (0^{0}C)	0.845
	Frozen fish (–28 to –30^{0}C)	0.440

The specific heat of fishes is calculated using the following two methods.

1. **Earle method:** This method is suitable, if at least moisture content of the fish is available. It can be used to find out the specific heat of fish both above and below its freezing point. Therefore, it is suitable to determine the specific heat of both unfrozen and frozen fishes as given below.

$$c_{fa} = m + 0.2\,d$$
$$c_{fb} = 0.5\,m + 0.2\,d$$

Where,

c_{fa} = Specific heat of fish above its freezing point, *i.e.*, for unfrozen fish (kcal/kg.^{0}C)

c_{fb} = Specific heat of fish below its freezing point, *i.e.*, for frozen fish (kcal/kg.^{0}C)

m = Mass fraction of moisture in the fish tissue (= moisture % ÷ 100)

d = Mass fraction of dry matter (fat + protein + ash) in the fish tissue (=1 − m)

> ***Mass fraction of a component*** *is its mass present in unit mass of the 'whole' (all the components). While expressing a component as percentage of the whole, 100 is assumed to be the mass of the whole, but while expressing a component as mass fraction of the whole, 1 is assumed to be the mass of the whole. For example, if the percentage of moisture in the tissue of a given fish is 80 %, it implies that 100 gram of the tissue containing all the components (moisture, protein, fat and ash) contains 80 gram of moisture. Mass fraction of moisture in the same tissue is m = M/100 = 80/100 = 0.8. It implies that 1 gram (unit mass) of the whole tissue contains 0.8 gram of moisture.*

2. **Zaitsev *et al.* method:** This method is suitable, if the proximate composition (*i.e.* moisture, fat, protein and ash content) of the fish is available. It can also be used to find out the specific heat of fish both above and below its freezing point. Therefore, it is suitable to determine the specific heat of both unfrozen and frozen fishes as given below.

$$c_{fa} = m + 0.5\,f + 0.36\,pa$$
$$c_{fb} = 0.5\,m + 0.5\,f + 0.36\,pa$$

Where,

c_{fa} = Specific heat of fish above its freezing point, *i.e.*, for unfrozen fish (kcal/kg. ^{0}C)

c_{fb} = Specific heat of fish below its freezing point, *i.e.*, for frozen fish (kcal/kg. ^{0}C)

m = Mass fraction of moisture in the fish tissue (= moisture % ÷ 100)

f = Mass fraction of fat in the fish tissue (= fat % ÷ 100)

pa = Mass fraction of protein and ash together= { (protein + ash) % ÷ 100}

» The difference in the specific heat of fish above its freezing point (c_{fa}) and below its freezing point (c_{fb}) is due to the difference in the specific heat of water, which is 1.00 kcal/kg.^{0}C above its freezing point (*i.e.* for liquid water) and 0.5 kcal/kg.^{0}C below its freezing point (*i.e.*, for ice). Therefore, in case of 'c_{fa}', the term is 'm' while in case of 'c_{fb}', the term is 0.5 m.

10. Latent heat of fusion:

» Latent heat of fusion of fish is the amount of heat that must be supplied to change unit mass of fish from its hard frozen solid state to its soft normal state. It is also the amount of heat that must be removed to change unit mass of fish from its normal soft state to its hard frozen solid state. It is expressed by the following equation.

$$L_f = \frac{Q}{M_f}$$

Where,

L_f = Latent heat of fusion of fish

Q = Quantity of heat to be supplied to change the fish from its hard frozen solid state to its soft normal state or

Quantity of heat to be removed to change the fish from its normal soft state to its hard frozen solid state.

M_f = Mass of fish

Table 1.6: The latent heat of fusion of some fishes

Fish/shellfish	Latent heat of fusion (kcal/kg)	Fish/shellfish	Latent heat of fusion (kcal/kg)
Haddock, cod	62.1	Salmon	51.1
Halibut	60.0	Tuna	55.9
Herring (Kippered)	55.9	Lobster (American)	63.1
Herring (Smoked)	51.1	Shrimp	66.2
Menhaden	49.4	Mackerel	45.4

» As only the water present in the fish tissue solidifies during its freezing, the latent heat of fusion of fish is always less than that of water (i.e. < 80 kcal/kg).

» Lean fishes, which contain more percentage of water, have higher values of latent heat of fusion, while fatty fishes, which contain less percentage of water, have lower values.

- Latent heat of fusion of fishes is low in the season when fat content of tissue is high. This is because; moisture content which counts for the latent heat of fusion is low when fat content is high. [In general, there exists an inverse relationship between water and fat content in the fish tissue and roughly speaking together they account for about 80% of the tissue weight.]
- Freezing of fish involves the lowering of its temperature leading to the crystallization of its tissue water only which has the latent heat of fusion of 80 kcal/kg.
- Therefore, the latent heat of fusion of fish is calculated using the following equation.

$$\mathbf{L_f = 80\ m}$$

Where,

L_f = Latent heat of fusion of fish
m = Mass fraction of moisture in fish tissue

11. Thermal conductivity (k) of fish:

- The thermal conductivity of fish is the amount of heat transferred by conduction through the body of fish per unit cross-section area per unit time per unit temperature gradient.

$$\mathbf{k_f} = \frac{Q}{A.T.\ \frac{\Delta t}{l}}$$

Where,

k_f = Thermal conductivity of fish
Q = Quantity of heat transferred through the fish body
A = Cross-section area of the fish body measured at right angle to the direction of heat flow
T = Time (duration) of heat transfer
$\frac{\Delta t}{l}$ = Temperature gradient (*i.e.* change in temperature per unit length of fish body)
$\Delta t = (t_1 - t_2)$ Where,
t_1 = Temperature at the hotter end of the fish body

t_2 = Temperature at the cooler end of the fish body
L = Length of the fish body

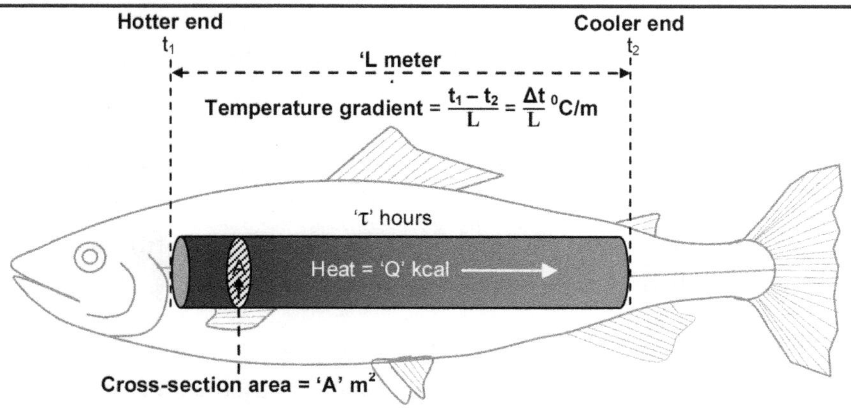

$$\textbf{Thermal conductivity of fish} = k_f = \frac{Q}{A\,.\,T\,.\,\frac{\Delta t}{L}}$$

Figure 1.8: Thermal conductivity of fish

Table 1.7: The thermal conductivity of some fishes at 30°C

Fish	Thermal conductivity (kcal/m.hr.°C)	Fish	Thermal conductivity (kcal/m.hr.°C)
Cod	0.482	White pomfret	0.390
Rohu	0.396	Black pomfret	0.380
Mackerel	0.395	Hilsa	0.387

» The thermal conductivity of fish varies from species to species depending on their chemical composition, because the thermal conductivity of different constituents of fish which contribute to the chemical composition is different as given below.

Constituent	Thermal conductivity (kcal/m.hr.°C)
Moisture	0.52
Dry matter	0.22

- Lean fishes which contain less fat (resulting in less dry matter) and more water, have higher thermal conductivity than fatty fishes which contain more fat (resulting in more dry matter) and less water.

- For a given species of fish, thermal conductivity is low in the season when fat content of tissue is high.

- The thermal conductivity of water is 0.52 kcal/m.hr.^{0}C, while that of ice is 2.07 kcal/m.hr.^{0}C (about four times). Therefore, there is little change in the thermal conductivity of fish between 0^{0}C and 30^{0}C. But when fish is frozen, its thermal conductivity sharply rises.

- During freezing, the tissue water changes to ice thereby increasing the thermal conductivity to about three times the original value. Thus, the thermal conductivity of frozen fish is greater than that of unfrozen fish above 0 ^{0}C as shown below. Hence, heat transfer is faster in frozen fish than in unfrozen chilled fish.

Fish	Temperature (^{0}C)	Thermal conductivity (kcal/m.hr.^{0}C)
Pike perch	–1	0.522
	–30	1.5

- In the frozen state, the orientation of muscle fibers is an important parameter and heat conduction is 15% to 30% greater along the muscle fibers than across the muscle fibers.

- The thermal conductivity of fish is calculated using the following equations.

$$\mathbf{k_{fa} = 0.52\,m + 0.22\,d}$$
$$\mathbf{k_{fb} = 2.07\,m + 0.22\,d}$$

Where,

k_{fa} = Thermal conductivity of fish above freezing point, *i.e.*, for unfrozen fish (kcal/m.hr.^{0}C)

k_{fb} = Thermal conductivity of fish below its freezing point, *i.e.*, for frozen fish (kcal/m.hr.^{0}C)

m = Mass fraction of moisture in the fish tissue (= moisture % ÷ 100)

f = Mass fraction of fat in the fish tissue (= fat % ÷ 100)

d = Mass fraction of dry matter (protein + fat + ash) = (1 – m)

12. Thermal diffusivity of fish:

» Thermal diffusivity of fish is defined as its thermal conductivity divided by the product of its density and specific heat. It is expressed by the following equation.

$$\alpha_f = \frac{k_f}{\rho_f \cdot c_f}$$

Where,

α_f = Thermal diffusivity of fish
k_f = Thermal conductivity of fish
ρ_f = Density of fish
c_f = Specific heat of fish

» The thermal diffusivity of fish is a measure of the rate at which a fish undergoes change in its temperature during heating or cooling as compared to thermal conductivity which is a measure of the rate at which heat is transferred through the fish.

» The thermal diffusivity of some fishes between 0^0C and 40^0C is given below.

Fish	Thermal diffusivity (m^2/hr)	Fish	Thermal diffusivity (m^2/hr)
Pike perch	0.000432	Bream	0.000415
Cod	0.000423	Sturgeon	0.000406

» From the equation of thermal diffusivity [$\alpha_f = k_f / (\rho_f \cdot c_f)$], it can be seen that thermal diffusivity of fish increases with increase in thermal conductivity and decrease in density and specific heat of fish.

» The thermal diffusivity of fish changes with season, due to the variation of its chemical composition with season.

» The thermal diffusivity of water is 0.000468 m^2/hr, while that of ice is 0.00396 m^2/hr (about nine times). This is the reason why water freezes faster than it melts. Moreover, due to this, there is little change in the thermal diffusivity of fish between 0^0C and 40^0C. But when it is frozen, its thermal diffusivity sharply rises.

» During freezing, the tissue water changes to ice thereby increasing the thermal diffusivity to about three times the original value. Thus, the thermal diffusivity of frozen fish is greater than that of unfrozen fish above 0⁰C. Hence, frozen fish undergoes change in temperature at a faster rate than unfrozen chilled fish.

Fish	Temperature (⁰C)	Thermal diffusivity (m^2/hr)
Pike perch	Unfrozen (0 to 40⁰C)	0.000432
	Frozen	0.0011-0.0016

» The thermal diffusivity of fish is calculated using the following two methods.

a. **Method I (Thermal data method):** This method uses the following equation, which involves other thermal data.

$$\alpha_f = \frac{k_f}{\rho_f \cdot c_f}$$

Where,

α_f = Thermal diffusivity of fish (m^2/hr)
k_f = Thermal conductivity of fish (kcal/m.hr.⁰C)
ρ_f = Density of fish (kg/m^3)
c_f = Specific heat of fish (kcal/kg.⁰C)

b. **Method II (Moisture method):** This method is used when better data are lacking to determine the thermal diffusivity provided the moisture content of the fish or fishery product is over 40% (Ashrae,1977).

$$\alpha_f = 0.000318 + (\alpha_w - 0.000318)\frac{M}{100}$$

Where,

α_f = Thermal diffusivity of fish (m^2/hr)
α_w = Thermal diffusivity of water at the desired temperature (m^2/hr)
M = Moisture percentage in the fish tissue

» In the processing of fish, thermal diffusivity, due to its influence on the rate at which a fish undergoes change in its temperature during heating or cooling, affects rate of chilling, freezing, thawing and high temperature processing (*e.g.*, cooking. frying, canning, pasteurization etc.) of fish.

- If the thermal diffusivity of a fish is high, it chills, freezes, thaws and gets heated up quickly due to faster rate of change of temperature.
- Due to the higher thermal diffusivity of ice (about nine times) than water, fish freezes faster than it thaws for the same temperature difference.
- During freezing or thawing of fish, it is only the water (moisture) present the fish tissue that undergoes a change in phase from liquid water to solid ice (during freezing) or from solid ice to liquid water (during thawing).
- During freezing of fish, the layers of ice formed at the surfaces diffuse heat from inside to outside at a faster rate and therefore fish freezes quickly.
- On the other hand, during thawing, the layer of water formed on the surfaces of fish diffuse heat from outside to inside slowly and therefore fish thaws slowly.

13. Enthalpy of fish:

- The enthalpy of fish at a definite temperature is its heat content per unit mass at that temperature with reference to a reference temperature datum of – 40⁰C where its heat content is assumed to be zero. It is expressed as follows.

$$h_{ft} = \frac{Q^{t}_{-40}}{M_f}$$

Where,

h_{ft} = Enthalpy of fish at a given temperature 't'
Q^{t}_{-40} = Heat content at the given temperature 't' with reference to a reference temperature datum of – 40⁰C where its heat content is assumed to be zero
M_f = Mass of the fish

- The reference temperature is taken as – 40⁰C where the heat content is arbitrarily designated as zero because, most of foods including fish and fishery products are commercially frozen to a low temperature of – 40⁰C.
- As the heat content of fish is assumed to be zero at – 40⁰C, the enthalpy of fish at a given temperature is the amount of heat to be added to or removed from unit mass of fish to bring it from – 40⁰C to that temperature.

» Therefore, the enthalpy of fish is calculated by summing up all the heat to be added to or removed from unit mass of fish to bring it from – 40⁰C to that temperature as given below.

$$\Delta h_{ft} = \frac{\Sigma^{t}_{-40} Q}{M_f}$$

Where,

h_{ft} = Enthalpy of fish at a given temperature 't'

Q = Sum of different amounts of heat to be added to or removed from fish to bring it from – 40⁰C to the given temperature 't'

M_f = Mass of the fish

Table 1.8: Enthalpy of fish at various temperatures

Temperature (⁰C)	Enthalpy (kcal/kg)	Temperature (⁰C)	Enthalpy (kcal/kg)	Temperature (⁰C)	Enthalpy (kcal/kg)
40	112.25	6	82.39	-12	15.84
36	108.73	4	80.65	-14	14.18
32	105.21	2	78.90	-16	12.69
28	101.69	0	77.16	-20	10.03
24	98.17	– 1	71.16	-24	7.67
20	94.65	– 2	42.16	-28	5.55
16	91.94	– 3	32.70	-32	3.60
14	88.39	– 4	28.05	-36	1.77
12	87.64	– 6	23.01	-40	0.00
10	85.89	– 8	19.99		
8	84.14	– 10	17.73		

» For example, the enthalpy of fish at a temperature 't ⁰C' which is above its freezing point is calculated by summing up all the following three quantities of heat to be added to unit mass of fish to bring it from – 40⁰C to that temperature.

Q_1= Quantity of heat added to raise the temperature of fish from – 40⁰C to its freezing point of –1⁰C

Q_2= Quantity of heat added to melt the frozen fish completely at its freezing point of – 1°C

Q_3= Quantity of heat added to raise the temperature of fish from – 1°C to t °C

$$\text{Enthalpy of fish at the temperature 't °C'} = \Delta h_{ft} = \frac{Q_1 + Q_2 + Q_3}{M_f}$$

» The experimentally determined enthalpy values of a typical lean fish at different temperatures are given below. The values are plotted to get the enthalpy curve (Figure 1.9).

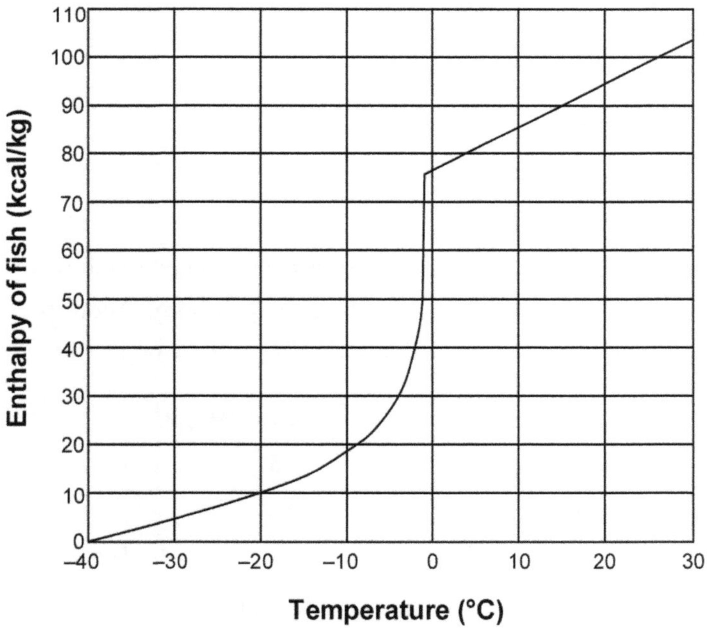

Figure 1.9: Enthalpy curve of fish

2 Chemical Characteristics of Fish

The chemical characteristics of fish are described in the following three ways.

I. Elemental composition

II. Molecular composition

III. Proximate composition

I. Elemental composition of fish: It is the percentage-wise expression for the relative proportion of different 'elements' that constitute the fish body. About 60 elements are present in the fish body. Most important among them are given in the following table.

Element	% of fish body weight
Oxygen	75
Hydrogen	10
Carbon	9.5
Nitrogen	2.5 – 3.0
Calcium	1.2 – 1.5
Phosphorous	0.6 – 0.8
Sulphur	0.3
Other elements	Traces

II. Molecular composition of fish: It is the percentage-wise expression for the relative proportion of different 'categories of molecules' that constitute the fish body. The different categories of molecules that constitute the fish body are as follows.

1.	Water
2.	Carbohydrates
3.	Proteins
4.	Lipids
5.	Vitamins
6.	Minerals
7.	Nucleic acids
8.	Non-protein nitrogenous substances

III. Proximate composition of fish: It is the percentage-wise expression for the relative proportion of 'the four major constituents' of the fish body such as moisture, crude protein, crude fat and ash. The four constituents which are included to give the proximate composition of fish are given below.

a. Moisture

b. Crude protein

c. Crude fat

d. Ash

In case of fish, the fifth constituent, fibre, which mainly refers to carbohydrates, is not considered as the carbohydrate content in fish is negligible. The general proximate composition of fish depicting the variation of its constituents and the respective percentage variations are given in the following Table 2.1.

Table 2.1: Variation in proximate composition of fish

Constituent	% by weight	Variation
Moisture	66-84	18.0
Crude protein	15-24	9.0
Crude fat	0.1-22	21.9
Ash	0.8-2	1.2

As moisture and proteins are the major constituents of its body, fish is considered a 'high-moisture, protein food'. Among the four constituents, the protein and ash contents do not show much variation while considerable variation is observed in crude fat and moisture content. In general, there exists an inverse relationship between moisture and crude fat content in the fish tissue and together they account for about 80% of the tissue weight as given below:

Crude fat + Moisture = 80%

Usually, dark-fleshed pelagic fishes are fatty containing 'more fat' and ' less water' while white demersal fishes are lean containing 'less fat' and 'more water' in their tissue. The factors which influence the proximate composition of fish are species, age, sex, season, habitat and feed (in case of cultured fish).

CHARACTERISTICS OF CONSTITUENTS OF FISH TISSUE

The characteristics of the following important constituents of fish tissue are described in details.

a. Moisture
b. Proteins
c. Lipids
d. Carbohydrates
e. Energy
f. Minerals
g. Vitamins
h. Nitrogenous extractives

A. Moisture (water):

» Among all the constituents of fish body, moisture constitutes the maximum of its body weight.

» It accounts for more than half of the weight of fish. Its content varies considerably from 66% to 84% with a percentage variation of about 18 % (Table 2.1).

» Some fishes like Bombay duck and some deep water fishes have exceptionally high moisture content above 90%.

» The water present in the fish tissue is called 'fish tissue water'.

» It is present in the fish tissue in four forms as given below.

i. Free water:

» It is the water present in the fish tissue so loosely, not being bound to any tissue constituent, that it is available as a solvent and freezes at a sub-zero temperature of -20^{0}C.

» Thus, this water is available in free for any chemical reaction.

» This is again of two types.

i. ***Excess water***: Its removal does not change any cellular behaviour.

ii. ***Metabolic water***: Its removal causes biologically inactive but not of lethal effect.

ii. Constitution water:

» This is of two types.

i. ***Vital water***: Its removal causes cell death.

ii. ***Ramanent freezable water***: It provides synergistic effect to vital water.

iii. Absorption water:

» It is more strongly bound to solid components.

» It cannot be frozen even at very low temperature.

» It can be removed by heating.

iv. Crystallization water:

» It is strongly bound to solid.

» It is unfreezable.

» It cannot be removed by freeze drying.

» It can be removed at very high temperature.

- The absorption and crystallization water together are called as ***Bound water***.

- It is the water present in the fish tissue which is bound so tightly to its hydrophilic (water-loving) tissue constituents that is not it available as a solvent nor does it freeze at a sub-zero temperature of - 20⁰C.
- Food containing higher quantities of bound water is more stable and less damage can happen during freezing.
- A very small portion of the tissue water in fish, amounting to about only 7% is present as bound water, leaving the rest, about 93%, as free water.
- The **moisture content** of fish muscle is responsible for its spoilage and microbial action.

Importance of Water in Fish Processing:

The water present in the fish tissue is same as normal water and therefore has little nutritional value. However, it has tremendous importance in the processing of fish.

Fish is a high-moisture food item as its moisture content is very high (66-84%). Such high moisture content makes the fish a highly perishable food commodity. The high moisture content is conducive for many spoilage processes like microbial and enzymatic activity. This is because microbes and enzymes require high moisture content for their activities. Therefore, some of the processing methods like drying, salting, and freeze-drying are based on removal of moisture from the fish tissue. In freezing preservation also the tissue water is removed in the sense that it is converted to ice so that it is not available for microbial or enzymatic activities.

When fish is exposed to any external action (freezing, heating, drying, pH change or osmotic pressure), the ratio between bound water and free water changes due to which the original properties of fresh fish alters which adversely affect its quality.

B. Proteins:

» There are a number of substances present in the fish tissue which contain nitrogen in them. These substances are collectively called 'nitrogenous substances'. Some of the important ones are proteins, free amino acids (FAA), urea, ammonia, trimethylamine oxide (TMAO), creatine, creatinine etc.

» The most important among the nitrogenous substances, which constitutes the bulk of it, is the proteins. The rest of the nitrogenous substances which are not part of protein are collectively called 'non-protein nitrogenous substances' (NPN substances). They include free amino acids (FAA), urea, ammonia, trimethylamine oxide (TMAO), creatine, creatinine etc.

» It is expressed as follows.

Nitrogenous substances = Proteins + NPN substances

» The protein content of fish is found out by first determining its nitrogen content and then multiplying it by a factor of 6.25. While determining the nitrogen content of fish, the value obtained is 'total nitrogen (TN)' which includes the nitrogen content of both proteins and NPN substances. Thus, total nitrogen is the sum of protein nitrogen (PN) and non-protein nitrogen (NPN). It is expressed as follows.

TN = PN + NPN

» Therefore, the value obtained by multiplying TN with 6.25 is called 'crude protein', *i.e.*, a rough estimate of proteins, in the sample as it includes both proteins and NPN substances. It is expressed as follows.

Crude protein = TN X 6.25

» If the 'true protein' content of a sample of fish has to be found out, then both TN and NPN are determined separately and PN is obtained by subtracting NPN from TN. True protein content is obtained by multiplying PN with 6.25. It is expressed as follows.

True protein = PN X 6.25 = (TN – NPN) X 6.25

» Thus, the protein content of fish can be expressed in two ways as follows.

i. Crude protein: It gives a rough estimate of proteins in the sample as in includes both proteins and NPN substances.

ii. True protein: It gives the exact content of proteins in the fish.

» The crude protein content of fish varies considerably from 15% to 24% with a percentage variation of about 9.0 %.

» Fish proteins are classified into the following three groups based on their location in the muscle cells and solubility in salt solution.

1. Sarcoplasmic proteins (SPP) or water-soluble proteins (WSP)

2. Myofibrillar proteins (MFP) or salt-soluble proteins (SSP)
3. Stroma proteins (SP) or insoluble proteins (IP)

1. Sarcoplasmic proteins (SPP) or water-soluble proteins (WSP):

» These are the proteins present in the sarcoplasm of the muscle cells. They are soluble in water and dilute solutions of neutral salts of low ionic strength of the order less than 0.15M.

» The sarcoplasmic protein is a large family of several proteins, which includes enzymes associated with the normal physiological processes of fish body, pigments, heme proteins (iron-containing proteins *viz.*, haemoglobin, the red pigment of blood and myoglobin, the red pigment of muscle), haemocyanin (a copper-containing protein that forms the blue pigment of blood in shellfishes), calcium-binding proteins (parvalbumins and calmodulins) and antifreeze proteins present in the sarcoplasm of fishes of the Arctic and Antarctic waters, (*e.g* Antarctic krill, *Euphausia superba*).

» The antifreeze proteins are chemically glycoproteins and are therefore called anti-freeze glycoproteins (AFGP). They lower the freezing point of fish tissue, thereby preventing the freezing of their tissue fluid at the sub-zero temperatures that prevail in these places. In absence of these antifreeze proteins, the fishes of these places would die due to the freezing of their body fluid at freezing temperature.

2. Myofibrillar proteins (MFP) or salt-soluble proteins (SSP):

» These are the proteins, which constitute the myofibrils of the muscle cells. They are soluble in solutions of neutral salts of high ionic strength of the order more than 0.5M.

» Myofibrillar proteins are classified into three groups as given below.

a. Contractile proteins:

- They are involved in contraction and relaxation of living muscle. They include the following two proteins.

  1. Myosin
  2. Actin

- In normal muscle of fish in living condition, myosin and actin remain separated from each other due to presence of ATP. However, after death, due to irreversible breakdown of ATP, myosin binds with actin irreversibly to form 'actomyosin'.

b. Regulatory proteins:

- They are not directly involved in muscle contraction and relaxation, but regulate these actions. They include mainly two proteins as given below.

  1. Tropomyosin
  2. Troponin

c. Structural proteins:

- They give support to the contractile and regulatory proteins thereby giving the myofibrils their structure. They include several proteins few of which are given below.

1. Titin (Connectin)	8. F-protein
2. Telethonin (T-cap)	9. α-Actinin
3. Nebulin	10. CapZ
4. Tropomodulin (Tmod)	11. M protein
5. C-proteins	12. Myomesin
6. X-protein	13. Skelemin
7. H-protein	14. Desmin

3. Stroma proteins (SP) or insoluble proteins (IP):

» These proteins constitute the connective tissue of fish muscle, which support the different structures of the fish body.

» Unlike SPP and MFP which are present inside the muscle cells, SP are present outside these cells. Thus, they are the components of the extracellular matrix of the cells. They are neither soluble in solutions of neutral salts nor in dilute acids or alkalis.

» The stroma proteins present in the fish muscle includes the following two types.

  a. Collagen
  b. Elastin

a. Collagen:

» It is the main structural protein of the various connective tissues of fishes.

- It is present in the form of elongated fibrils in fibrous tissues such as ligaments, skin and swim bladder.
- Tendons found in terrestrial animals are absent in fish. Tendon is different than ligament in the sense that tendons bind muscle to bone while ligaments bind bone to bone.
- Collagen is also abundant in corneas, cartilage, bones, blood vessels, gut, inter-vertebral discs and the dentin in teeth. In muscle tissue of fish, it is present in the endomysium, perimysium, epimysium and myocommata.
- Different fish species contain varying amounts of collagen in the body tissues, which influences the textural properties of their muscle.
- When connective tissue in fish meat containing a large proportion of collagen is heated by moist heat, say by boiling or steaming, for a long time, the collagen changes to a water soluble protein gelatin. Gelatin is a mixture of polypeptides.
- In raw fish, it is difficult to remove the meat from the fish body. However, when it is heated, the meat can be removed easily from body. This is caused by the gelatinization of collagen present in the myocommata, which holds the muscle tissues together, into gelatin.

b. Elastin:

- It is a protein found mainly in the arteries, intestine, skin, heart, notochord, elastic cartilage and swim bladder of fishes.
- It is elastic in nature, unlike collagen which is strong and inelastic.
- It is responsible for allowing tissues in the body to "snap back" to their original shape after being stretched or contracted just like a rubber band.
- For example, it helps in the contraction and expansion of the blood vessels allowing flow of blood.
- Elastin is very resistant to moist heat due to which cooking does not affect the elastin in connective tissue.
- It is different from collagen in that, collagen provides rigidity, while elastin gives stretchability. Collagen is the framework, giving strength and foundation, while elastin allows returning to the shape collagen gives a tissue after stretching or changing.

Comparison of SPP, MFP and SP of fish and terrestrial animals:

Sarcoplasmic protein:

- The sarcoplasmic proteins of fish muscle are more or less similar to those of terrestrial animals.
- Their proportion in total proteins is slightly lower in fish meat (26-30%) than that in terrestrial animal meat (35-40%) (Table 2.2).
- Their content in fish meat varies with fish species, but is generally higher in pelagic fishes and lower in demersal fishes.

Myofibrillar protein:

- Fish contains a higher proportion of myofibrillar proteins (65%) than terrestrial animals (40%) (Table 2.2).
- Due to higher content of myofibrillar proteins, fish meat is considered an excellent base material for many fabricated functional foods.

Stroma protein:

- In fish, stroma proteins account for about 3% of total proteins in teleosts and for about 10% in elasmobranchs.
- Terrestrial animals contain a very high proportion of stroma proteins contributing to about 17% of the total proteins (Table 2.2).
- This is the reason for the soft texture and easy digestibility of fish meat as compared to terrestrial animal meat.

Table 2.2: Comparison of SPP, MFP and SP

		Type of meat	
		Fish meat	**Terrestrial animal meat**
Percentage of different meat proteins	Sarcoplasmic proteins	26-30	35-40
	Myofibrillar proteins	65	40
	Stroma proteins	3 (In teleosts) 10 (In elasmobranchs)	17

Shellfish protein:

» The muscle tissue of shellfish usually ranges lower than most of fish (22% to 27%).

Crustaceans – 9.4% to 15.3%.
Mollusks - Mussels – 12%
Squids – 9.8%
Gastropods – 9.9%

» The body liquid of oysters is quite rich in organics, albumins, peptides and mucin.

» Crustacean proteins frequently are bound to sugars forming glucoproteins.

Lobster protein contains 2.2% sugars.

Crab protein contains 2.8% sugars.

» Shellfish muscle tissue is characterized by its high content of non-protein nitrogen (NPN).

» Both methionine and threonine amino acids were found to be lower (30% to 40% less) in both crustaceans and molluscan shellfish as compared to most fish, both marine and fresh water species.

» Crustacean muscles show a higher value of acidic amino acid, glutamic acid and aspartic acids and a lower value of valine, isoleucine, threonine and lysine when compared with fish muscle.

Nutritional Importance of Fish Proteins:

1. Essential amino acids:

» With respect to amino acids, fish proteins contain all the eight essential amino acids (histidine, isoleucine, luecine, lysine, methionine, phenylalanine, threonine, valine) and, like milk, eggs and mammalian meat proteins, have very high biological value.

» Cereals are usually low in lysine and/or the sulphur containing amino acids (methionine and cysteine), whereas fish protein is an excellent source of these amino acids.

» Moreover, fish proteins are superior to plant and meat proteins with respect to biological value (BV) and net protein utilization (NPU) (Sankar and Nair, 2001).

Essential amino acids:

- *Out of the 20 amino acids that constitute the proteins, 10 are essential amino acids and the rest 10 are non-essential amino acids.*
- *An amino acid is categorized as an 'essential amino acid' for an organism, if it is so essential for its normal health that it must be ingested by the organism, as it cannot be synthesized by its own body.*
- *Non-essential amino acids are the amino acids which can be synthesised in the body of a living organism.*
- *The 10 essential amino acids are phenylalanine, valine, threonine, tryptophan, isoleucine, methionine, histidine, arginine, leucine and lysine.*
- *The common mnemonic used to remember these EAAs is PVT TIM HALL. Another less commonly used one is MATT VIL PHLy.*
- *However, the number of EAAs found in literature varies from 8 to 11 as described below.*

a) *8 EAAs: Two of the above ten amino acids, i.e., arginine and histidine are considered semi-essential as they are required by children.*

b) *9 EAAs: One of the above ten amino acids, i.e., arginine is considered semi-essential as it is required by children.*

c) *10 EAAs: All the 10 amino acids as given above as found in most literature (PVT TIM HALL)*

11 EAAs: In addition to the above 10 amino acids, tyrosine is considered as an EAA if the diet is deficient in phenylalnine as tyrosine is produced from phenylalanine.

» Therefore, in cereal based diets, a supplement of fish can, therefore, enhance the biological value significantly (Huss, 1998) and help in eradicating protein malnutrition.

Table 2.3: Essential amino acids (%) in various proteins (Source: Braekkan, 1976)

Amino acids	Fish	Milk	Beef	Eggs
Lysine	8.8	8.1	9.3	6.8
Tryptophan	1.0	1.6	1.1	1.9
Histidine	2.0	2.6	3.8	2.2
Phenylalanine	3.9	5.3	4.5	5.4
Leucine	8.4	10.2	8.2	8.4
Isoleucine	6.0	7.2	5.2	7.1
Threonine	4.6	4.4	4.2	5.5
Methionine-cystine	4.0	4.3	2.9	3.3
Valine	6.0	7.6	5.0	8.1

» The amino acid composition of fish varies with species and fishing season.

» The content of several amino acids like arginine, histidine, tryptophan, tyrosine, methionine and phenylalanine is lower during spawning, which means that the value of the fish as food is lower at that time.

Table 2.4: Amino acid composition of the muscle protein of some fishes and shellfishes (g of amino acid /100g protein) (Source: Gopakumar, 1997)

Species → **Amino acids ↓**	**Catla** (*Catla catla*)	**Rohu** (*Labeo rohita*)	**Mrigal** (*Cirrhinus mrigala*)	**Tilapia** (*Tilapia mosambica*)	**Oil sardine** (*Sardinella longiceps*)	**Seer fish** (*Scomberomorus guttatus*)	**Shark** (*Scoliodon spp.*)
Aspartic acid	10.25	9.32	10.24	12.91	10.05	7.16	9.23
Threonine	5.61	4.76	3.99	5.32	4.66	3.66	5.07
Serine	3.31	3.27	3.75	4.05	3.91	2.89	3.18
Glutamic acid	17.42	12.98	14.11	17.05	14.33	14.13	12.91
Proline	1.92	4.13	2.63	4.07	1.93	1.83	3.26
Glycine	9.01	3.84	3.72	6.68	3.77	3.16	6.64
Alanine	5.44	6.49	5.87	7.36	4.88	4.75	5.46
Valine	6.31	4.24	4.87	5.81	4.73	1.01	5.27
Cysteine	0.96	2.06	1.24	0.84	1.63	4.63	0.42
Methionine	2.10	2.05	2.42	2.97	2.80	2.29	1.38
Isoleucine	4.38	5.66	4.18	6.58	3.29	7.08	6.17
Leucine	9.19	8.25	7.97	9.83	5.97	12.78	10.17
Tyrosine	4.32	3.22	3.07	1.74	3.41	1.56	2.74
Phenylalanine	0.00	3.70	3.78	3.10	3.93	4.18	3.31
Histidine	5.25	6.17	3.78	2.53	5.57	3.69	5.23
Lysine	8.30	12.22	13.12	15.76	10.64	9.16	12.83
Arginine	5.82	3.60	5.46	5.62	6.47	6.19	4.78
Tryptophan	1.18	1.28	0.00	0.00	1.89	0.00	0.00

Species → **Amino acids ↓**	**Catfish** (*Arius maculates*)	**Silverbelly** (*Leiognathus linedatus*)	**Prawn** (*Penaeus indicus*)	**Mussel** (*Perna viridis*)	**Cuttlefish** (*Sepia spp.*)	**Lobster** (*Perulus sewelli*)
Aspartic acid	7.73	11.78	10.02	9.56	7.68	6.42
Threonine	3.43	4.28	3.20	4.00	2.83	2.57
Serine	2.80	2.68	3.33	3.63	2.65	2.69
Glutamic acid	12.40	17.03	17.73	14.96	8.99	9.18
Proline	2.07	3.99	7.53	4.31	5.96	5.70
Glycine	3.90	4.92	7.75	9.32	2.44	3.64
Alanine	6.03	6.46	6.86	5.10	3.36	4.36
Valine	4.21	5.89	3.48	4.74	1.36	1.27
Cysteine	0.66	1.35	0.74	2.43	0.65	2.32
Methionine	1.94	2.48	3.87	1.75	1.17	1.63

Isoleucine	7.52	4.90	3.10	5.88	2.57	3.61
Leucine	5.78	7.51	7.24	7.94	5.23	4.79
Tyrosine	1.54	1.63	1.83	4.15	0.49	3.16
Phenylalanine	2.16	5.36	3.33	4.67	0.63	3.16
Histidine	2.42	2.51	2.26	4.46	3.01	1.89
Lysine	7.83	12.25	8.51	10.46	7.24	6.21
Arginine	5.84	6.49	2.05	7.26	5.18	1.60
Tryptophan	0.00	1.02	0.00	0.00	0.00	0.00

2. **Easily digestible**: Due to their easy digestibility, fish proteins are superior to plant and meat proteins with respect to biological value (BV) and net protein utilization (NPU). Therefore, in diets based mainly on cereals, a supplement of fish can raise the biological value significantly and help in eradicating protein malnutrition.

Importance of Proteins in Fish Processing:

1. **Organoleptic quality:** Proteins do not have much influence in the taste and flavour of fish.
2. **Functional properties:** With respect to their functional properties, fish proteins are much better than other plant and animal proteins. Functional properties of proteins are the properties which influence the eating quality of the fish. They include solubility, viscosity, water holding capacity, fat holding capacity, emulsifying capacity, emulsion stability, gelation, foaming etc. The better functional properties of fish proteins as compared to animal proteins is due to its higher content of myofibrillar proteins (65%) than that of animal proteins (40%).
3. **Enzyme activity:** All the enzymes present in fish, like those in all other organisms, are proteins. In living fish, they help in all the metabolic processes. After death, they go on hydrolyzing the fish tissue components. Such hydrolysis of tissue components of fish by its own enzymes is called 'autolysis'. The enzymes associated with autolysis can be categorized into two groups as given below.
   a. **Digestive or gut enzymes:** These enzymes are synthesised within the secretory cells of the digestive system (e.g. those of pancreas, intestine etc.) and are stored within these cells as inactive precursors called 'pro-enzymes' or 'zymogens'. For digestion of food, they are secreted into the gut where they are converted to their active enzyme form. Therefore, they are 'extracellular enzymes'. After death these enzymes hydrolyse the gut wall and hydrolyse the muscle tissue of the belly cavity adjacent to the gut. These enzymes include pepsin, trypsin and chymotrypsin.

b. **Muscle and internal organ enzymes:** These enzymes are synthesised within the cells of the muscles and internal organs (liver, kidney, spleen and gonads) and are stored within these cells confined within their lysosomes. Such confinement does not allow these enzymes to hydrolyse the cytoplasm and other cytoplasmic organelles. Thus, these enzymes are 'intracellular enzymes'. Whenever, the activity of these enzymes is required, the membrane of the lysosomes is ruptured and the enzymes within them are released out on the substrates to be hydrolysed. After death of fish, when oxygen supply to the cells stops, the membranes of the lysosomes get damaged thereby releasing the enzymes into the cytoplasm where they hydrolyse the cellular components such as proteins (most prominent), carbohydrates, fats and nucleic acids. The most important among these enzymes are the proteases such as cathepsins, calpains and collagenase. Among them, the most important is cathepsins which are responsible for the very fast rate of muscle softening of fish after death. It is a group of proteases which hydrolyse most of the muscle proteins. Out of the different types of cathepsins identified (cathepsin A, B, C, D, E, H, L etc.), cathepsin A, B, C, D and L are present in fish muscles. The concentration of cathepsins in fish muscle is more than that in other animal muscles. This is one of the reasons why fish softens and spoils more quickly than other animals. However, the concentration of cathepsins in the muscles is much less than that in the internal organs. The second group of proteases called calpains or calcium activated factor (CAF) hydrolyses the myosin, troponin and Z line proteins leading to postmortem tenderisation of fish muscles. The third group of proteases is collagenase, which hydrolyses the protein collagen of the connective tissue.

4. **Muscle softening:** Besides enzymes being proteins, the most important tissue component they hydrolyse during autolysis is also proteins. Such hydrolysis of proteins of fish tissue by its own enzymes is called 'proteolysis'. As proteins are the most important structural components of fish tissues, their hydrolysis results in gradual softening of its tissues, particularly of the muscle tissues. Autolysis is manifested only as softening of the muscle tissue due to which autolysis and proteolysis are often used synonymously. Extensive muscle softening is indicative of onset of spoilage. That is why; softness of muscle is very often used by consumers in determining the freshness of fish. Such extensive muscle softening may occur, for example, when fish is stored at higher ambient temperature.

'Tenderisation', which is usually done in case of terrestrial animals in order to improve their texture by making the muscles soft, tender and pliable, is not done in case of fishes. As meat of terrestrial animal is tough and difficult to eat even after long cooking, they are subjected to the process of tenderization by keeping them at a cooler temperature after slaughter. During the process the enzymes present in their body hydrolyse the muscle proteins which makes the meat soft and pliable, easy to be eaten. Such tenderization is not done in case of fish because of two reasons as given below.

a. **High enzyme activity of fish enzymes**: The enzymes of fish tissue are about ten times more active than those of the terrestrial animals.

b. **Soft texture of fish muscle:** The muscle tissue of fish has soft texture due to its low content of connective tissue (Collagen and elastin).

Because of these two reasons, fish muscle become very soft in a short period of time from death making further tenderization or softening undesirable.

5. **Nutrient for microbial activity:** Autolysis, though manifest as softening of fish tissue, does not, in any way, lead to direct spoilage of fish as it does not produce any off-odour off-flavour substance that distracts the consumers. However, the hydrolysed end products of autolysis, particularly those produced by hydrolysis of proteins (*e.g.*, the free amino acids) are simpler compounds which become a nutrient-rich medium for microbial activity. The microbes present in the fish tissue grow profusely utilising these simpler compounds resulting in the production of off-odour and off-flavour substances as their metabolic end-products. The accumulation of these off-odour off-flavour substance render the fish spoiled as they distract the consumers.

6. **Protein denaturation:** Fish proteins have lower thermal stability as compared to other plant and animal proteins. As a result, they are more susceptible to denaturation. In fresh fish, denaturation of proteins occurs due to rough handling or even, storage at high ambient temperature, *e.g.*, when fish is kept in sun on the deck. During processing of fish, denaturation can occur due to exposure to high temperature (canning and frying), low temperature (freezing), chemicals (salting) and due to dehydration (drying) as well as acidification (pickling).

7. **Texture of fish:** As proteins are the main structural components of fish body, they are responsible for the its texture. Therefore, any undesirable change in its proteins results in the deterioration of its texture. The two important protein changes responsible for the texture deterioration of fish are given below.

   a. **Protein denaturation:** It refers to the change in the native three-dimensional conformation (*i.e.*, structure) of proteins caused by the alteration of their secondary, tertiary or quaternary structure without affecting the primary structure (*i.e.*, without breaking the amino acid chain).

b. **Protein hydrolysis (Proteolysis):** It refers to the hydrolysis of the proteins resulting in the alteration of their primary structure caused by the breakdown of their amino acid chains.

C. Lipids:

» The lipid or fat content of fish shows the highest variation from 0.1% to 22% with a percentage variation of about 21.9%.

» Lipids which remain as solid or semi-solid at room temperature are called 'fats' and occur predominantly in animals as 'animal fat'.

» Lipids which remain as liquid are called oils and are found mainly in plants as 'vegetable oil'.

» However, lipids present in fish are also largely oils as they remain as liquid at room temperature after their extraction form fish. Therefore, fish fat is usually called 'fish oil'.

» The crude fat in fish and fish products can be determined by extracting fat from a dried sample with anhydrous ethyl ether or petroleum ether. This method is referred to as Soxhlet method.

Grouping of fishes based on fat content:

Based on the average content of fat in the flesh, fishes are grouped into three categories as follows (Table 2.5).

1. **Lean fishes:** In these fishes, the fat content of flesh is less than 2%.
2. **Medium-fatty fishes (Semi-fatty fishes):** These fishes contain fat between 2% and 5% in their flesh.
3. **Fatty fishes:** Their fat content of flesh exceeds 5%.

Table 2.5: Categories of fishes based on fat content

Lean fishes			Medium fatty fishes	Fatty fishes
Pink perch	Ray	Catla	Tuna	Oil sardine
Croaker	Shark	Rohu	Pomfret	Mackerel
Lizard fish	Silver belly	Mrigal	Common carp	Grey mullet
Bombay duck	Snapper	Cat fish	Milk fish	Eel
Caranx	Sole	Murrel	Mullet	Herring
Jew fish	White bait	Tilapia	Pearl spot	Pilchard
Cod	Haddock	Flounder	Mussel	Salmon
Hake	Perch	Pollack		Trout
Prawn	Sole	Whiting		
	Lobster	Squid		

Grouping of fishes based on fat and protein content:

- Based on the average content of fat and proteins in the flesh, fishes are grouped into five categories as follows (Figure 2.1).

a. **Low oil, high protein fishes:** These fishes contain less than 5% fat and 15-20% proteins.

b. **Medium oil, high protein fishes:** These fishes contain 5-15% fat and 15-20% proteins.

c. **High oil, low protein fishes:** These fishes contain more than 15% fat and less than 15% proteins.

d. **Low oil, very high protein fishes:** These fishes contain less than 5% fat and more than 20% proteins.

e. **Low oil and low protein fishes:** These fishes contain less than 5% fat and less than 15% proteins.

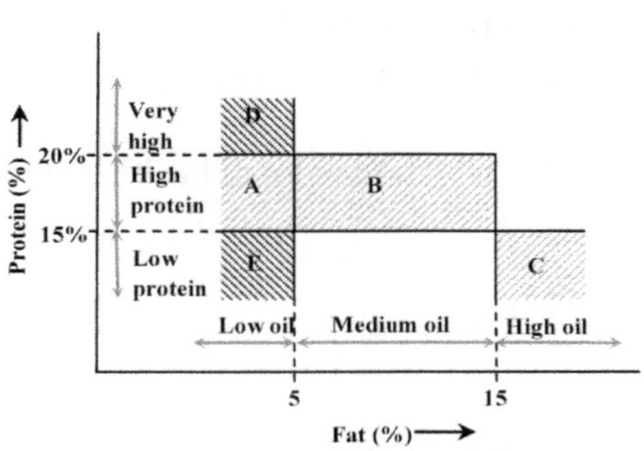

A: Low oil, high protein fishes (Fat: <5%, Protein: 15 to 20%)

B. Medium oil, high protein fishes (Fat: 5 to15%, Protein: 15 to 20%)

C. High oil, low protein fishes (Fat: >15%, Protein: <15%)

D. Low oil, very high protein fishes (Fat: <5%, Protein: >20%)

E. Low oil & low protein fishes (Fat: <5%, Protein: 15%)

Figure 2.1: Grouping of fish based on fat and protein content

COMPONENTS OF FISH LIPIDS

» Among the different types of lipids, the following seven components are very important in fish.

1. Fatty acids
2. Fats
3. Phospholipids
4. Cholesterol
5. Hydrocarbons
6. Fat soluble vitamins
7. Pigments

1. Fatty acids:

» Fatty acids in their free form are present in the fish tissue only in traces. The bulk of it, however, is present in its esterified form as fats and phospholipids.

» Usually the total fatty acids present in the tissues (both free and esterified forms as fats and phospholipids) are taken into consideration while determining the quality of tissue lipids.

» In this regard, it is important to note that, the fatty acids present in terrestrial animals rarely contain more than two double bonds and are therefore mostly saturated.

» In contrast, fish oils contain upto 40% long chain (C14 to C22) fatty acids which are highly unsaturated containing three to six double bonds. Therefore, these fatty acids are called 'polyunsaturated fatty acids (PUFA)' *i.e.* containing more than two double bonds.

» As the number of double bonds is very high (usually three to six) they are called 'highly unsaturated fatty acids' (HUFA). However, the term HUFA is not in common usage.

» Moreover, most of these fatty acids are ω-3 PUFA which have lot of health benefits. Thus, fish is said to be a rich source of ω-3 PUFA.

» In contrast, vegetable oil is also rich in PUFA, but most of them are ω-6 PUFA which do not have much health benefit.

» Marine fishes contain more PUFA with 4 to 6 double bonds (about 88%) than freshwater fishes (about 70%).

Fatty acid composition of fish lipids:

» The important fatty acids present in fish lipids are given below.

Carbon number	Common name	IUPAC name
Saturated fatty acids		
C8:0	Caprilic acid	Octanoic acid
C10:0	Capric acid	Decanoic acid
C12:0	Lauric acid	Dodecanoic acid
C14:0	Myristic acid	Tetradecanoic acid
C16:0	Palmitic acid	Hexadecanoic acid
C18:0	Stearic acid	Octadecanoic acid
C20:0	Arachidic acid	Eicosanoic acid
C22:0	Behenic acid	Docosanoic acid
C24:0	Lignoceric acid	Tetracosanoic acid
Monounsaturated fatty acids		
C14:1, ω-5	Myristoleic acid	*Cis*-9-tetradecenoic acid
C16:1, ω-7	Palmitoleic acid (Zoomaric acid or Physetoleic acid)	*Cis*-9-hexadecenoic acid
C18:1, ω-9	Oleic acid	*Cis*-9-octadecenoic acid
C18:1, ω-7	Vaccenic acid	*Cis*-11-octadecenoic acid
C20: 1, ω-9	Gondoic acid	*Cis*-11-eicosenoic acid
C20: 1, ω-11	Gadoleic acid	*Cis*-9-eicosenoic acid
C22:1, ω-9	Erucic acid	*Cis*-13-docosenoic acid
C22:1, ω -11	Cetoleic acid	*Cis*-11-docosenoic acid
C24:1, ω-9	Nervonic acid	*Cis*-15-tetracosenoic acid
Polyunsaturated fatty acids (PUFA)		
C18: 2, ω-6	Linoleic acid	All-*cis*-9,12-octadecadienoic acid
C18: 3, ω-3	α-Linolenic acid (ALA)*	All-*cis*-9,12,15-octadecatrienoic acid
C20: 4, ω-6	Arachidonic acid	All-*cis*-5,8,11,14-eicosatetraenoic acid
C20: 5, ω-3 (EPA)	Timnodonic acid	All-*cis*-5,8,11,14,17-eicosapentaenoic acid
C22: 5, ω-3 (DPA)	Clupanodonic acid	All-*cis*-7,10,13,16,19-docosapentaenoic acid
C22: 6, ω-3 (DHA)	Cervonic acid	All-*cis*-4,7,10,13,16,19-docosahexaenoic acid

* γ-Linolenic acid (GLA) : C18:3 ω-6

2. Fats:

- Fats are the storage form of lipids and are therefore called 'depot fats'.
- Among the naturally occurring fats, usually animal fats are saturated fats as they are mostly made of saturated fatty acids. Therefore, they are mostly present in solid form and are called fats, *e.g.*, animal fats.
- In contrast, plant fats are unsaturated fats as they are mostly made of unsaturated fatty acids. Hence, they are mostly present in liquid form and are called oils, *e.g.*, vegetable oils.
- However, though fish is an animal, its fats are mostly unsaturated as they are made of unsaturated fatty acids. So, they are mostly present in liquid form and are called '***fish oils***'.

Distribution of fats in the fish body:

- Usually, fat is deposited in the fish body in three places viz. a) Livers, b) Muscle and c) Adipose tissues.
- The distribution of fats in the body of fish varies among species, based on which they are categories into the following three groups.

a. Lean fishes:

- In these species, the fats are stored only in their livers.
- Such fishes include demersal (bottom-dwelling) fishes like lizard fish, pink perch, croaker, cod, saithe and hake.

b. Fatty fishes:

- They store fats in special cells called adipocytes (adipose cells or fat cells) distributed in different parts of the body other than liver.
- Fats are usually stored as fat droplets in the cytoplasm of the adipocytes almost filling the cell volume.
- The adipocytes have a phospholipid membrane surrounded by a rather weak collagen network. These cells are found in large number under the skin (subcutaneous depot) and in the abdominal cavity (visceral depot).
- In most fishes, the depot fats include only fats (triglycerides or triacylglycerol).

» In few fishes, the depot fats are waxes.

» Fatty fishes include pelagic fishes like oil sardine, mackerel, herring, sturgeon, salmon and sprat.

» In these fishes fat is stored in the adipocytes located in the skeletal muscle tissue between myocommata and the muscle cells; in the subcutaneous layer; in the connective tissue near the bones (along the vertebral column and at the base of the fins and head bones) or in the abdominal cavity in the shape of fatty deposits (adipose tissues) around the viscera.

c. Medium fatty fishes (Semi-fatty fishes):

Adipocytes:

Adipose, fat, cytes: cells. 'Cytes' is used as a suffix to mean different types of cells

1. Osteocytes: Bone cells
2. Chondrocytes: Cartilage cells
3. Myocytes: Muscle cells (Cardiac myocytes: cardiac muscle cells; skeletal myocytes: Skeletal muscle cells)
4. Erythrocytes: Red blood cells (RBC)

» They store fats in limited parts of their body tissues only or in lower quantities than typical fatty fishes.

» They are categorised as demersal fish though they spend only some time at the bottom. Such fishes include barracuda, mullet and shark.

Grouping of fishes into lean, semi-fatty or fatty as mentioned above, however, is complicated as it needs determination of the fat content of different tissues. Therefore, a more practical way is to determine the fat content of the muscle tissue alone and group them as lean (<2%), medium fatty (2-5%) or fatty (>5%) species as described previously.

3. Phospholipids:

» Phospholipids present in fish, like those in other organisms, are the important structural component of biological membranes (*i.e.*, unit membrane) and are therefore called 'membrane lipids' or 'structural lipids'.

- The important phospholipids present in fish are given below.
    a. Phosphatidyl choline (Lecithin)
    b. Phosphatidyl ethanolamine (Cephalin)
    c. Phosphatidyl serine
    d. Phosphatidyl inositol
    e. Lysophosphatidyl choline (Lysolecithin)
    f. Lysophosphatidyl ethanolamine(Lysocephalin)
    g. Diphosphatidyl glycerol (Cardiolipin)
    h. Phosphatidal ethanolamine (Plasmalogens)
    i. Sphingophosphatidyl choline (Sphingomyelin)

[Note the difference between ***phosphotidyl*** *in case of* ***cephalin*** *and* ***phosphotidal*** *in case of plasmalogen]*

4. Cholesterol:

- Cholesterol is present in all fishes as it is an integral component of the cell membrane of all animal cells.
- Plant cells and bacteria cells mostly do not have cholesterol. Therefore, plants and plant products have negligible quantity of cholesterol. It is interesting to note that vegetable oils do not have cholesterol at all or have in traces.
- Cholesterol is present in fish tissue in two forms i.e. a) Free cholesterol and b) Cholestryl ester in which cholesterol is esterified to a fatty acid.
- The total content of cholesterol in fish issue varies from 45 mg/100 g to 150 mg/100 g of which about 23 mg/100 g to 92 mg/100 g is free cholesterol.
- Generally, the cholesterol content of fin fishes, fatty or lean, is low whereas it is high in most of the shellfishes (shrimp, prawn, squid, octopus).
- Freshwater fishes contain more cholesterol than marine fishes.
- Based on the content of cholesterol and fat, fishes can be grouped into three categories.

a. Low-cholesterol low-fat species

b. Low-cholesterol high-fat species

c. High-cholesterol low-fat species

» However, there is no species which has 'high cholesterol and high fat' content.

5. Hydrocarbon:

» The only hydrocarbon found in fish is ***squalene*** which is present only in the liver of shark.

6. Fat soluble vitamins:

» They include vitamins A, D, E and K.

» Cod liver oil and shark liver oil are rich sources of these vitamins.

7. Pigments:

» The important pigments found in fishes are given below.

a. Haemoglobin:

- It is found in the red blood cells present in the blood of fishes.

b. Myoglobin:

- It is found in the muscle cells of fishes.

c. Haemocyanin:

- It is present in the haemolymph (blood) of invertebrates Ex. Shrimp, crab.
- It is not bound to blood cells. Instead, it is suspended directly in the haemolymph.
- It gives blue coloration to the blood.

d. Chlorophyll:

- It is present in sardine which imparts greenish shade to its oil.

e. Lutein, astacene, astaxanthine and tetraxanthine:

- These are present in several fishes imparting pale yellow to reddish colour to them.
- The red colour meat of salmon is due to presence of astaxanthine in its muscle tissues.
- Crustaceans contain astacene and astaxanthine in their muscle tissue bound to proteins resulting in a blue or blue-grey colour.
- When they are heated by boiling or frying, the bond between the proteins and astaxanthine is broken as the proteins get denatured. The free astaxanthine, which has orange-red colour, imparted the orange-red colour to the fried or boiled prawn.

Nutritional Importance of Fish Lipids:

1. Control of cardiovascular diseases (CVDs):

» Fish fat is beneficial to human health as it is rich in ω-3 PUFA. Most important among which are eicosapentaenoic acid (EPA: C20: 5, ω-3) and docosahexaenoic acid (DHA: C22: 6, ω-3).

» The ω-3 fatty acids in fish arise from the phytoplankton in the food chain that it eats.

» These ω-3 PUFA have hypocholesteromic effects in that they reduce the cholesterol level in the body thereby reducing the chance of cardiovascular diseases (CVDs).

» The main cause of CVDs is the high level of cholesterol in the body which results in the formation of atherosclerotic plaques (atheroma) inside the lumen of the blood vessels. As a result the inner diameter of the blood vessels decreases resulting in increased blood pressure. High level of blood cholesterol also leads to hardening of the blood vessels thereby reducing their elastic nature to contract and expand. All these lead to increased blood pressure. Moreover, the formation of atheroma and its subsequent growth leads to blockage of blood vessels.

» Cholesterol remains in the body in two forms such as high density lipoprotein cholesterol (HDL-cholesterol) and low density lipoprotein cholesterol (LDL-cholesterol). HDL-cholesterol reduces the chance of atheroma formation due to which it is called 'good cholesterol'. In contrast, LDL-cholesterol increases the chance of atheroma formation and is therefore, called 'bad cholesterol'.

» It is important to note that, though high level of cholesterol in our body is the cause of CVDs, cholesterol is also required by our body for its normal metabolism.

» High intake of saturated fats and trans-fats increases the cholesterol level in the body while high intake of unsaturated fats decreases it. Therefore, in order to overcome CVDs, one should avoid taking saturated fats and take more of unsaturated fats. Preference should be given to those fats which are rich in PUFA, particularly those rich in ω-3 PUFAs.

» Among the PUFAs, the two most important which are remarkable for their effect in reducing cholesterol level in the body are EPA and DHA. Fish oil is a rich source of these two PUFAs.

» Thus, in order to overcome CVDs, it is advisable to

- Avoid saturated fats,
- Avoid trans-fats,
- Take more of unsaturated fats rich in PUFA, particularly those rich in ω-3 PUFA such as fish oils, and
- Control the level of dietary cholesterol so that it remains in the normal range and does not become excessively high.

» Hence, one should avoid foods which are rich in cholesterol such as egg yolk, animal fats (e.g., butter) and shellfishes.

» Thus, intake of recommended amounts of DHA and EPA in the form of dietary fish or fish oil supplements lowers cholesterol level and reduces the risk of CVDs.

» In contrast to freshwater fishes, marine fishes have better nutritional value with respect to PUFA content as they contain higher percentage of PUFA with 4-6 double bonds (about 88%) against about 70% in freshwater fishes.

» In contrast to fish fats, which are rich in ω-3 PUFA, animal fat is mostly saturated for which its consumption may help elevate CVDs as it increases cholesterol level in the body.

» Vegetable oils are rich in PUFA, but these are mostly of ω-6 type for which they are not as effective as fish oil in controlling CVDs.

2. **Essential fatty acids:**

   » Fish oil also contains good amount of 'essential fatty acids' which are sometimes known as vitamin F.

   » Only two fatty acids are known to be essential for humans which include linoleic acid (an ω-6 fatty acid) and α-linolenic acid (alpha linolenic acid, ALA, an ω-3 fatty acid)

3. **Brain development and function:**

   » DHA, which is found in fish oil in abundant quantity, is essential for the growth and functional development of the brain in infants.

   » It is also required for maintenance of normal brain function in adults. The inclusion of plentiful DHA in the diet improves learning ability,

4. **Control of other diseases:**

   » DHA can also reduce the risk of several diseases associated with the nervous system such as Alzheimer's disease and Parkinson's disease.

   » Many other diseases such as rheumatoid arthritis and even cancer appear to be reduced by eating fish oil.

5. **Low cholesterol:**

   » Excluding shellfishes, most of the fin fishes, lean or fatty, have low cholesterol level due to which they can be consumed safely without increase the risk of CVDs.

6. **Hydrocarbons:**

   » The hydrocarbon, squalene, found in shark liver has several pharmaceutical and industrial uses.

7. **Fat soluble vitamins:**

   » The fat soluble vitamins such as vitamins A, D, E and K present in fish liver oil, such as cod liver and shark liver oil, are required for different metabolic processes.

Importance of Lipids in Fish Processing:

1. Organoleptic quality:

- » Fat content is one of the main criteria for judging the organoleptic quality of any fish species.
- » Its content in the fish tissue influences the sensation of the cooked flesh in the mouth of the eater.
- » A well-fed fatty fish tastes very smooth and succulent (juicy), although the sensation is produced by oil, not water.
- » After spawning, when the fat is at its lowest level, the main sensation is of dryness or fibrousness and the taste is disappointing.

2. Lipid deterioration:

- » Although lipids are responsible for the good taste of fish, they undergo deterioration during processing and storage leading to the loss of quality of fish and fish products.
- » Fish lipids undergo the following four types of deterioration.

  a. Hydrolytic rancidity (due to hydrolysis of lipids)

  b. Oxidative rancidity (due to lipid autoxidation)

  c. Flavour reversion (change of flavour)

  d. Polymerisation (formation of lipid polymers)

a. Hydrolytic rancidity:

- » It is the rancidity resulting from the accumulation of free fatty acids (FFA) produced by hydrolysis of tissue fats by lipolytic body enzymes (lipases) during postmortem autolysis and by lipolytic bacteria during postmortem microbial putrefaction of fish tissue.

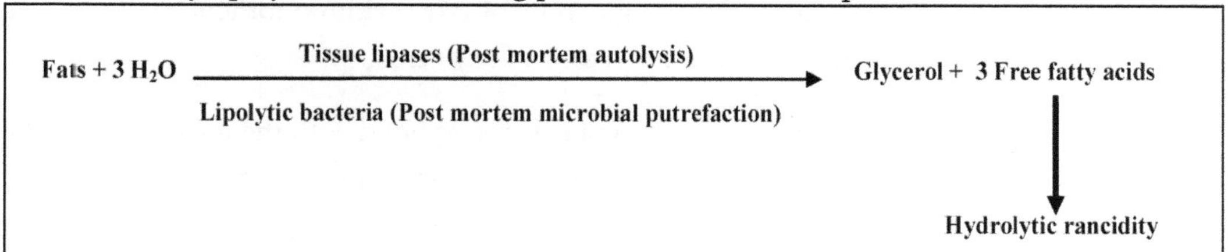

Figure 2.2: Hydrolytic rancidity

b. Oxidative rancidity:

- » The oxidative rancidity develops in fish after death or in fishery products during processing and storage due to accumulation of certain end products of lipid autoxidation.
- » Lipid autoxidation is the process of addition of atmospheric oxygen to the unsaturated fatty acids of the lipids at their of points of unsaturation, *i.e.*, to the double bonds, and subsequent reactions leading to the production and accumulation of end products which impart a typical rancid flavour called 'oxidative rancidity'.
- » As fish lipids are rich in unsaturated fatty acids, they are highly susceptible to autoxidation when exposed to atmospheric oxygen leading to development of rancid flavour.

c. Flavour reversion:

- » Lipids containing the fatty acid 'linolenic acid' undergo a type of deterioration called 'flavour reversion' when exposed to air.
- » Flavour reversion is the development of a particular type of oxidised off-flavour, which develops at comparatively low levels of lipid oxidation.
- » This off-flavour is described as "greasy', "beany", "fishy" or "painty". It is produced with considerably less oxygen than auto-oxidation.
- » As fish oil contains linolenic acid, though at a small level of less than 2% of total lipids, it is susceptible to flavour reversion.

d. Polymerisation:

- » It is a type of lipid deterioration, which occurs only when lipids are exposed to high temperature processing such as canning and frying.
- » In the process of polymerisation, single units of fatty acid monomers combine in a chain manner to form a chain or polymer of fatty acids.
- » The unsaturated fatty acid of one fat molecule forms bond with the unsaturated fatty acid of another fat molecule at the point of unsaturation of both the fatty acids (i.e., points where there are double bonds) (Figure 2.3).
- » It is particularly prominent in frying oil in which, due to polymerization, the size

of the lipid molecules increases leading to increase in the viscosity of the oil which greatly enhances its foaming.

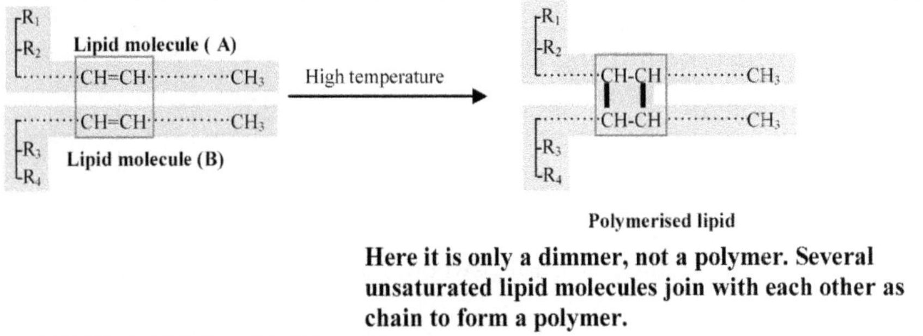

Figure 2.3: Polymerisation of lipids (A simplified representation)

D. Carbohydrates:

- The carbohydrate contains carbon (C), hydrogen (H) and oxygen (O) and can be represented by the general formula $C_x(H_2O)_y$. In the carbohydrate such as sugar, starch, cellulose etc, H and O are always in the proportion in which they occur in water i.e. 2:1.
- Fish contains far less carbohydrate than foods of plant origin. The small amounts that are present can be ignored so far as nutritive value is concerned.
- The major carbohydrate in fish muscle is known as glycogen or, animal starch.
- A fish muscle may contain 0.1 to 1.0% carbohydrate.
- Mollusks have a higher content, typically in the range of 1% to 7%.
- This quantity can vary seasonally, and in any case, declines rapidly during the stress and struggle associated with capture, and post mortem.

E. Energy:

- Calories are needed to provide energy for proper functioning of the body. The number of calories in a food depends on the amount of energy the food provides.

Food component	Energy density	
	kJ/g	kcal/g
Fat	37	9
Ethanol (drinking alcohol)	29	7
Proteins	17	4
Carbohydrates	17	4
Organic acids	13	3
Polyols (sugar alcohols, sweeteners)	10	2.4
Fiber	8	2

» Fish and fish products are a cheap source of protein for body building and to provide energy.

» The amount of energy used by a normal man can be measured in the form of heat produced and also by measuring the oxygen used and the carbon dioxide formed or, by measuring the energy content of the food eaten.

F. Minerals:

» The mineral content in fish tissue varies very little from 0.8% to 2% with a percentage variation of about 1.2%.

» The total mineral content of fish is expressed as 'total ash content' which is left over after complete combustion of fish tissue.

» The important minerals that are present in the fish tissue are given below.

Minerals	mg/100 g fish tissue*	Minerals	mg/100 g fish tissue*
Sodium	72	Chlorine	197
Potassium	278	Silicon	4
Calcium	79	Manganese	0.82
Magnesium	38	Zinc	0.96
Phosphorous	190	Copper	0.20
Sulphur	191	Arsenic	0.37
Iron	1.55	Iodine	0.15
* The values given are average of the mineral contents of several species of fishes.			

» Although 15 minerals are required for human body, about 5 to 6 are necessary for normal life. Minerals have 3 main functions.

- As constituents of the bones and teeth. These include calcium, phosphorus and magnesium.
- As soluble salt, which help to control the composition of body fluids and cells. These include sodium and chlorine in the fluids outside the cells and potassium, magnesium and phosphorus inside the cells.
- As essential adjuncts to many enzymes and other proteins viz. haemoglobin, that are necessary for the release and utilization of energy. Iron (Fe) and phosphorus and most of the other elements act in this way.

Nutritional Importance of Minerals in Fish:

» Minerals are inorganic substances required by all organisms. Therefore, all of them require an adequate supply of these minerals.

» The requirement for minerals varies among species and individuals.

» However, all of these minerals are required to be present in certain specific forms for their use by an organism. For example, phosphorous (P) in elemental form is extremely toxic, while in the form of phosphate ion (PO_4^{3-}) is an essential mineral.

» In fish, most of the minerals are present in their ionic forms either as anions (Cl^-) or cations (Na^+, K^+, Ca^{++}, Mg^{++}, Fe^{+++}, Mn^{++}, Zn^{++}, Cu^{++}).

» Few of them are also present in their organically bound forms, being bonded in an organic compound (P, S, Si, As, I).

» Although fish cannot be the only source of an essential mineral, it provides a well-balanced supply of minerals in a readily usable form.

» The nutritional importance of the minerals present in fish is given below.

1. Sodium, potassium, calcium, magnesium and chlorine:

» These maintain the ionic balance between the extracellular and intracellular fluid.

» Sodium (Na^+) and calcium (Ca^{++}) ions are maintained at higher concentrations in the extracellular fluid while potassium (K^+) and magnesium (Mg^{++}) ions are maintained at higher concentrations in the intracellular fluid.

» For different activities, these ions move across the cell membranes.

» Chlorine, present as negatively charged chloride ions (Cl^-), maintains the electrical neutrality by passing across the cell membrane against the movement of these positively charged ions (Na^+, K^+, Ca^{++} and Mg^{++}).

» Since high levels of sodium in the body causes high blood pressure, one should consume mainly foods naturally low in sodium. Fresh fish is one of the best choices for curbing sodium intake since all fish varieties are low in sodium.

» About 99% of the calcium present in human body is found in bones and teeth while the rest 1% is found in cells, blood, and other body fluids.

» The most important function of calcium in human body is formation and maintenance of bone and teeth.

» Therefore, dietary intake of calcium is essential for the normal growth of children as well as for women during their reproductive years to minimize subsequent osteoporosis. While milk is the major source of calcium, fish can also be a useful source.

» Calcium and phosphorus in the skeleton comprise 70% to 80% of the minerals.

» While, fish meat alone, excluding its bones, is not a good source of calcium, fish can be a very good source in the following two conditions.

  a. When very small fishes are eaten along with their bones as their bones are not hard enough to cause any oral injury or throat entanglement.

  b. When canned fish with lot of bones is eaten in which, due to the severe thermal processing at a very high temperature for a very long time, the bones become very soft to be eaten along with the meat.

2. Magnesuim and manganese:

» These are important as co-factors for several enzymatic reactions in the human body.

3. Phosphorous:

» About 85% of the phosphorous present in the body is found in the bones and teeth.

» Phosphorous, together with calcium, provides structure and strength to bones and teeth.

» Calcium and phosphorous in the bone and teeth comprise 70-80% of the minerals in the body.

4. Sulphur:

- Sulphur is present in the body mainly as the component of the sulphur-containing amino acids namely methionine and cysteine.
- Formation of disulphide bonds between pairs of these amino acids in a polypeptide chain is one of the factors which give proteins their three-dimensional structural conformation.

5. Iron:

- Iron is a component of heme-proteins like haemoglobin and myoglobin.
- Haemoglobin is found in the erythrocytes (red blood cells, RBCs) present in the blood which binds oxygen and transport it to different parts of body via blood circulation.
- Myoglobin is present in the muscle cells which help in the storage of oxygen in the muscle.
- Two-third of the iron present in the body is found in the haemoglobin.
- Iron is also a component of non-heme proteins like transferrin and ferritin.
- Transferrin is an iron-binding blood plasma protein that controls the level of free iron in biological fluids.
- Ferritin is an intracellular protein that stores and release iron in a controlled fashion.
- Deficiency of iron in the body is one of the reasons of anaemia, the lowered ability of blood to carry oxygen.

6. Silicon:

- It is mainly required for the proper development and functioning of bones, cartilages, connective tissues, skin, hair and nails.

7. Zinc:

- It mainly helps in proper functioning of the immune system.

8. Copper:

- Copper has a major role in the synthesis of haemoglobin.

- A copper-protein complex known as caeruloplasmin (ferrodoxin I), which contains six copper atoms per molecule, plays an important role in the transport of iron in transferrin for haemoglobin synthesis.

9. Arsenic:

- Though it is essential for the body, its biological functions in not clearly known.
- It is more known for its toxicity (arsenic toxicity) caused through arsenic-rich drinking water.

10. Iodine:

- Iodine is required for the proper functioning of the thyroid gland which produces two hormones, triiodthyronine or T3 and thyroxine or T4.
- Iodine is an important component of these two hormones. Therefore, iodine is essential to produce them.
- These hormones have extensive roles in the body.
- They act on every type of cells raising their metabolic rate.
- Deficiency of iodine leads to a disease called 'goitre' which is still found in many inland areas where the availability of marine fish is poor, because marine fish is a good source of iodine.
- To prevent goiter, it is recommended to take iodised salt instead of normal evaporated salt.

Importance of minerals in fish on its processing:

- The minerals present in fish have little role in its processing. However, during processing the mineral content, particularly the content of sodium (Na^+) and chlorine (Cl^-) increases. This is because, salt (NaCl) is added in several processing methods as a preservative (in salting preservation), for taste (in most of the processed products) and for gel formation (in gelled products like kamaboko, fish sausage etc.)

G. Vitamins:

- Different types of vitamins are present in fish in very small quantities. Even, the total quantity of vitamins in fish is so small that it is not taken into account in analysis of proximate composition.

» It is important to note that fish contains all the 13 vitamins, though their quantities are species specific and can furthermore, vary with season and among different parts of the fish body.

» Depending on solubility, vitamins present in fish are categorized into two groups i.e.

1. Fat soluble vitamins: Vitamins A, D, E and K
2. Water soluble vitamins: Vitamin B complex (eight B vitamins) and vitamin C.

Table 2.6: The quantity of vitamins in the muscle tissue of fatty fishes

	Vitamins	Quantity of vitamin (mg/g)		Vitamins	Quantity of vitamin (mg/g)
1	Vitamin A	0.025	8	Vitamin B_5 (Pantothenic acid)	0.50
2	Vitamin D	0.015	9	Vitamin B_6 (Pyridoxine)	0.50
3	Vitamin E	0.012	10	Vitamin B_7 (Biotin)	0.005
4	Vitamin K	Traces	11	Vitamin B_9 (Folic acid)	0.08
5	Vitamin B_1 (Thiamine)	0.05	12	Vitamin B_{12} (Cyanocobalamin)	0.001
6	Vitamin B_2 (Riboflavin)	0.12	13	Vitamin C	3.0
7	Vitamin B_3 (Niacin)	3.0			

» Fish is an important source of the fat soluble vitamins A, D, E and K, as it often contains them in higher concentrations than in land animals.

» As they are fat soluble, they remain in those parts of the fish body where fat gets deposited.

» Therefore, in fatty fishes in which the fat gets deposited in the muscle tissues, these vitamins remain in the muscle tissue.

» In lean fishes where the fat gets deposited in the livers, they remain in the livers.

» When oil is extracted from fish, these vitamins get extracted into the oil.

- Usually liver oils (cod liver oil and shark liver oil) are rich in vitamins A and D. Therefore, these liver oils are used for pharmaceutical purpose to overcome the deficiency of vitamins A and D.
- Among the water soluble vitamins, fish is a poor source of vitamin C. Like all other animals, it is concentrated in the adrenal cortex which is not usually eaten.
- In contrast, fish contains modest amounts of all the eight vitamins of the vitamin B complex.
- Fish liver is a rich source of these vitamins.
- Another important source is fish roes.
- In several species of fishes, the content of vitamin B1 (thiamine) decreases drastically after their death as they contain the enzyme thiaminase which destroys the thiamine present in their body. Cooking of fish destroys this enzyme.
- Therefore, animals fed with raw fish alone face thiamine deficiency diseases.

Nutritional Importance of Vitamins in Fish:

The nutritional importance of the vitamins present in fish is given below.

1. Vitamin A:

- It is required for normal vision and seeing in dim light.
- Its deficiency causes xerophthalmia (dry eyes) and night blindness (nyctalopia).
- Liver oils of cod and shark are rich sources of vitamin A.
- Muscle tissue of fatty fishes contains modest amounts of Vitamin A in contrast to those of lean fishes.

2. Vitamin D:

- It is required for proper development of bones as it helps in the absorption of calcium from the food into the body.
- Its deficiency causes osteomalacia in adults and rickets in children, both of which result in defective bone formation.

- Liver oils of cod and shark are rich sources of vitamin D. Muscle tissue of fatty fishes contain average amounts of Vitamin D in contrast to those of lean fishes.
- Though liver oils of both cod and shark contain both vitamin A and D, shark liver is an excellent source of vitamin A while cod liver oil is rich in vitamin D.

(Remember: 'sh**A**rk' for vitamin A and 'co**D**' for vitamin D.)

3. Vitamin E:

- It is the most common natural antioxidant.
- It prevents lipid auto-oxidation by acting as a free radical scavenger.
- It protects tissue lipids from auto-oxidation in spite of their contact with oxygen that enter into the body via respiration. In particular, it protects the membrane lipids from oxidative damage as it gets incorporated into the lipids of the membrane because it is fat soluble.
- Thus, lipids present in fish tissue, though rich in unsaturated fatty acids which are highly susceptible to auto-oxidation, do not undergo oxidative deterioration due to presence of vitamin E.
- Deficiency of vitamin E causes neurological problems due to poor nerve conduction. Although it occurs mainly in plant products, animals also contain it and among animals, fish is a good source, particularly fish liver oil is a very rich source.

4. Vitamin K:

- It helps in clotting of blood, thereby preventing excessive bleeding from cuts.
- Deficiency of vitamin K usually does not occur in adults, as it is synthesised by the normal intestinal bacterial flora and absorbed into the body.
- However, in newborn infants, as the intestinal tracks are sterile and it takes weeks for the formation and stabilization of a normal bacterial flora. For this, they are given a single dose of vitamin K.
- Deficiency occurs in adults due to continuous use of antibiotics which destroy the normal intestinal flora. Though fish contains vitamin K, green leafy vegetables are its main source.

5. Vitamin B complex:

» All the members of the vitamin B complex, such as B_1, B_2, B_3, B_5, B_6, B_7, B_9 and B12 are coenzymes which help enzymes in their work as catalysts in metabolism.

» The specific nutritional role of these vitamins is given below.

i. **Vitamin B_1 (Thiamine)**: It helps all the cells of the body to change carbohydrates into energy. Its deficiency causes the disease 'beri beri' characterised by weight loss, body weakness and pain, brain damage, irregular heart rate, heart failure, and death if left untreated.

ii. **Vitamin B_2 (Riboflavine)**: It is present in flavoproteins in the forms of flavine mononucleotide (FMN) or flavine adenine dinucleotide (FAD). Both FMN and FAD are electron carriers in electron transport chain (ETC). Deficiency of vitamin B_2 results in stamatitis including painful red tongue with sore throat, chapped and fissured lips and inflammation of the corners of the mouth.

iii. **Vitamin B_3 (Niacin)**:It is a precursor of the coenzymes nicotinamide adenine dinucleotide (NAD) and nicotinamide adenine dinucleotide phospahate (NADP) both of which are coenzymes for many dehydrogenases participating in many hydrogen transfer processes. Deficiency of vitamin B_3 results in the disease pellagra, which is characterized mainly by diarrhoea, dermatitis, and dementia (the three D's).

iv. **Vitamin B_5 (Pantothenic acid)**: It helps in the synthesis of coenzyme A (CoA) which is used for the synthesis of many other metabolic compounds such as acetyl CoA and succinyl CoA during energy production. Deficiency of vitamin B_5 results mainly in irritability, fatigue, and apathy due to impaired energy production.

v. **Vitamin B_6 (Pyridoxine)**: It mainly helps in the metabolisms of macronutrients such as glucose, amino acids and lipids as well as in the synthesis of haemoglobin. Deficiency mainly causes roughening of skin and sore tongue.

vi. **vi. Vitamin B_7 (Biotin)**: It is a coenzyme for several types of carboxylase enzymes which are responsible for transfer of CO_2 in metabolic processes. Therefore, it is essential for the metabolism of glucose, amino acids and fatty acids. Deficiency causes hair loss and dermatitis in the form of a scaly, red rash around the eyes, nose, mouth, and genital area.

vii. **Vitamin B_9 (Folic acid)**: It helps in the synthesis of the nucleic acids (DNA and RNA) as it is required for the synthesis of the nucleosides such as adenosine, guanosine and thymidine. Therefore, it is required for the production and maintenance of new cells.

Its deficiency limits cell division thereby leading to megaloblastic anemia which is characterised by large immature RBCs. As a result, the oxygen-carrying capacity of blood becomes low which causes fatigue and weakness. It deficiency in pregnant women results in neural tube defect (NTD), in which the neural tube in the developing embryo does not close completely. In infants and children, its deficiency slows growth rate.

viii. Vitamin B_{12} (Cyanocobalamin): It is required for the metabolism of nucleic acids, amino acids and fatty acids. Its deficiency results in pernicious anaemia which is a type of megaloblastic anaemia characterised by decreased gastric production of HCl and deficiency of intrinsic factor (IF), a substance normally secreted by the parietal cells of the gastric mucosa that is essential for vitamin B_{12} absorption in the ileum.

6. Vitamin C:

» It is a coenzyme for several enzymes most of which are associated with the synthesis of collagen.

» Its deficiency results in a disease called scurvy whose symptoms are weakness, bleeding gums, joint pain and reddish spots on the skin.

Importance of vitamins in fish on its processing:

» Few of the vitamins present in fish have certain role in its processing as given below.

1. Vitamin E which is an antioxidant, controls development of rancidity caused by autoxidation of the tissue lipid.

2. Vitamin C also controls rancidity by acting as a synergist for antioxidants. It also controls browning reactions.

Effects of processing on vitamins:

» Unlike other constituents of fish such as moisture, proteins, lipid and ash which have certain importance on its processing, vitamins have little importance as described above. However, different processing methods have notable effects on them as given below.

1. The fat soluble vitamins, being insoluble in water, are not washed away during washing of fish in water. However, when fat is removed from fish, most of these vitamins are lost into the extracted fat (fish oil). Vitamin A is stable under ordinary cooking conditions, but is destroyed by contact with products of lipid auto-oxidation

found in rancid fat. As unsaturated fats are more prone to autoxidation, higher is the content of unsaturated fats, greater is the chance of destruction of vitamin A. Loss of vitamin A is more prominent in dry fish as its fat is more susceptible to autoxidation.

2. The water soluble vitamins get washed away during washing of fish as well as during leaching out of tissue water in different processing steps.
3. Most of the vitamins of the vitamin B complex, particularly thiamine, are destroyed by heat. Destruction of thiamine is considerable if heat treatment is severe as in case of canning.
4. Vitamin C is destroyed by heat and contact with air as it gets oxidised easily.

H. Nitrogenous extractives:

» A number of low molecular weight substances in fish are water soluble components. These substances amount to approximately 2% of the fresh weight. Besides some of the individual compounds are also present in fish.

» These extractives are essential to the taste and smell of the fish.

» The ratios of the individual compounds differ markedly with species and such variations are considered to be largely responsible for differences in flavor.

» A distinction is made between N-extractives and N-free extractives in fish.

» The N-containing extractives can be defined as the water-soluble, low molecular weight, nitrogen- containing compounds of non-protein nature.

» These are the non-protein nitrogenous (NPN) substances that get extracted from the tissue into an aqueous solution. It is obtained by subtracting the protein nitrogen (PN *i.e.*, content of nitrogen in the proteins of the tissue) form the total nitrogen (TN *i.e.*, content of total nitrogen in the tissue). It is expressed as follows.

$$\mathbf{NPN = TN - PN}$$

Where,

NPN = Non-protein nitrogen
TN = Total nitrogen
PN = Protein nitrogen

- As NPN substances do not include proteins which are the most abundant of all the nitrogenous substance present in the fish tissue, the percentage of NPN substances in the total nitrogenous (TN) substances is very less.
- NPN compounds account for a relatively high % of the Total Nitrogen in the muscles of some aquatic animals, 10-20% in teleosts, about 20% in crustaceans and molluscs, and 30 – 40% in elasmobranchs.

Category of fish	NPN (as percentage of TN)
Teleosts	10
Crustaceans and molluscs	20
Elasmobranchs	> 30

- In contrast, NPN compounds in land animals usually represent not more than 10% of the total nitrogen.

The important components of NPN substances are given below.

1. Nitrogenous bases
2. Free amino acid
3. Guanidine derivatives
4. Dipeptides
5. Purine derivatives
6. Pyrimidine derivatives
7. Urea

1. Nitrogenous bases:

- These are the NPN substances having 'basic' nature due to which their presence in the tissue raises its pH.
- These are the nitrogenous bases which are volatile in nature.
- Though they may be present in small quantities in live fish, a large amount of these substances are formed in the dead fish during spoilage.

» As they are volatile, immediately after their formation, they volatilize into their gaseous state and escape into the surrounding atmosphere.

» Upon reaching the consumers' nose, they produce the sensation of odour. As all of these volatile bases have off-odour, their formation in the fish tissue after death results in the off-odour of spoiling fish.

» The volatile bases include the following

a. Trimethylamine (TMA)

b. Dimethylamine (DMA)

c. Methylamine (MA)

d. Ammonia (NH_3)

2. Nitrogen free extractives:

» The N-free extractives are low molecular weight components which do not contain nitrogen, e.g. glucose and sugar phosphate, pyruvic and lactic acid.

Table 2.7: Major differences in muscle extractives

Compound in mg/100 g wet weight	Fish Cod	Fish Herring	Fish Shark species	Crustaceans Lobster	Poutry Leg muscle	Mammalian muscle
1) Total extractives	1200	1200	3000	5500	1200	3500
2) Total free amino-acids	75	300	100	3000	440	350
a. Arginine	<10	<10	<10	750	<20	<10
b. Glycine	20	20	20	100-1000	<20	<10
c. Glutamic acid	<10	<10	<10	270	55	36
d. Histidine	<1.0	86	<1.0	-	<10	<10
e. Proline	<1.0	<1.0	<1.0	750	<10	<10
3) Creatine	400	400	300	0	-	550
4) Betaine	0	0	150	100	-	-
5) Trimethylamine oxide (TMAO)	350	250	500-1000	100	0	0
6) Anserine	150	0	0	0	280	150
7) Carnosine	0	0	0	0	180	200
8) Urea	0	0	2000	-	-	35

*It should be noted that the unit in this table refers to the total molecular weight of the compound

- » TMAO constitutes a characteristic and important part of the NPN-fraction in marine species.
- » This component is found in all marine fish species in quantities from 1 to 5 % of the muscle tissue (dry weight) but is virtually absent from freshwater species and from terrestrial organisms.
- » TMAO is formed by biosynthesis in certain zooplankton species.
- » These organisms possess an enzyme (TMA mono-oxygenase) which oxidizes TMA to TMAO.
- » TMA is commonly found in marine plants as are many other methylated amines (monomethylamine and dimethylamine).
- » Plankton-eating fish may obtain their TMAO from feeding on these zooplankton (exogenous origin).
- » Certain fish species are also able to synthesize TMAO from TMA, but this synthesis is regarded as being of minor importance.
- » The TMA-oxidase system is found in the microsomes of the cells and is dependent on the presence of Nicotinamide ademine denucleotide phosphate (NADPH):

$$(CH_3)_3N + NADPH + H^+ + O_2 \rightarrow (CH_3)_3NO + NADP^+ + H_2O$$

- » The amount of TMAO in the muscle tissue depends on the species, season, fishing ground, etc.
- » In general, the highest amount is found in elasmobranchs and squid (75-250 mg N/100 g); cod have somewhat less (60-120 mg N/100 g) while flatfish and pelagic fish have the least.
- » Pelagic fish (sardines, tuna, mackerel) have their highest concentration of TMAO in the dark muscle while demersal, white-fleshed fish have a much higher content in the white muscle.
- » In elasmobranchs, TMAO seems to play a role in osmoregulation, and it has been shown that a transfer of small rays to a mixture of fresh and sea water (1: 1) will result in a 50 % reduction of intracellular TMAO.
- » The role of TMAO in teleosts is more uncertain.

» Several hypotheses for the role of TMAO have been proposed:
  - TMAO is essentially a waste product, a detoxified form of TMA
  - TMAO is an osmoregulator
  - TMAO functions as an "anti-freeze"
  - TMAO has no significant function. It is accumulated in the muscle when the fish is fed a TMAO-containing diet

» TMAO together with high amounts of taurine could have additional effects, at least in fresh water fish.

» Quantitatively, the main component of the NPN-fraction is creatine.

» In resting fish, most of the creatine is phosphorylated and supplies energy for muscular contraction.

» The NPN-fraction also contains a fair amount of free amino-acids. These constitute 630 mg/ 100 g light muscle in mackerel (*Scomber scombrus*), 350-420 mg/ 100 g in herring (*Clupea harengus*) and 310-370 mg/100 g in capelin (*Mallotus villosus*).

» The relative importance of the different amino- acids varies with species.

» Taurine, alanine, glycine and imidazole-containing amino-acids seem to dominate in most fish.

» Of the imidazole-containing amino-acids, histidine has attracted much attention because it can be decarboxylated microbiologically to histamine.

» Active, dark-fleshed species such as tuna and mackerel have a high content of histidine.

» The amount of nucleotides and nucleotide fragments in dead fish depends on the state of the fish.

» The volatile base nitrogen is below 20 mg% of the wet muscle in the fishes, molluscs and in the crustaceans excepting the prawns.

» The quantity of volatile base nitrogen ranges from 9 to 24 mg% of the wet muscle.

» The volatile base nitrogen in fresh fish muscle is mainly ammonia which has its origin in the deamination of the muscle adenylic acid, a reaction which is associated with muscular work and also with processes leading to the denaturation of muscle protein.

- The AAN (Alpha Amino Nitrogen) in the teleosts is only about 17 to 80 mg% of the total non-protein nitrogen while in the invertebrates, it accounts for over 40% of the NPN and crustaceans upto 280 mg%.
- The AAN content depends on environmental condition of fish.
- The higher values of AAN are mainly due to the increased level of basic free amino acids present in them.
- The urine of fishes contains only a small quantity of AAN, 14% of the nitrogen excreted by invertebrates has been found to be amino acid nitrogen.

IV. Loss of nutritional value of fish during processing and preservation:

1. **Exudation**: This is also called as Drip. During thawing, 4% protein loss is observed through drip.
2. **Cooking at 60^0C:** Approximately 20% to 30% weight loss occurs due to exudation of fluids from fish body.
3. **Canning**: No loss from nutritional point of view. But loss of Vitamin B1 and C is observed.
4. **Salting and smoking:** 1-5% nutritional loss is seen during salting and 8-30% loss during smoking is observed.
5. **Freezing**: Fat rancidity, desiccation, denaturation of protein, freezer burn and rusting occurs during freezing which result to 5-6% loss of nutrients.

V. Metabolism in Fish:

- "Metabolism" is the word used to cover the system of chemical processes that keep something alive. Fish obtains its energy through food for building and maintaining the body parts needed to function. There are different metabolic processes through which the chemical energy of the nutrients is converted for utilization by the organism.
- Metabolism itself relies on three main things:
  1. Respiration and nutrition to supply metabolites (the products it uses, built out of both inorganic and organic matter)
  2. Osmoregulation for a stable working environment
  3. Excretion to get rid of all the poisons and other waste products produced as side-effects

- In fishes, the metabolism covers two processes: catabolism and anabolism. Catabolism is the process of breaking down metabolites to produce active energy, while anabolism uses those same products to build new body tissue for growth, maintenance, and reproduction.
- The metabolism can work at different speeds, depending on environmental conditions, and is controlled by hormones produced in the fish's body. The metabolic rate can change with a variety of factors:
  - Size – bigger fish have relatively slower metabolic rates
  - Age – young fish grow more but don't need the reproductive side yet
  - Activity – busy fish need a faster rate
  - Condition – fish in poor condition need more tissue maintenance
  - Environment – temperature, oxygen levels and salinity all affect the rate
- Food is taken into the gut, broken down into small units and transported by the blood to sites in the body where it is required to build up new tissues for growth, for replacement of worn out tissues or to meet the demands of the reproductive system. Simultaneously some compounds are converted to energy and those which are no longer required are broken down in the tissue.
- The energy in fish is mainly used for (a) muscle activity, (b) synthesis of cell materials and (c) heat production. During metabolic processes, a large number of chemical reactions take place.
- Catalysts help these reactions to take place in right direction and at the desired speed. But these catalysts are able to change the substances in the reaction without themselves being changed. They are also called as enzymes. Enzymes are biological catalysts and are special proteins. All the chemical reactions in the metabolism make their own specific enzyme.
- If everything is normal in a fish's environment, it produces energy by oxidation. This requires a constant supply of enough oxygen. If there isn't enough, the fish will produce energy in white muscle tissue using "glycolysis". Adrenaline stimulates the tissue and causes glycogen to be converted to glucose and energy without the need for oxygen. Unfortunately, this also produces poisonous lactate, so glycolysis can only be sustained for short periods. Oxygen and energy will also be needed to break down the lactate, so it's a sort of "oxygen debt" in times of emergency.

- If the fish's environment is low-stress, stable, free of disease and supplied with everything required, excess energy can be used for growth and reproduction. In general, only the excess is used for these purposes, so good growth and active reproductive behavior is positive signs that favourable living conditions are being maintained.
- At the other end of the process, waste products generated by using metabolites are excreted from the fish's body. All waste is toxic, whether produced in energy creation or tissue growth and maintenance. Most of this waste consists of carbon dioxide and ammonia (both of which are ejected through the gills by diffusion), water and some larger molecules like purine, which eventually becomes urea and is removed with water by the kidney.

3 Biological Characteristics of Fish

- Fish is a cold blooded aquatic vertebrate which breaths by means of pharyngeal gills, propels and balances by means of fins.
- However, the term '*fish*' is used broadly to mean all the aquatic organisms including fish, shellfish, aquatic mammals and seaweeds.
- All the products obtained from them are called '***aquatic products***'.
- The term '***seafood***' includes all the food items prepared from sea animals and plants.
- The term, '***marine products***', includes both seafood and inedible products prepared from marine plants and animals, excluding those from the fresh water organisms.
- The common aquatic organism important in fish processing can be classified into the following groups.

i. Fishes: It includes the true fishes as per the definition of fish given above. They are very often called **finfishes,** as they possess fins. It has two important groups.

a. Elasmobranches: All cartilaginous fishes

b. Teleosts: It includes all the bony fishes. Based on habitat they are of the following three groups.

i. Marine teleosts

ii. Fresh water teleosts

iii. Brackish water teleosts

ii. Shellfishes: It includes invertebrates with external hard shell. It has two important groups

a. **Molluscs:** They have calcarious shell (made of calcium carbonate) and include three groups.

i. **Gastropods**: These have single piece of shell.

ii. **Bivalves:** These have a shell made of two halves held together by an adductor muscle.

iii. **Cephalopods:** These have shell appendages on the head without any external shell.

Examples: Squid & cuttlefish (Have internal pen called cuttle shell)
Octopus (Has no shell at all)

b. **Crustaceans**: They have chitinous shell (made of chitin).

iii. **Marine mammals:** These include the mammals inhabiting the seas. They mainly have two groups.

1. **Mysticeti** (Baleen/whalebone whales)
2. **Odontoceti** (Toothed whales)

iv. **Seaweeds:** These are the aquatic plants growing in the seas and brackish water lakes.

Fishes	**Elasmo branches**		Sharks	Scoliodon laticaudus
			Rays	Himantura bleekeri
			Skates	
	Teleosts	**Marine**	Oil sardine	*Sardinella longiceps*
			Indian mackerel	*Restrelliger kanagurta*
			Pomfrets	*Pampus sp.*
			Seer fish	*Scomberomorus sp.*
			Tuna	*Thunnus sp*
			Ribbon fishes	*Trichiurus sp.*
			Anchovies	*Anchoviella sp.*
			Marine cat fishes	*Tachysurus sp.*
			Lactarius	*Lactarius lactarius*
			Flat fishes	*Pleuronectes platessa*
			Soles	*Cynogossus macrolepidotus*
			Silver bellies	*Stolephorus indicus*
			Scianids	*Sciaena aneus*
			Carangids	*Caranx melampygus*

Fishes		**Fresh water**	Catla	*Catla catla*
			Rohu	*Labeo rohita*
			Mrigal	*Cirrhinus mrigala*
			Common carp	*Cyprinus carpio*
			Grass carp	*Ctenopharyngodon idella*
			Silver carp	*Hypopthalmicthys mollitrix*
		Brackish water	Bhetki	*Lates calcarifer*
			Mullets	*Mugil sp.*
Shellfishes	**Molluscs**	**Gastropods**	Whelk	*Babylonia canaliculate*
			Winkle	*Littorina littorea*
		Bivalves	Oyster	*Crassostrea madrasensis*
			Mussel	*Perna viridis*
			Cockle	*Cerastoderma edule*
			Scallop	*Pectine sps*
		Cepholopods	Squid	*Loligo duvaucelii*
			Cuttlefish	*Sepia sp.*
			Octopus	*Octopus* sp.
	Crustaceans	Prawns (shrimps)		*Penaeus Mondon*
		Lobsters		*Panulirus homarus*
		Crabs		*Portunus sanguinolentus*
		Crawfish		*Procambarus clarkii*
Marine mammals		**Mysticeti** (Baleen/ whalebone whales)	Blue whale	*Balaenoptera musculus*
			Fin whale	*Balaenoptera physalus*
			Hump back whale	*Megaptera novaeangliae*
			Sei whale	*Balaenoptera borealis*
		Odontoceti (Toothed whales)	Sperm whale	*Physeter macrocephalus*
Seaweeds			Gracilaria	*Gracillaria edulis*
			Sargassum	*Sargassum muticum*
			Gelidium	*Gelidium spinosum*
			Entermorpha	*Enteromorpha prolifera*
			Ulva	*Ulva fasciata*
			Turbinaria	*Turbinaria ornata*
			Caulerpa	*Caulerpa racemosa*

1. Anatomy in fish processing:

Anatomy has certain role in fish processing. The important structures of fish body and their importance in fish processing are described below.

I. Eye:

- The lens of the eyes of fishes is brilliantly transparent.
- The eyes of a freshly killed fish are brilliantly black.
- When fish starts to spoil, it gradually looks cloudy or milky. The cloudiness goes on increasing with the extent of spoilage. Therefore, the transparency of eye is used as an indicator of freshness of fish.

II. Gills:

- The gills harbour a high load of bacteria because of the large amount of water passing over them for respiration.
- After death, these bacteria grow leading to a gradual change in the colour of the gills from bright red to dark brown.
- Therefore, the colour of the gills is used as an indicator of the freshness of fish.
- Moreover, due to their high microbial load, very often gills are removed immediately after harvesting in larger fishes. This process is called '*gilling*' or '*degilling*'.

III. Skin:

- It harbours a high load of bacteria due to its contact with water.
- Due to its impermeability to microbes, they cannot enter into the underlying muscle and internal organs. Therefore, the muscle and internal organs are sterile.
- Soon after harvesting, the skin is removed as some consumers do not like to eat it.
- Removal of skin also reduces the microbial load.
- The process of removal of skin of fish is called '*skinning*' or '*deskinning*'.
- Skin of shark is tough and is used for manufacture of 'shark leather'.

IV. Gut:

- After death of fish, the gut enzymes remain active and hydrolyse the gut wall as well as the flesh close to it.
- Therefore, the flesh of the belly area loses its firm-elastic texture and becomes soft-flabby which is undesirable.
- Besides, the gut has the highest load of microbes in fish body as high load of bacteria enter into the gut along with the food.
- After death, these microbes start spoiling the fish.
- Therefore, in most fishes, gut is usually removed after harvesting. This process is called '*gutting*' or '*degutting*'.
- In case of shrimps and prawns, this process is called '*deveining*' as their guts present dorsally and look like vein.
- The gut of few fishes is used for the preparation of 'surgical sutures'.

V. Internal organs:

- These are the organs which remain inside the belly cavity or visceral cavity. Therefore they are also called '*visceral organs*'.
- These organs along with the gut remain in the visceral cavity and are collectively called '*viscera*'.
- During dressing of fish for further processing, the gut along with all the visceral organs are removed by the process called '*evisceration*'.
- During the process of gutting, actually the whole viscera are removed. Therefore, the terms 'gutting' and 'evisceration' are used synonymously.
- The internal organs include the following.

a. Livers:

- The liver of fish is generally of 1% to 7% of body weight. In most fishes it varies between 1-2% but in case of shark, cod, it varies between 4-7% of body weight.
- The liver is considered as fat depot. The liver oil is rich in Vitamin A and D or, some content of sterol related hydrocarbons.

» In lean fishes, excess fat in the body is deposited in the livers, *e.g.* shark and cod. Therefore, livers of shark and cod are usually used to extract liver oil, which are rich source of vitamins A and D.

» On the other hand, in fatty fishes, excess fat is deposited in the flesh as body fat and just under the skin as subcutaneous fat. This fat is extracted and used as 'fish body oil' (or fish oil, as it is commonly called) for several purposes.

» The high fat content in flesh and under skin helps them in their buoyancy for remaining in the pelagic layer.

» Shark also contains considerable quantity of *squalene* (a hydrocarbon) in its liver.

» The density of squalene is lower (0.86) than that of water, for which it helps in the buoyancy of shark. It is extracted from shark liver and has industrial use.

b. Gall bladder:

» It is usually found somewhere within the lobes of the liver.

» It stores bile secreted by liver.

» During dressing of fish, it should be removed carefully without rupturing. This is because, bile has a bitter taste and if the gall bladder is ruptured, the bile leaks out and comes in contact with the flesh imparting a bitter taste to it. Even repeated washing with water cannot completely remove this bile.

c. Pancreas:

» In case of cartilaginous fish, the pancreas is compact and performs both as an endocrine as well as an exocrine gland.

» The endocrine function is performed by the β-cells of the 'islet of Langerhans' present in the pancreas which secret the hormone **insulin**. This hormone is converts excess glucose in the blood to glycogen for its storage in the liver.

» The exocrine function is performed by the rest of the pancreas which secrets some of the digestive enzymes that pass into the intestine. These enzymes help in the digestion of the food.

» In most of the teleosts, the pancreas is diffused and is not visible. It remains scattered throughout the belly cavity.

» In few species, the cells of the Islet of Langerhans are not present in the pancreas, rather are present as separate large visible bodies called 'Brockmann bodies' or 'principal islets'. In these fishes, the exocrine pancreas is partially embedded in the liver.

d. Kidney:

» Kidney of fish lies in the visceral cavity beneath the backbone.

» It is a long reddish brown structure looking like a clot of stale blood.

e. Gonads:

» They are commonly called 'fish roes' and are used as food.

» The ripening ovary of a female fish is called 'hard roe' and the testes of a male one 'soft roe'.

» The ovaries of mullets, fish roes, have high demand as food and fetch good market price.

» The ovary of some fishes like sturgeon, salmon and trout is called '**cavier**' which is one of the costliest fish product used for human consumption.

» The milt or chitterlings (*i.e.*, sperm) that flows out from the testes is generally not eaten, rather it is used for manufacture of *protamine*, which is used in combination with insulin to control diabetes. It prolongs the effectiveness of the insulin injected to the diabetics.

f. Air bladder:

» Swim bladder or air bladder is white thick walled organ filled with gas due to which it looks like a balloon.

» Most fish have swim bladder except some fishes like sharks.

» Air bladder from some fishes is used to prepare 'fish maws' and 'isinglass' which have industrial applications.

g. Others:

» The others internal organs include heart, spleen and mesentery.

» Heart helps in blood circulation.

» The spleen consists of a large number of distinct dark red lobes of various shapes and sizes scattered in the adipose (fatty) tissues surrounding the gut.

» Spleen helps in the fragmentation of the worn out RBC and therefore called the graveyard of RBC.

» Mesentery is the membranous 'visceral peritoneum' which holds all the visceral organs in place. It is an extension of the membranous 'parietal peritoneum' which lines the visceral cavity.

» In some fishes, the parietal peritoneum is black due to which its presence distracts the consumers. Therefore, after evisceration, the black peritoneum is removed completely.

VI. Skeletal system:

» The skeletal system of fish supports the body structures and helps in its movement.

» The fish skeletal muscle provides around 50% of total body weight of fish.

» It includes the following components.

1. Exoskeleton: It is present external to the fish body and includes the following.

a. Scales:

- Most of the fin fishes have scales attached to the skin.
- As scales are inedible, they are removed from the fish body before it is used as food.
- The process of removal of scales of fish is called 'scaling' or 'descaling'.
- Scales of some fishes are used for preparation of 'pearl essence'.

b. Shells:

- Most of the shellfishes have a protective layer of shell on their body surface.
- Shells are of two types as follows.

i. Chitinous shell:

- These are found on the body of crustaceans.
- As chitinous in nature (*i.e.* made of chitin), these shells are used for preparation of 'chitin' and 'chitosan', which have several industrial and pharmaceutical uses.

ii. Calcareous shell:

- These are found on the body of most of the molluscs.
- As calcareous in nature (*i.e.* made of calcium carbonate, $CaCO_3$), these shells are used for preparation of lime (CaO) which has several industrial, agricultural and pharmaceutical uses.

2. Endoskeleton:

- It is present inside the fish body completely covered by flesh or skin.
- Endoskeleton is made of cartilages in elasmobranches and of bones in teleosts.
- From nutritional point of view, the bone of fish is a rich source of the mineral 'calcium' as it contains calcium phosphate.
- The bone of fish obtained from fish processing industries as waste, is also used to prepare 'fish bone meal' which has agricultural uses.
- The endoskeleton of fish is made of two parts.

a. Axial skeleton:

- It is the skeleton, which remains along the main axis (axial) of fish body.
- It includes two components.

i. Vertebral column:

- The vertebral column or backbone runs from head to tail and is composed of a series of segments called vertebrae.
- The vertebrae of the trunk region have lateral processes those bear the ribs.
- Usually there are also a corresponding number of false ribs or pin bones extending more or less horizontally into the muscle tissue.
- These pin bones cause a great deal of trouble during consumption as food.

ii. Skull:

- It is the structure, which encloses the brain.
- It is also called cranium (brain box).

b. Appendicular skeleton:

- It is the skeleton, which supports the appendages of fish.
- The appendages of fish are fins.
- The appendicular skeleton includes the following.

i. Skeleton of the fins:

- It consists of the parallel 'fin rays' and 'basal radials' from which arise the fin rays.
- As fins are usually not eaten, during dressing of fish, these are removed by the process called 'finning' or 'definning'.
- The fins of sharks are used to extract the 'shark fin rays', which is used to prepare 'shark fin ray soup'.

ii. Gridles:

- They hold the skeleton of the paired fins and include 'pectoral girdles' and 'pelvic girdles'.

VII. Cardiovascular system:

» The cardiovascular system (cardio: heart; vascular: blood vessels) is the system, which circulates blood in the body.

» This system is of interest in fish processing, since larger size fishes which are to be filleted, should be bled immediately after capture to remove most of their blood.

» It is done to get attractive fillets with uniform white colour.

» If kept without bleeding for a long time after capture, the red colour blood oozes into the white tissues of the fillets imparting it a reddish colour which, with passage of time, undergoes undesirable discoloration to a dull brown colour.

» However, smaller fishes are not bled as it is time-consuming.

VIII. Muscular system:

» The muscular system consists of the muscle tissues present in the fish body, which constitutes the bulk of the edible portion.

» Muscle is different from meat. The 'muscle' of living fish undergoes several changes after death (post mortem changes) to become 'meat'.

» The muscle is made of cells called 'muscle cells' or 'myocytes'.

» Very often, they are much longer in comparison to their diameter, for which they look like fibers. That is why; they are very often called 'muscle fibers' or 'myofibers'.

» There are three types of muscle in fish as follows.

1. Smooth muscle - Un-striated muscle
2. Cardiac muscle - Striated muscle
3. Skeletal muscle

1. Smooth muscle:

- It constitutes the smooth parts of the fish body such as the gut, veins, arteries and capillaries.
- It also constitutes the flesh of molluscs.
- It is 'unstriated muscle', as it does not have striated (banded) appearance under microscope.
- It is made of cells called 'smooth muscle cells'/ 'smooth myocytes' or, 'smooth muscle fibers' / 'smooth myofibers'

2. Cardiac muscle:

- It constitutes the muscles of the heart.
- It is 'striated muscle', as it has striated (banded) appearance under microscope due to the presence of alternate dark bands and light bands.
- It is made of cells called 'cardiac muscle cells'/ 'cardiac myocytes' or, 'cardiac muscle fibers' / 'cardiac myofibers'

» The smooth muscle and cardiac muscle constitute a small part of the body weight and are of little importance as food.

» It is the skeletal muscle which contributes to the bulk of the fish body and is used as food.

» It is also the skeletal muscle which undergoes contraction during rigor mortis (stiffening of body after death).

» Therefore, it has great importance in fish processing and is dealt with in details below.

3. Skeletal muscle:

- It is the muscle anchored to the skeleton of fish.
- The skeletal muscle consists of sections containing contractile proteins surrounded by connective tissue that is attached in turn to the skeleton and to the skin. The connective tissue consists of sheets of collagen known as myocommata.
- Similar to cardiac muscle, it is also 'striated muscle', as it has striated (banded) appearance under microscope due to the presence of alternate dark bands and light bands.
- It is made of cells called 'skeletal muscle cells'/ 'skeletal myocytes' or, 'skeletal muscle fibers' / 'skeletal myofibers'.
- Skeletal muscle fibers, depending on their colour and utilisation for fast or slow movement are of three types as given below.

a) White fibers (Fast fibers):

- These are white colour muscle cells used for 'quick' and 'fast' activities (*e.g.*, fast swimming to escape from a predator).
- Energy is produced in these cells quickly by a short anaerobic 'glycolytic path' [Glucose (no air) $\rightarrow$ 2 Lactic acid + 2 ATP].
- As yield of ATP is less, (i.e., 2 ATP/glucose), more glucose is stored in these cells.

b) Red fibers (Slow fibers):

- These are red colour muscle cells, used for 'slow' and 'continuous' activities (*e.g.*, continuous slow swimming on the surface of water, in search of food).
- Energy is produced in these cells continuously by a long aerobic 'oxidative path' [Glucose + $O_2 \rightarrow 6\ CO_2 + 6\ H_2O$ + 38 ATP].

- As yield of ATP is more, (*i.e.*, 38 ATP/glucose), less glucose is stored in these cells.

c) Pink fibers (Intermediate fibers):

- These are pink colour muscle cells, used for intermediate type of movement.
- Energy is produced in these cells both by glycolytic and oxidative paths.
- In fish, fibers of different types are confined to distinct locations thereby resulting in mainly two types of skeletal muscles with distinct colour.
- With respect to their use as food, they also have different taste, texture & stability.
- The two types of skeletal muscles in fish are as follows.
  a. White muscle (White meat or fast muscle)
  b. Dark muscle (Dark meat or slow muscle)

i. White muscle (White meat or fast muscle):

» It is white in colour as it is made of white fibers or fast fibers.

» In most fishes, it constitutes the bulk of the skeletal muscle.

» It is located 'deep' in the body and is capable of fast contraction (Figure 3.7).

» It remains 'inactive' most of the times and is recruited only when rapid bursts of activities are required *e.g.*, when a fish swims fast to escape from a predator.

» In most of the 'demersal fishes' which remain 'inactive' as they don't attempt to float, almost the whole skeletal muscle is made of white muscle with very little or no dark muscle.

ii. Dark muscle (Dark meat or slow muscle):

» It is brown or dark-red in colour as it is made of red fibers or slow fibers.

» It constitutes a small fraction of the muscle tissue in some fishes.

» It is found 'superficially' just under the skin beneath the lateral line region running roughly parallel to the long axis of the fish body (Figure 3.1).

» It is used for slow continuous action over long periods without rest.

- It is involved in slow cruising for which power requirement is very low *e.g.*, slow continuous swimming.
- In most of the pelagic fishes which have to swim more or less continuously so as to keep them afloat in the surface layer of the sea, dark muscle constitutes a large proportion of their skeletal muscle (upto 48% of body weight).
- In cross section, usually dark muscle has a flat triangular shape with the blunt tip pointing medially.
- However, it can have more complicated shapes.
- Different fishes have different quantities and shapes of dark muscle.

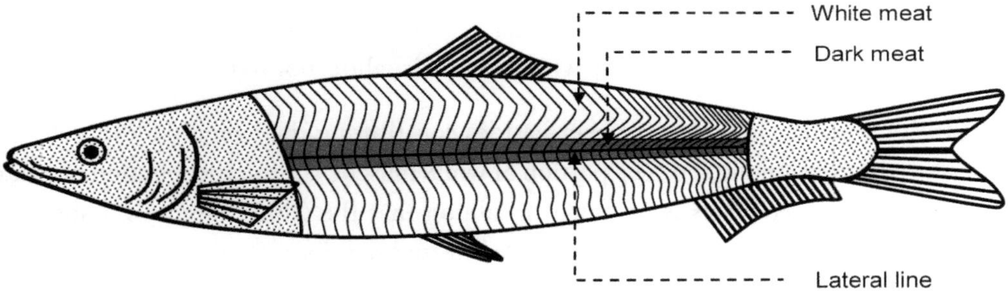

Figure 3.1: Longitudinal view of fish
(A portion of skin removed to show white and dark muscle)

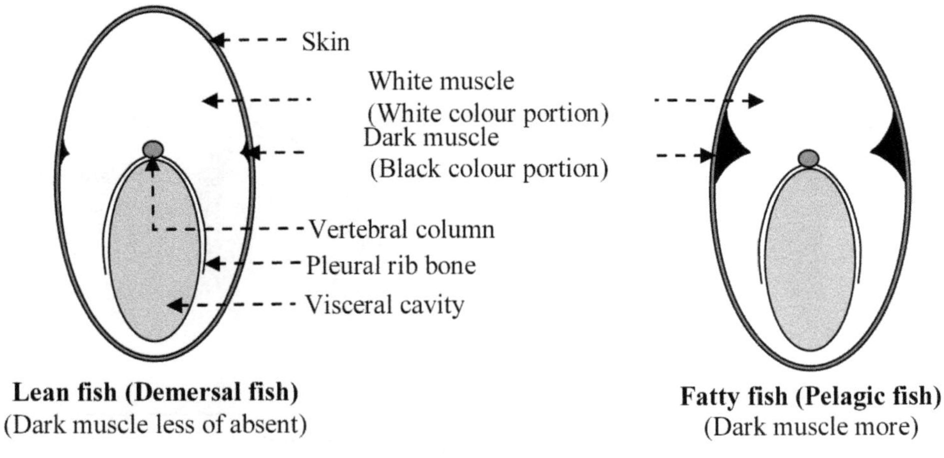

Figure 3.2: Transverse-section view of fish

» The white muscle and dark muscle have distinct colour, texture, taste and stability. Most of the times, the dark muscle is not desired due to its distracting dark colour, poor texture and bad taste. It also has low stability as it becomes rancid quickly due to its high lipid content. Therefore, sometimes, dark meat is removed and rejected. It is used for preparation of pet foods.

» Besides the white fibers and red fibers, in fish, between the dark and white muscles, there is a group of pink fibers.

» In some fishes like tuna, in addition to the superficial dark muscle, there is another deep-seated dark muscle (Figure 3.2). These two types of dark muscles have distinct properties. Moreover, in most tuna species, the white muscle is red due to its very high content of the red coloured muscle pigment 'myoglobin'. Therefore, their meat has the appearance of red meat similar to that of terrestrial animals. However, in albacore tuna, the colour of the white muscle is white, not red.

» In some salmonids, the red fibers, besides constituting the lateral dark muscle, are also scattered among the deep portion of white muscle. The interpenetration of white muscle with red fibers is so much in salmon that the white muscle is designated 'mosaic muscle', meaning a mosaic of white and red fibers (Figure 3.3).

» The reddish meat colour found in some species of salmon and sea trout does not originate from the red muscle pigment, 'myoglobin', rather it is due to the red carotinoid pigment, 'astaxanthine' present in the muscle of these fishes. This pigment comes to the muscle through the food they consume.

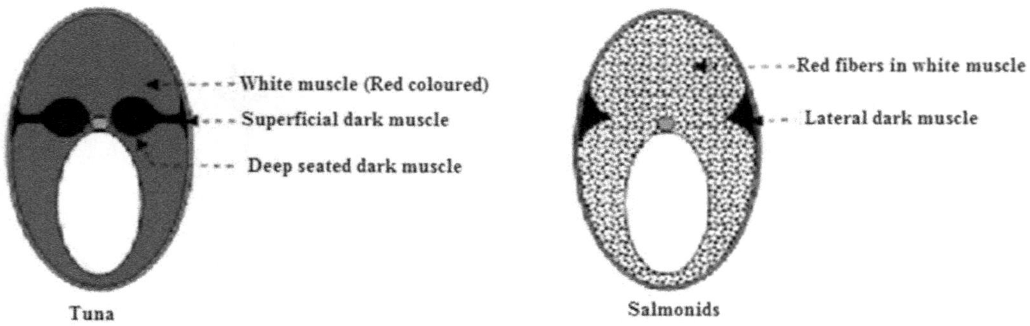

Figure 3.3: Dark muscles in Tuna and Salmonids

2. Structure of fish muscle:

» The structure of fish muscle usually refers to the structure of the skeletal muscle of fish. It is the muscle anchored to the skeleton of fish. It extends along the longitudinal axis of the fish body covering both the sides of the vertebral column.

» The muscle mass of each side is called a '**fillet**'. A fillet is a piece of flesh obtained by cutting the fish body from head to tail along the side of the vertebral column. Thus, two fillets can be obtained from each fish.

» The fillets have an almost oblong triangular shape, with the widest portion remaining towards the head-end and gradually tapering towards the tail-end.

» Fillet muscle provides around 50% of the total weight of fish and constitutes the edible portion of fish body. Therefore, very often, the fish fillets are cut out and used for icing and freezing preservation.

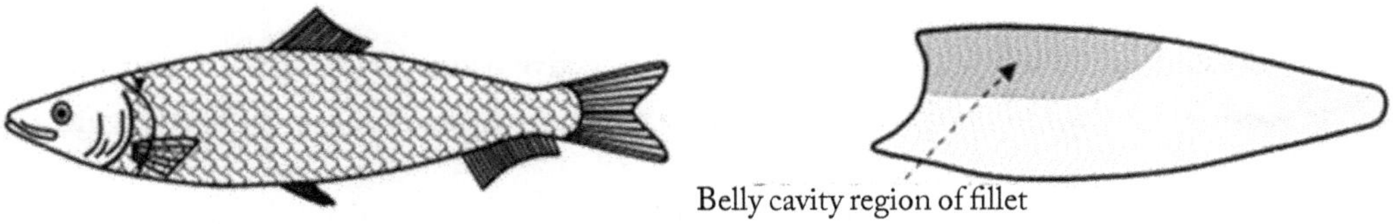

Figure 3.4: Fish and its fillet

a. Myotome:

» The muscle is made of several transverse muscle segments called 'myotomes' or 'myomeres' (Figure 3.5).

» They are plough-shaped and look like the letter 'm' turned through 90° to right so that its base remains towards the head.

» The myotomes are separated from each other by connective tissue sheets called 'myocommata' or 'myosepta'.

» The numbers of myotomes is usually identical to the number of vertebrae.

» Each myotome is made of large number of elongated cells called 'muscle cells'. They are the building units of the muscle.

» The muscle cells are very long as compared to their diameter for which they look like fibers. That is why, very often they are called 'muscle fibers' or 'myofibers'.

» Each muscle fiber extends between two myocommata.

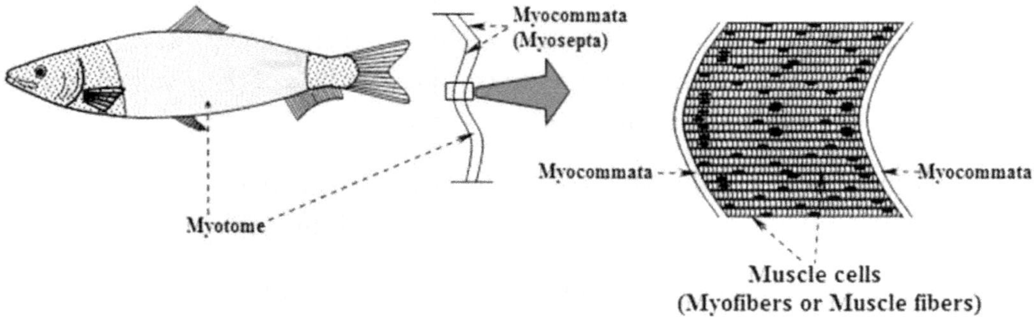

Figure 3.5: Structure of myotome

» The skeletal muscle is supported by four connective tissue systems such as endomysium, perimysium, epimysium and myocommata which are made of collagen fibers (Figure 3.6).

» Each muscle cell is surrounding by a casing called 'endomysium'.

» A group of muscle cells form fasciculi, which is surrounded by another casing called 'perimysium'.

» A group of fasciculi form the muscle tissue which is surrounded by a casing called 'epimysium'.

» This muscle tissue fits into the myotomes and is anchored on both ends to myocommata.

» The myocommata are anchored to the skeleton and skin.

» Each muscle cell is confined by the cell membrane called 'sarcolemma' (Figure 3.6).

» Each cell is almost completely filled with large number of parallel strands running along the entire length of the cell. These strands are called 'myofibrils'.

» The myofibrils have a diameter of about 0.001 mm.

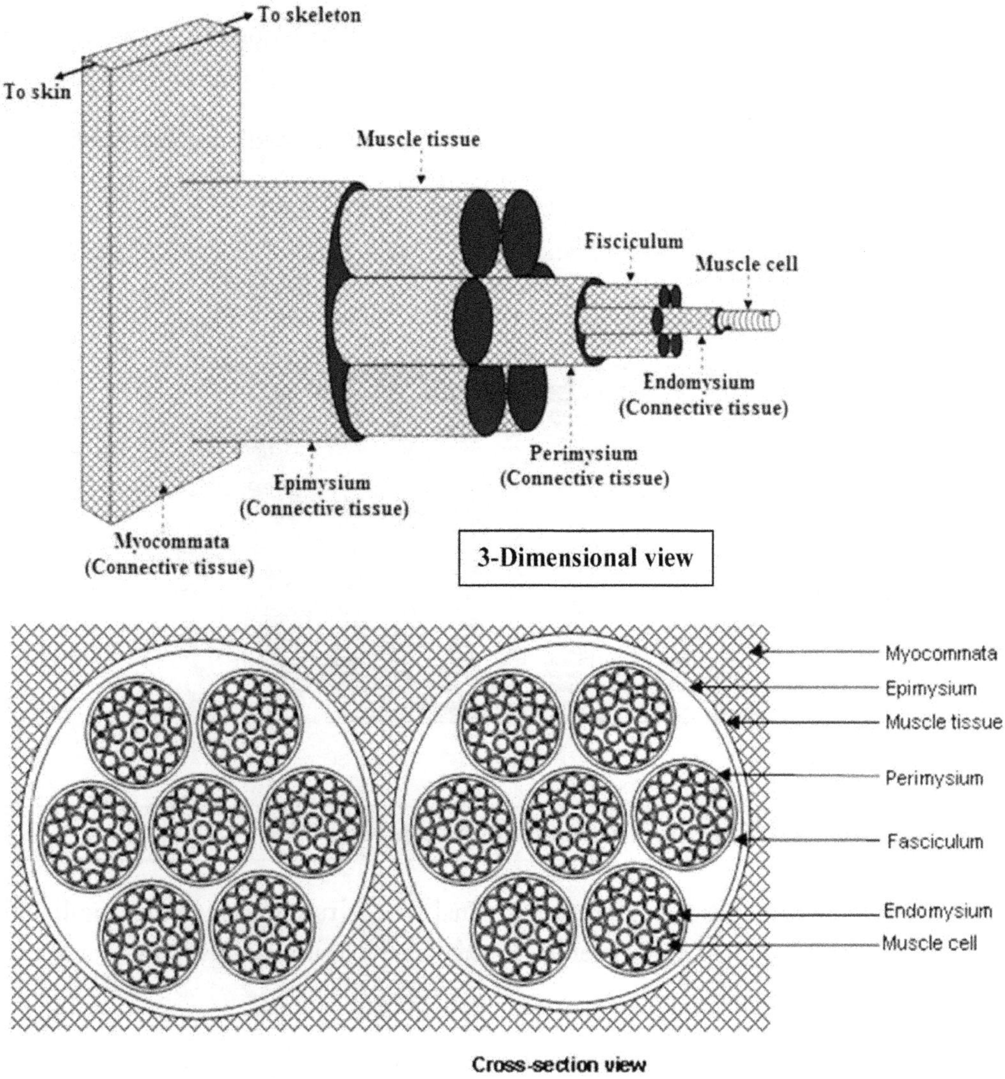

Figure 3.6: Connective tissues of fish muscle

b. Sarcoplam:

- The cytoplasm of muscle cell is called 'sarcoplasm'.
- As the muscle fibers are almost completely filled with myofibrils, there is only a peripheral 'sarcoplasm'.

» Muscle cells are multinucleated and the nuclei are located in the peripheral sarcoplasm.

» In the sarcoplasm numerous mitochondria, glogi bodies, ribosomes, lyosomes, glycogen granules, fat droplets, myoglobin and many enzymes are also found.

» Endoplasmic reticulum is also found in the muscle cell and is called 'sarcoplasmic reticulum'.

» The diameter of a muscle cell varies between 0.1 to 0.2 mm.

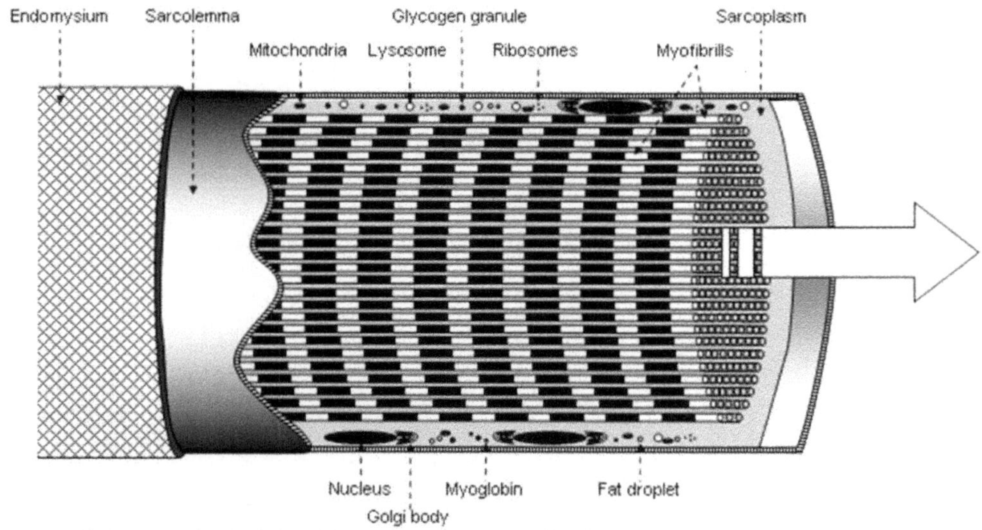

Figure 3.7: Structure of muscle cell of fish

» The sarcoplasmic reticulum has 'longitudinal tubules' running along the length of the muscle cell and 'transverse cisternae' formed by the terminals of several longitudinal tubules (Figure 3.8).

» The transverse cisternae run on both the sides of the Z-lines of the myofibrils.

» Another tubular system is formed by the invagination of the sarcolemma inward to the myofibrils along the Z-lines. This system is called the 'T-system' or 'T-tubule' (T= transverse).

» The two transverse cisternae of the sarcoplasmic reticulum remaining on the both the sides of a Z line and the central T-tubule running along the Z line constitute a system called a 'triad' (triad=a combination of three).

» The triads form a loose network, which almost covers each myofibril.

» The triads function in the excitation and relaxation of the muscle.

» In a relaxed muscle, the transverse cisternae of the sarcoplasmic reticulum contain high concentration of calcium ions (Ca^{++}) (Figure 3.7). As the calcium ions are very small and are permeable through the membrane of the sarcoplasmic reticulum, they continuously diffuse freely from the cisternae of the sarcoplasmic reticulum to the sarcoplasm. Therefore, the high concentration is maintained by ATP-fuelled Ca-pumps located on the membrane of sarcoplasmic reticulum which continuously pump Ca^{++} from the sarcoplasm into the cisternae using ATP.

» After death of a fish, due to disappearance of ATP (fuel), the Ca-pumps cease to function due to which Ca^{++} ions freely diffuse out from the cisternae into the sarcoplasm resulting in the muscle contraction during post mortem rigor mortis (stiffening of muscle after death).

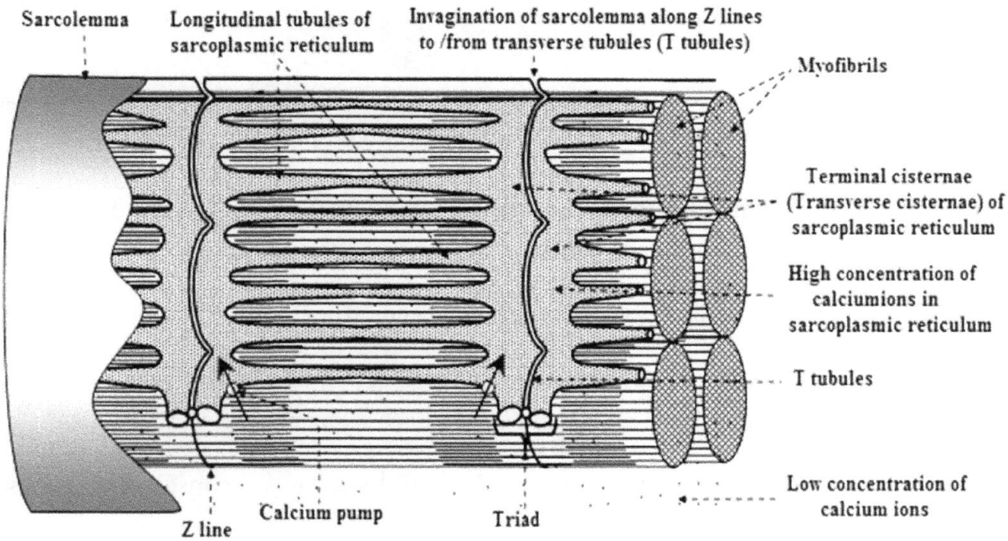

Figure 3.8: Sarcoplasmic reticulum of fish muscle cell

» The muscle cell is almost completely filled with the myofibrils.

» The myofibrils have striated appearance resulting from the alternate 'light bands' and 'dark bands'.

» The striations on the myofibrils are in register, *i.e.*, similarly oriented across all the myofibrils of a muscle cell (Figure 3.9). That means, the dark and light bands of each myofibril coincides with of the dark and light bands of every other myofibril in the muscle cell.

» This gives a striated appearance to the whole muscle cell.

» The muscle cells are also aligned synchronously to give the whole skeletal and cardiac muscle a striated appearance under microscope.

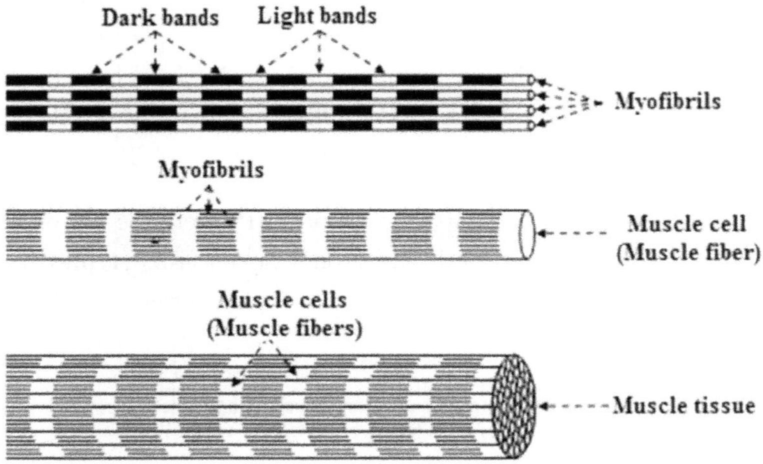

Figure 3.9: Striated appearance of myofibrils, muscle fiber and muscle tissue

» The dark bands are also called 'anisotropic bands' or 'A bands' (Figure 3.10). These bands look 'dark' in ordinary light.

» The light bands are called 'isotropic bands' or 'I bands'. They look 'light' in ordinary light.

» Each myofibril is divided into several transverse segments by vertical membranous structures called Z lines located at the middle of I bands. Each segment is called a 'sarcomere'.

» Thus, a sarcomere is the segment of a myofibril between two consecutive Z lines.

» The sarcomere is the smallest functional unit of a muscle fiber.

» It is the contraction of the sarcomeres which results in the contraction of the myofibrils leading ultimately to the contraction of the individual muscle fibers and contraction of the muscle as a whole.

» The sarcomeres are approximately 2.3 μm long.

» The Z lines are made of the protein 'α-actinin'.

» Each myofibril is made of several filaments which orient along the length of the myofibril (Figure 3.11). These are called 'myofilaments'.

- There are two types of 'myofilaments'.
- The thick myofilaments are called 'myosin filaments' and the thin myofilaments are called 'actin filaments'.
- The thick filaments are longer and thicker than the thin filaments.
- Actin filament is of about 7 nm in diameter and myosin filament is of about 15 nm in diameter.
- The thin filaments are anchored to the Z lines at right angle to it so that they project in opposite directions and run along the myofibril.
- Two adjacent opposite-facing actin filaments, anchored to a Z line, do not remain in the same horizontal line.
- As Z line is a zig-zag vertical line, the opposing actin filaments are anchored at the alternate pointed tips extending in the direction of the tips.
- The thick filaments remain in the center of the sarcomere (*i.e.*, between two Z lines) and run along the myofibril.
- Each thick filament extends between two pairs of opposing actin filaments in a sarcomere.
- The thick filaments impart dark appearance to the myofibrils. Therefore, the segments of the myofibril in which the thick filaments are present constitute the dark bands or A bands.
- The segments of the myofibril present between alternate dark bands have light appearance and are therefore constitute the light band or I band.
- The actin filaments are present in the I bands and extend beyond it into the A bands as they protrude among the thick filaments towards the center of the sarcomeres.
- Thus, there is a zone called 'overlap zone' where the thin filaments overlap the thick filaments.
- Therefore, A band contains thick filaments as well as the overlapping portions of thin filaments from both the sides whereas I bands contain only thin filaments.
- As the thin filaments protrude from both the sides of the Z lines, each I bands is bisected by a Z line.

» At the center of the A band is a portion where the actin filaments from both the sides cannot reach. This portion, therefore, is slightly less dark than the normal dark A band and is called the 'H zone'.

» At the center of the H zone is a line called 'M line'. The M line is made of a pair of proteins called 'M-line proteins'.

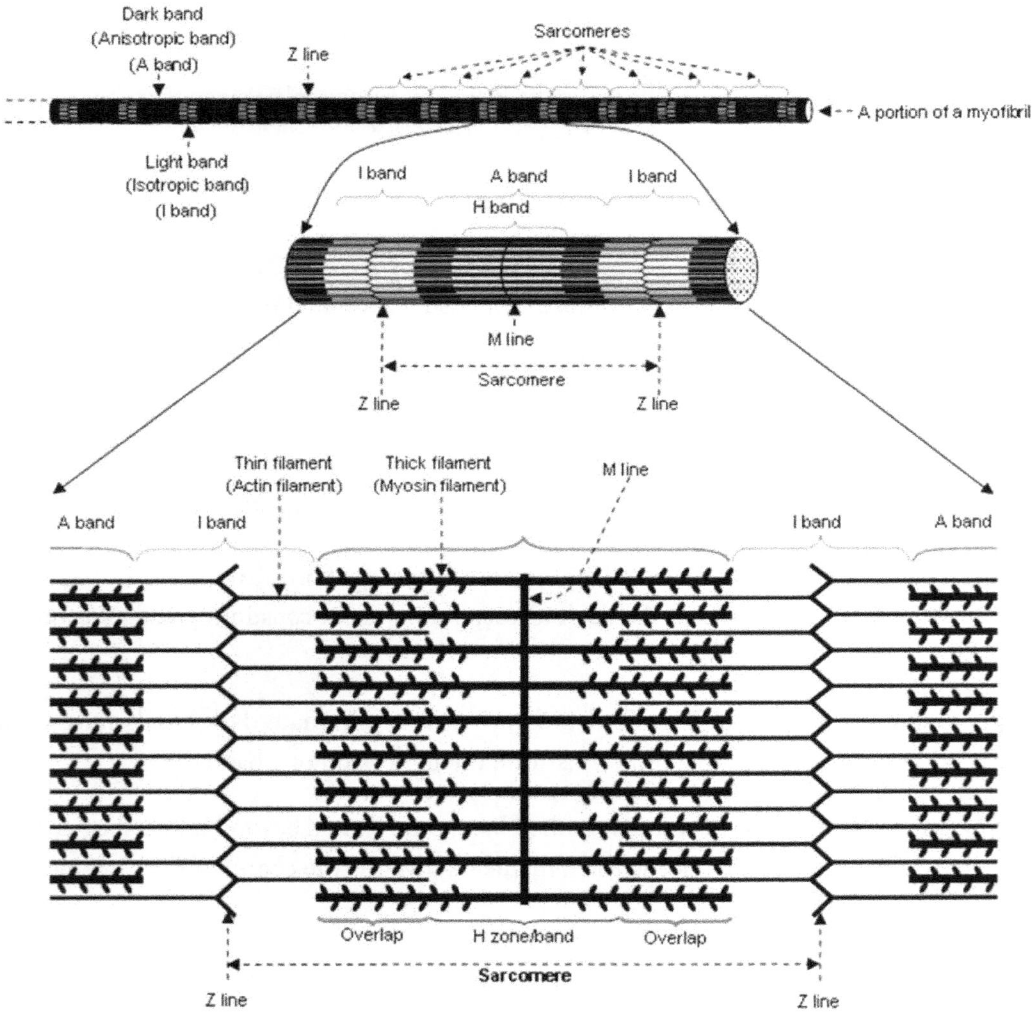

Figure 3.10: A sarcomere

» The cross sections of a myofibril at different positions show that the thick filaments are present as a bundle in each H zone and the thin filaments are present as a bundle in each I band (Figure 3.11).

» Both the bundles are present in the overlap zone where the thick filaments overlap the thin filaments.

» All the filaments are arranged in a hexagonal array.

» Each thick filament is surrounded by six thick filaments while each thin filament is surrounded by three thin filaments.

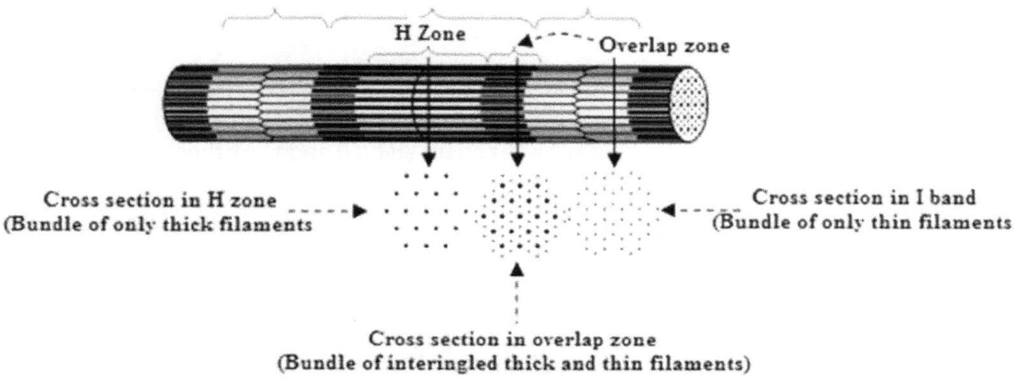

Figure 3.11: Arrangement of thick and thin filaments in the myofibrils

» Each 'thick filament' is made of a bundle of slender protein molecules called 'myosin' (Figure 3.12).

» Each myosin molecule is highly asymmetrical and consists of three parts *viz.*, a tail, a neck and a pair of heads.

» The tail is fibrous (fiber or filament-like) and is very long. It remains along the length of the thick filament.

» The neck is also fibrous, but is very short. It connects to the tail and to the pair of heads.

» As both tail and neck are fibrous, very often the myosin molecule is described to consist of two parts, viz, a fibrous tail and a pair of globular heads.

» The neck protrudes from the thick filament thereby protruding the heads from the filament.

» As it has high flexibility, it acts as a hinge through which the heads can move forward and backward during muscle contraction.

- The heads are globular and are pear-shaped. They protrude from the thick filament in a regular helical array. They protrude away from the center of the thick filament.
- Midway along the length of a thick filament is a 'bare zone' devoid of the protruding heads. This zone contains only the myosin tails bundled together.
- Fish myosin, similar to that of other vertebrates, is a hexameric protein (*i.e.* containing six component proteins) consisting of two heavy chains, called myosin heavy chains (MHC) and two pairs light chains, called myosin light chains (MLC) (Figure 3.12 & 3.13).
- Each chain is a polypeptide and is called a 'chain' as each one is a chain of amino acids.
- One pair of light chain is regulatory light chain (RLC) and the other pair is essential light chain (ELC).
- Each myosin heavy chain (MHC) consists of a heavy meromyosin (HMM) head portion and a light meromyosin (LMM) tail portion.
- Each HMM, in turn, consists of myosin sub-fragment 1 (S1) or myosin head and myosin sub-fragment 2 (S2) or myosin neck.
- The LMM and S2 constitute the tail or rod of the myosin molecule and the S1 constitutes the myosin head.
- The light chains are located on the myosin heads.
- Each head has one each from the two pairs of light chains, *i.e.*, each head has one ELC and one RLC.
- The RLC remains behind the ELC in the head. Both the light chains are wrapped around the heads.
- Myosin head has two binding sites; (i) actin binding site where it binds to the actin filament and (ii) ATP binding site where it binds to an ATP.
- Myosin head has Ca^{++} ATPase activity in that. In presence of Ca^{++}, it can dissociate ATP to ADP and Pi with liberation of energy.
- The tail of each MHC forms a coiled right-handed helix.
- The two helices of the two myosin heavy chains coil around each other as a left-handed coiled-coil to give a double helix (left-handed super helix).

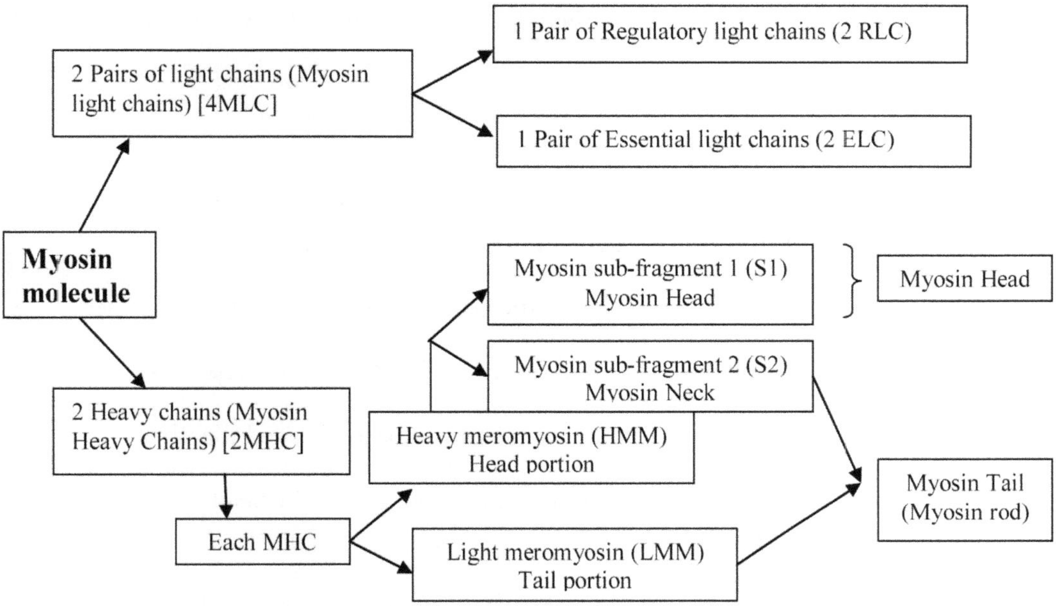

Figure 3.12: Components of a myosin molecule

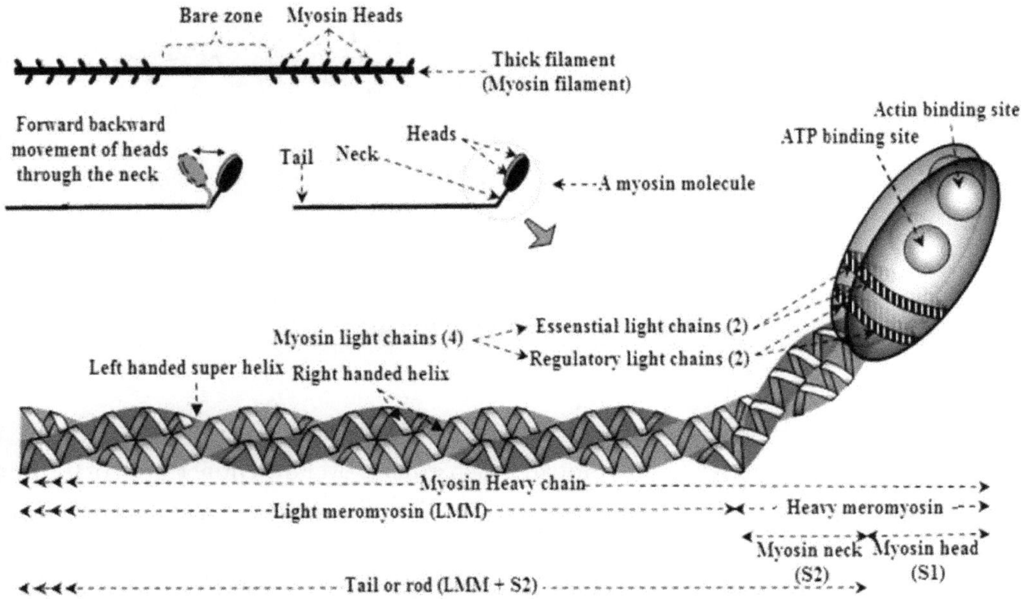

Figure 3.13: A myosin

- Each 'thin filament' consists of three types of proteins, *viz.*, F-actin, tropomyosin and troponin.

F-actin:

- The F-actin (filamentous actin) is a filamentous protein which extends over the entire length of the thin filament and forms its backbone.
- Each F-actin is made of two filamentous strands twisted around each other.
- Each strand is a polymer of several G-actin (globular actin) molecules.
- Therefore, each F-actin looks like two chains of beads twisted together.
- Each G-actin has a myosin-binding site where a myosin head can bind during muscle contraction.
- The thin filament has two ends, named 'barbed end' and 'free end'. The end which is anchored in the Z line is called the 'barbed end' and that which remains free pointing towards the myosin filament is called 'free end'.

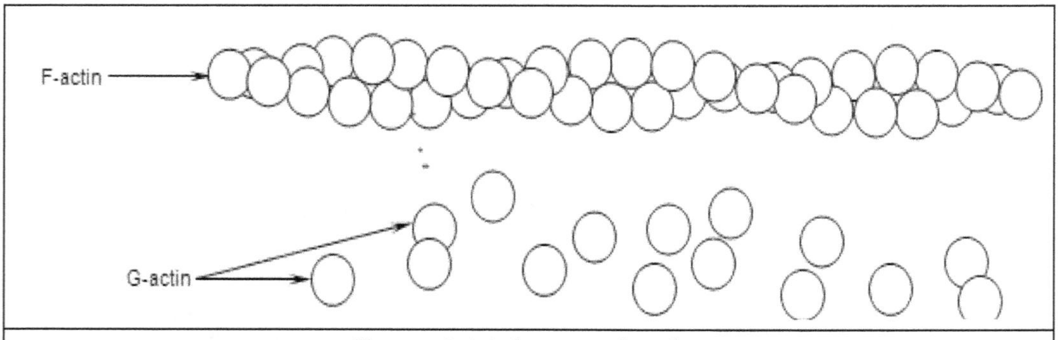

Figure 3.14: Actin molecules

Tropomyosin:

- Tropomyosin (Tm) is a fibrous protein which extends over the entire length of the thin filament remaining in the two grooves of the F-actin twisted strand (Figure 3.16).
- Thus, there are two long tropomyosin strands; one for each groove.
- Each long tropomyosin strand is made of short tropomyosin strands arranged end-to-end so that they appear like a long twisted strand.

» Each short tropomyosin strand makes a tropomyosin molecule and is made of two subunit strands twisted around each other. Each subunit strand has a helical structure.

» Thus, in each F-actin groove, there is a long tropomyosin twisted strand which is made of short tropomyosin twisted strands (tropomyosin molecules) arranged end-to-end. Each short tropomyosin strand extends over seven G-actin monomers.

» The tropomyosin molecules cover the myosin-binding sites on the G-actin molecules thereby preventing the binding of myosin heads to the actin filament (Figure 3.16).

» For binding of the myosin heads to the actin filament, as required during muscle contraction, the tropomyosin molecules have to be displaced so as to expose the myosin-binding sites on the G-actin molecules. This takes place with the help of troponin molecules.

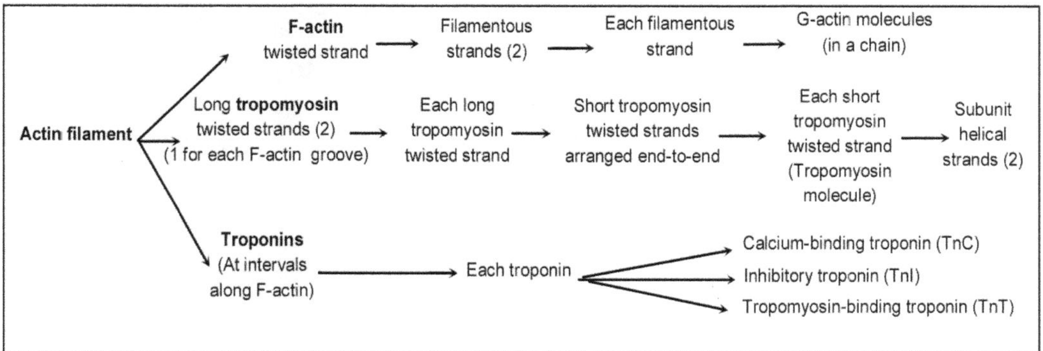

Figure 3.15: Components of an actin filament

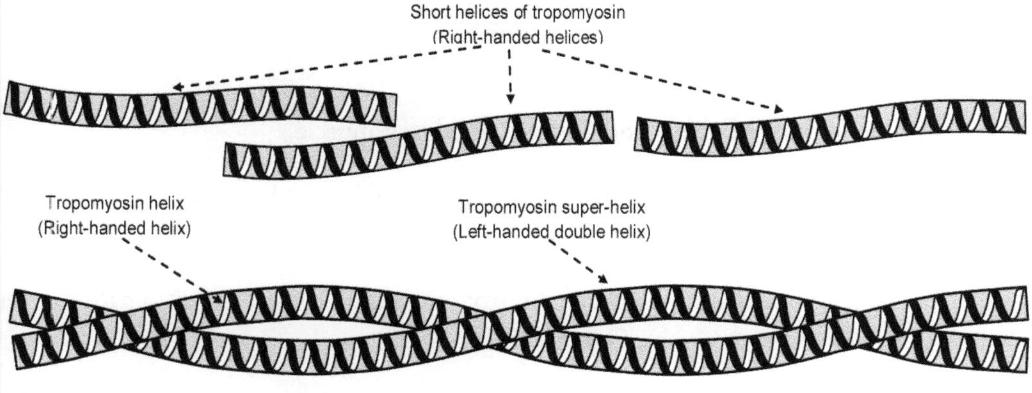

Figure 3.16: Tropomyosin molecules

Troponin (Tn):

- Troponin is a globular protein, which remains in the filament at a regular interval of 7 G-actin monomers (Figure 3.17). Each troponin has three subunits as given below.

1. TnC:

- It is the calcium-binding troponin (C= calcium) (TnC), which can bind two calcium ions (Ca^{++}).
- It is bonded to inhibitory troponin (TnI) only both in relaxed muscle and during muscle contraction.
- Binding of Ca^{++} ions to TnC results in a change in the three-dimensional conformation of the whole troponin molecule.

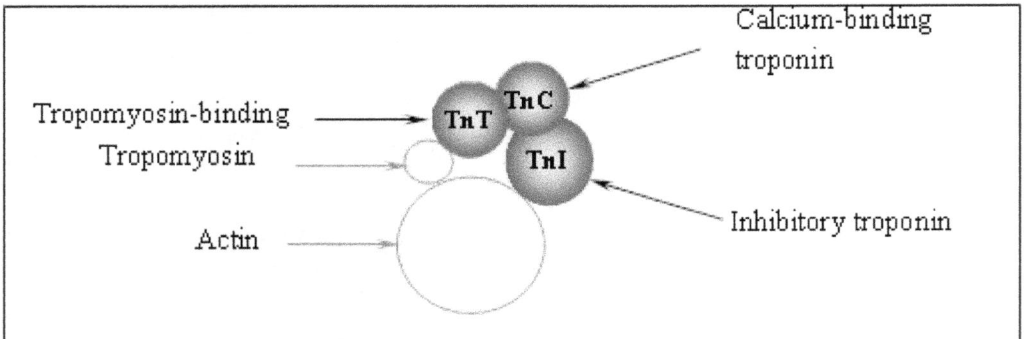

Figure 3.17: A troponin molecule

2. TnI:

- It is the inhibitory troponin (I= inhibitory), which inhabits contact of actin with myosin.
- It is bonded to TnC, TnT and G-actin in relaxed muscle.
- It binds to the G-actin in such a way that it inhibits the binding of myosin head to actin.
- This is accomplished as it pushes the tropomyosin (T) filament via TnT, to cover the myosin-binding site on G-actin (Figure 3.18).
- During muscle contraction, it is delinked from G-actin.

3. TnT:

- It is the tropomyosin-binding troponin (T=tropomyosin).
- It is bonded to TnI and tropomyosin (Tm) both in relaxed muscle and during muscle contraction (Figure 3.18).

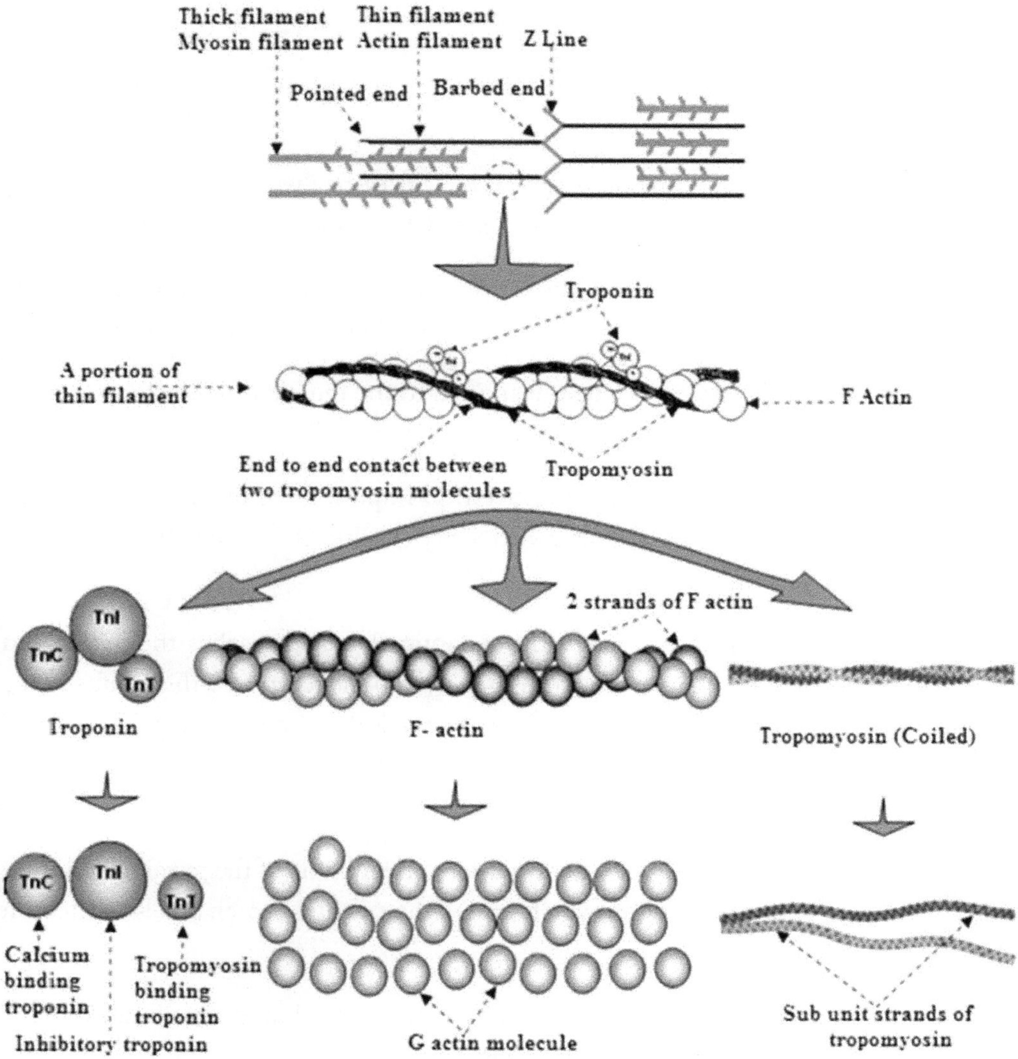

Figure 3.18: The filament

- For muscle contraction, a nerve ***impulse*** is created in the brain which is transmitted to the muscle cells by the long nerve cells connected to them (Figure 3.20).
- The junction of the muscle cell and nerve cell is called 'myo-neural junction' or 'neuro-muscular junction'.
- The terminal portion of the nerve cell which is dilated and remains very close to the surface of the muscle cell is called 'axon terminal'.
- The sarcolemma of the muscle cell at the myo-neural junction is highly folded and is called 'motor end plate'.
- The nerve cell and muscle cell are not in contact with each other as there is a narrow space between the axon terminal and the motor end plate. This space is called 'synaptic cleft'.
- The impulse created in the brain is transmitted as electrical impulse by alternate depolarisation and repolarisation of the cell membrane of the nerve cells.
- This electrical impulse cannot be transmitted to the muscle cell directly as there is no direct contact between the nerve cell and muscle cell.
- Therefore, at the axon terminal, the electrical impulse is converted to a chemical impulse and a neurotransmitter called acetylcholine is secreted from the axon terminal to the synaptic cleft.
- The acetylcholine passes through the synaptic cleft and reaches the motor end plate of the muscle cell where it binds to acetylcholine receptors. This results in the depolarisation of the sarcolemma of the muscle cells thereby converting the chemical impulse to electrical impulse. This electrical impulse spreads into the muscle cell by the spreading network of T-tubules.
- The T-tubules pass the impulse to the adjacent membrane of the terminal cisternae of the sarcoplasmic reticulum which causes the calcium ion channels located on their membrane to open.
- As a result, calcium ions stored in the sarcoplasmic reticulum diffuse freely into the sarcoplasm through the calcium ion channels.
- These calcium ions, upon reaching the myofibrils of the muscle cells, bind to the TnC (2 Ca^{++} per TnC) of the troponin resulting in the change in the three-dimensional conformation (quaternary structure) of the globular protein troponin as a whole (Figure 3.20).

» Due to this, the linkage among TnC, TnI and TnT changes and TnI gets delinked from the G-actin.

» As a result, the tropomyosin molecules, which are bonded to TnT subunits, moves deeper into the actin groove exposing the myosin-binding site on the G-actin molecules.

» Now, the myosin heads projecting from the myosin filaments (thick filaments) towards the actin filaments (thin filaments), bind to the G-actin molecules at the exposed myosin-binding sites. The bond formed is called 'actin-myosin cross-bridge'.

» Utilising ATP, the myosin necks bend moving the heads towards the center of the sarcomere.

» As the G-actin molecules are bonded to the myosin heads, the actin filaments also moves towards the center of the sarcomere. Thus, the actin filaments slide past the myosin filaments.

» As the actin filaments from both the sides move towards the center of the sarcomeres, the Z lines which anchor them also move from both sides towards the center of the sarcomere. At the end of the single movement of each myosin head, an ATP gets attached to it resulting in the breakdown of the cross-bridge.

» The head gets detached from the G-actin and comes back to its original position.

» Now, the attached ATP is broken down by the Ca^{++} ATPase present in the myosin head to ADP and Pi with liberation of energy.

» Using this energy the myosin head again forms cross-bridge with G-actin and bends the myosin neck moving the heads towards the center of the sarcomere.

» Thus, the series of myosin heads act like levers and make repeated oar-like movement to slide the thin filaments over the thick filaments.

» This leads to gradual movement of the Z lines towards the center of the sarcomeres.

» As a result, the sarcomere shortens leading to shortening of the myofibrils and muscle contraction.

» Such theory of muscle contraction is called '***sliding filament cross-bridge theory***'.

» With the development of cryo-electron microscopy, high precision x-ray crystallography and advanced immunochemical techniques, finer structures associated with muscle have been elucidated.

» These structures include the following.

1. Titin (Connectin)	8. F-protein
2. Telethonin (T-cap)	9. α-Actinin
3. Nebulin	10. Cap Z
4. Tropomodulin (Tmod)	11. M protein
5. C-proteins	12. Myomesin
6. X-protein	13. Skelemin
7. H-protein	14. Desmin

» Two proteins (titin and nebulin) also contribute to sarcomere structure and stability.

» Titin is an extremely large protein (3000 kd), and single titin molecules extend from the M line to the Z disc. These long molecules of titin are thought to act like springs that keep the myosin filaments centered in the sarcomere and maintain the resting tension that allows a muscle to snap back if overextended.

» Nebulin filaments are associated with actin and are thought to regulate the assembly of actin filaments by acting as rulers that determine their length.

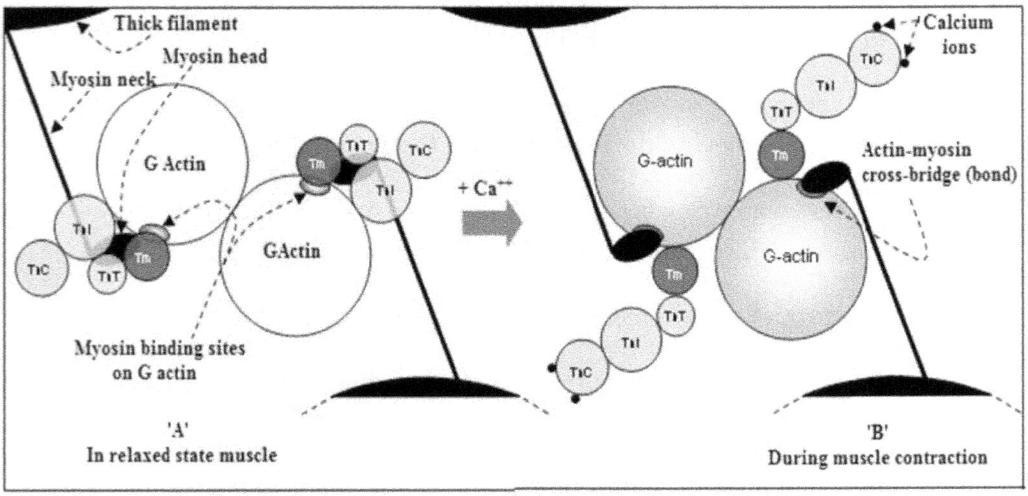

Figure 3.19: Thin filament (Cross sectional view)

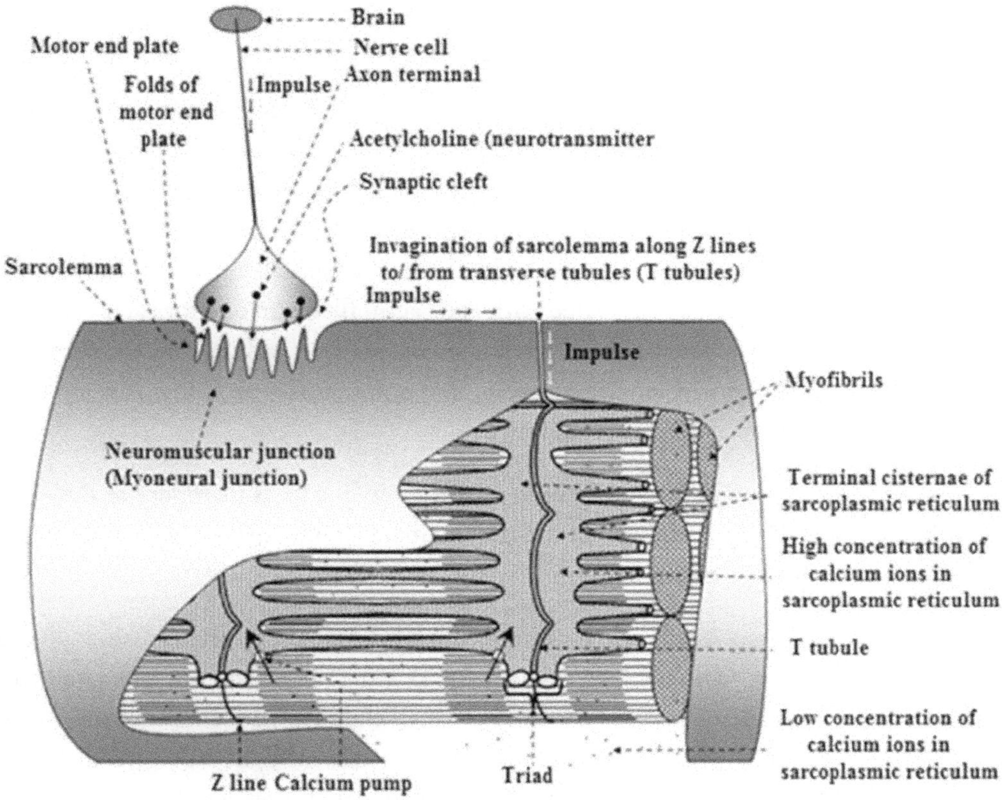

Figure 3.20: Impulse transmission for muscle contraction

1. **Titin (Connectin):** It is a filamentous protein extending from the Z-line to the M-line in the sarcomere, thus stretching over half the sarcomere. It is the third most abundant muscle protein, next to myosin and actin. It is the largest protein known so far. Therefore, it is called a 'giant protein' and is named titin (Greek: titan: a giant deity, or anything of great size). Among the various forms of titin (isoforms), the largest one is made of a chain of 34350 amino acid residues. It connects the thick filaments to the Z-line. Therefore, before being named as titin, it was called connectin. The elastic nature of muscle fibers is due to the elastic nature of titin. As it's I band portion is very extensible, it has the capacity to elongate or coil. Therefore, it acts as an 'elastic' which binds the myosin filament to the Z-line. Besides, a titin filament serves as the supporting structure on which the myosin molecules are held by adhesion to form the myosin filaments

2. **Telethonin (T-cap)**: It is a protein of the Z-line which anchors the titin to the Z-line thereby preventing it from being slipped out from the Z-line. As it acts like a cap, it is

a cap protein and is also call 'T-cap' (titin-cap). It is named so as it anchors the end of titin to Z-line (*tele*: end).

3. **Nebulin:** It is a giant protein and named so because of its nebulous (unclear) nature. It forms a filament-like structure that extends equal length along the thin filament. It is fully integrated into the structure of the actin filament. The nebulin binds actin with high affinity thereby forming actin-nebulin complex. Its N terminus is found near the pointed end while its C terminus is embedded near the barbed end of the thin filament. Two nebulin molecules are associated with each actin filament, one for each face of the filament. There is close association between nebulin and the thin filament. The correlation of its size to thin filament lengths suggests that nebulin may function as a 'molecular ruler' to determine thin filament length.

4. **Tropomodulin (Tmod):** It is a cap protein similar to T-cap. It is present at the end of nebulin and caps the pointed end of the thin filament. It is associated with determining the length of the thin filament.

5. **C-proteins:** It is a myosin-binding protein which binds together the myosin molecules to form the myosin filament. Thus, it is a component of the myosin filament which plays a role in its assembly (*i.e.*, in the assembly of the myosin filaments from myosin molecules). As it is a myosin-binding protein, it is also named MyBP-C[1] (myosin-binding protein C). As it is not a part of the myosin molecule, it is considered as a non-myosin protein. It has high affinity for myosin molecules. It is located in seven stripes (5th to 11th) along each half A-band of the myofibrils. MyBP-C has three isoforms, *viz*, cardiac MyBP-C (cMyBP-C) and fast (white) skeletal MyBP-C (wMyBP-C) and slow (red) skeletal MyBP-C (rMyBP-C). Slow (red) skeletal MyBP-C was originally described as X-protein (MyBP-X) as given below.

> MyBP-C[1] : *A group of non-myosin proteins present in the myosin filament are called 'myosin binding proteins (MyBP). They are of four types such as C-protein, H-protein, F-protein and X-protein. These are also called MyBP-C, MyBP-H, MyBP-F and MyBP-X proteins respectively. These proteins are detected on the myosin filament as a series of 11 transverse stripes in the central region of each half of the A-band. The proteins of the* 1st, 2nd *and* 4th *stripes from the M-line have not yet been identified. The protein of the* 3rd *stripe is H-protein while the proteins of the rest seven stripes (from* 5th *to* 11th*) are C-proteins.*

6. **X-protein:** It is the second type of non-myosin myosin binding protein (MyBP) found only in slow (red) skeletal muscle and is named MyBP-X. Due to its resemblance to MyBP-C, it is now considered as an isoform of MyBP-C.

7. **H-protein:** It is the third type of non-myosin myosin binding protein (MyBP) present in the myosin filament. It is the protein of the 3rd stripe present in the central region of each half of the A band. It is a low molecular weight MyBP-C with which it has sequence homology.

8. **F-protein:** It is the fourth type of non-myosin myosin binding protein (MyBP) present in the myosin filament. It is located on the S2 (myosin neck) of myosin molecules. It has phosphofructokinase activity.

9. α-Actinin: It is the major constituent of the Z-lines of myofibrils. Each α-actinin is an anti-parallel rod-shaped dimer (made of two monomers) with one actin-binding site at each end of the rod.

10. **CapZ:** It is a protein found in the Z-line which caps the barbed end of a thin filament. It prevents monomer (G actin) association at this end.

11. **M protein:** It is a protein present in the M line of the myofibrils. Its central region corresponds to the middle of the m line (i.e. M1 line) with its terminal regions organised along neighbouring thick filaments.

12. **Myomesin:** It is a flexible segmented rod present in the M line. It binds myosin to titin thereby linking titin molecules to the thick filament.

13. **Skelemin:** It is a protein present in the periphery of the M line that links the thick filaments with the intermediate[2] filaments which link adjacent myofibrills.

14. **Desmin :** It is a protein which makes the intermediate filaments that encircle Z lines and link adjacent myofibrils.

[2] *In the muscle cells the myofibrils are anchored to an interconnecting cytoskeleton of intermediate filaments. These are called intermediate filaments as their diameter is intermediate between those of the thick and thin filaments of the myofibrils. They encircle the myofibrils at the Z lines as two closely spaced rings and at the M lines as a single ring. From the peripheries of these rings emerge transverse filaments connecting the myofibrils to each other at neighbouring Z and M lines respectively and to the costameres at the sarcolemma. Costameres are a group of proteins located below the sarcolemma which form a junctional complex connecting the myofibrils at the Z lines with the sarcolemma as well as with the extracellular matrix of collagen present as the connective tissues. Longitudinal filaments also extend between consecutive Z lines.*

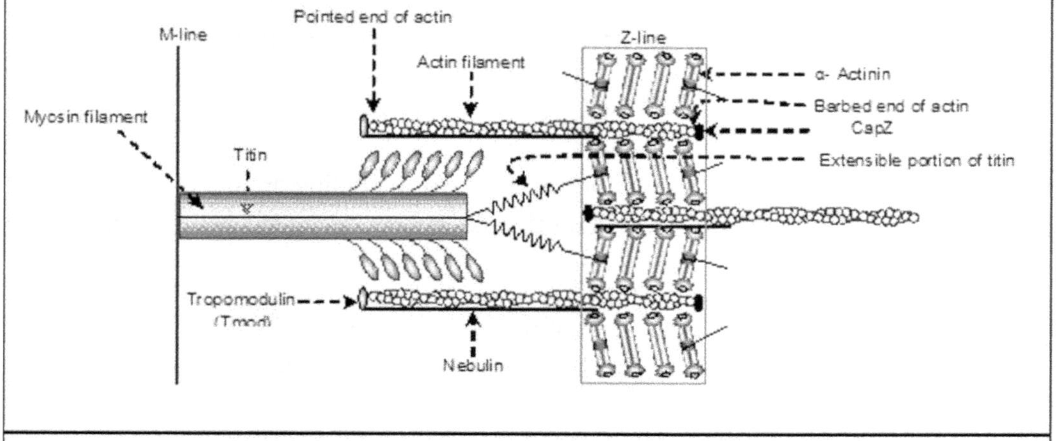

Figure 3.21: Titin, nebulin, tropomodulin (Tmod), CapZ

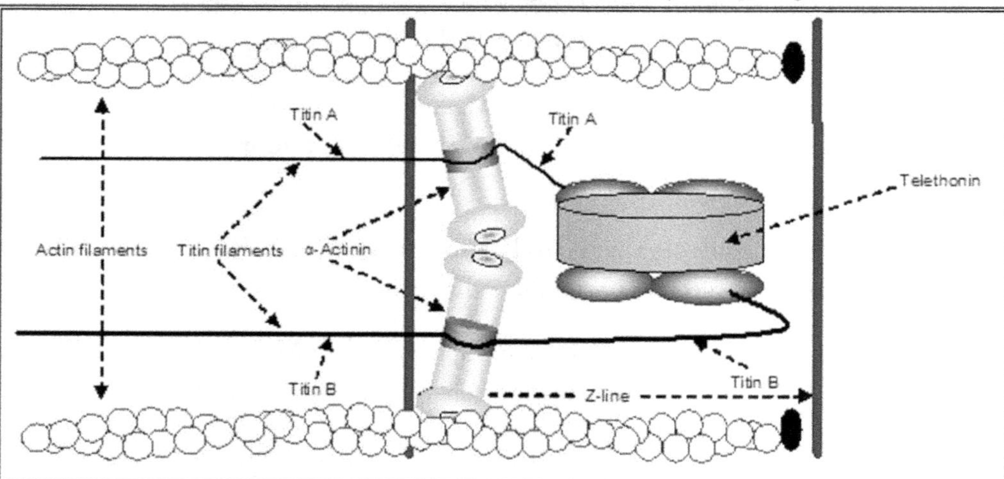

Figure 3.22: α-Actinin and telethonin (T-cap)

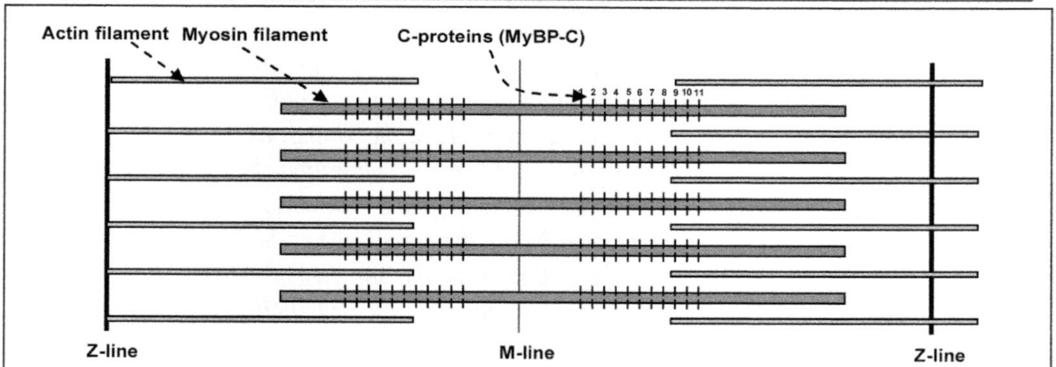

Figure 3.23: Myosin binding proteins (11 stripes)

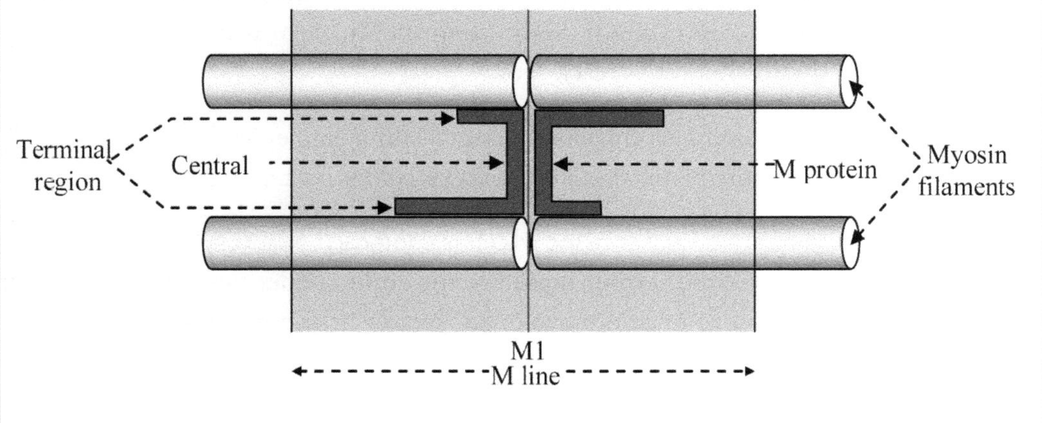

Figure 3.24: M protein

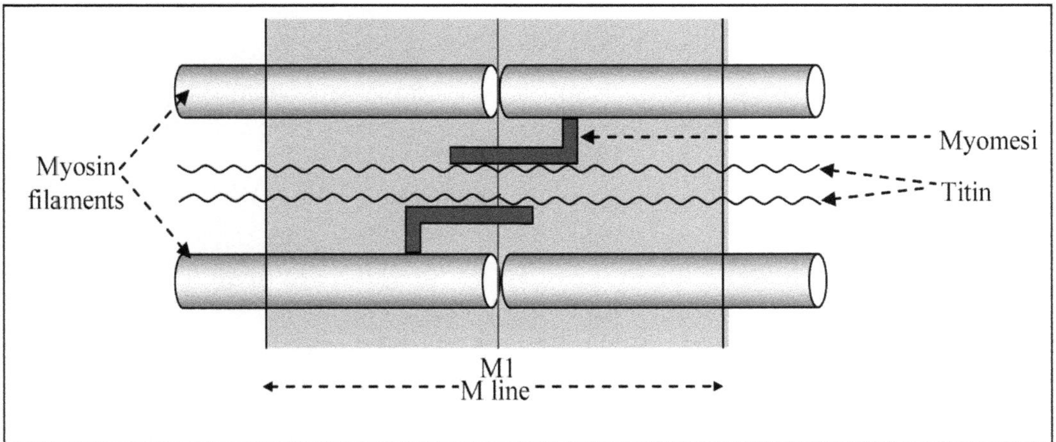

Figure 3.25: Myomesin

3. Muscle of crustaceans:

- The crustaceans differ substantially from the fish.
- The outermost layer, called the **Shell**, is an exoskeleton which supports and protects the softer parts of the body.
- The shell consists of modified protein and a polysaccharide (glucosamine) known as chitin, and may be calcified.
- This shell is shed at intervals and replaced by a new larger shell to accommodate the growth of the soft parts of the body.

- This moulting process (**Ecdysis**) requires profound physiological changes in the animal which also influence its eating and processing quality, in particular its susceptibility to a discolouration known as black spot.
- The muscles of crustaceans are bound externally by a pigmented layer called skin. It is a heavily pigmented layer that may easily discolour post mortem.
- The fat in fish is stored in fat tissue which is composed of special fat cells. In fatty fish species like sardine, herring, mackerel, the fat lies under the skin in the lining of the gut and also in muscle tissue. Lean species of fish carry most of their fat depots in the liver.

4 Post Mortem Changes in Fish and Safety Hazards

Fish has been a food item since ancient times. It is a lucrative nutritionally rich, cheap source of animal protein. At present it is gaining additional importance among the affluent people due to their increasing awareness about its unique nutritional properties as 'Health food'. Although advancement in technology of fishing and aquaculture has resulted in increasing the production of fish, its effective utilization has not yet met with the expected success, particularly in the tropical developing countries. A large portion of the fish catch is either discarded or sold at a low price due to the fast deterioration of its quality. Fish is a highly perishable food commodity as its quality starts deteriorating immediately after death. The lion's share of the total catch is consumed fresh in which the post-mortem changes have tremendous influence on quality. Moreover, the quality of processed fishery products depends largely on the quality of the raw material which in turn depends on the post-mortem changes.

Generally, food safety, organoleptic features, nutritional quality and aptitude to industrial transformation, and acceptability of fish as food are the important quality attributes of fish flesh. Freshness of fish is closely linked to sensorial changes and texture properties. Along with ante mortem muscle biochemistry, postmortem biochemical processes are directly linked to final quality attributes. The understanding of postmortem mechanisms is a pre-requisite for an accurate control of the quality of commercialized fish by the identification of objective indicators. Postmortem tenderization is one of the most important quality attributes of fish muscle. Loss of freshness is due to a complex combination of biochemical, chemical and physical processes, and is followed by muscle spoilage due to microbiological contamination. Autolytic modifications include protease action on proteins and connective tissue, and also fat hydrolysis. Tenderization is enzymatic in nature; physio-chemical conditions (pH, osmotic pressure) may modulate the proteolytic action of endogenous enzymes.

A. SENSORY EVALUATION

The sensory analysis of fresh fish was evaluated on the basis of general appearance (skin color and mucus), stiffness of muscle, consistency of belly cavity, color and shape of eyes, color, odor and mucus of gills, color and adherence to vertebral column and integrity and brightness of peritoneum (Table 4.1). But, in markets and landing sites it is done by assessing the appearance, texture and odour. The sensory attributes for fish are listed in Table below. The characteristic sensory changes in fish post mortem vary considerably depending on fish species and storage method. The scoring systems are based upon changes taking place during storage. The appearance of fish stored under chilled condition without ice does not change as much as for iced fish, but the fish spoil more rapidly and an evaluation of cooked flavour will be necessary. Knowledge of the time /temperature history of the fish should therefore be essential at landing. The suggested scale is numbered from 0 to 3, where 3 is the best quality.

If quality criteria of chilled fish during storage are needed, sensory assessment of the cooked fish can be conducted. Some of the attributes for cooked fish and shellfish are mentioned in Table below. A characteristic pattern of the deterioration of fish stored in ice can be found and divided into the following four phases:

» **Phase 1:** The fish is very fresh and has a sweet, seaweedy and delicate taste. The taste can be very slightly metallic. In fresh fish, the sweet taste is maximized 2-3 days after catching.

» **Phase 2:** There is a loss of the characteristic odour and taste. The flesh becomes neutral but has no off-flavours. The texture is still pleasant.

» **Phase 3:** There is sign of spoilage and a range of volatile, unpleasant-smelling substances produced depending on the fish species and type of spoilage (aerobic, anaerobic). One of the volatile compounds may be trimethylamine (TMA) derived from the bacterial reduction of trimethyl-aminoxide (TMAO). TMA has a very characteristic "fishy" smell. At the beginning of the phase the off-flavour may be slightly sour, fruity and slightly bitter, especially in fatty fish. During the later stages sickly sweet, cabbage-like, ammoniacal, sulphurous and rancid smells develop. The texture becomes either soft and watery or tough and dry.

» **Phase 4** The fish can be characterized as spoiled and putrid.

Table 4.1: Freshness ratings: Council Regulation (EEC) No. 103/76 OJ No. L20 (28 January 1976) (EEC, 1976)

Criteria				
	Marks			
Part of fish inspected	**3**	**2**	**1**	**0**
Appearance				
Skin	Bright, iridescent pigmentation, No discoloration Aqueous, transparent, mucus	Pigmentation bright but not lustrous Slightly cloudy mucus	Pigmentation in the process of becoming discoloured and dull Milky mucus	Dull pigmentation Opaque mucus
Eye	Convex (bulging) Transparent cornea Black, bright pupil	Convex and slightly sunken Slightly opalescent cornea Black, dull pupil	Flat Opalescent cornea Opaque pupil	Concave in the centre Milky cornea Grey pupil
Gills	Bright colour No mucus	Less coloured Slight traces of clear mucus	Becoming discoloured Opaque mucus	Yellowish Milky mucus
Flesh (cut from abdomen)	Bluish, translucent, smooth, shining No change in original colour	Velvety, waxy, dull Colour slightly changed	Slightly opaque	Opaque
Colour (along vertebral column)	Uncoloured	Slightly pink	Pink	Red
Organs	Kidneys and residues of other organs should be bright red, as should the blood inside the aorta	Kidneys and residues of other organs should be dull red; blood becoming discoloured	Kidneys and residues of other organs and blood should be pale red	Kidneys and residues of other organs and should be brownish in colour
Condition				
Flesh	Firm and elastic Smooth surface	Less elastic	Slightly soft (flaccid), less elastic Waxy (velvety) and dull surface	Soft (flaccid) Scales easily detached from skin, surface rather wrinkled, inclining to ealy
Vertebral column	Breaks instead of coming away	Sticks	Sticks slightly	Does not stick
Peritoneum	Sticks completely to flesh	Sticks	Sticks slightly	Does not stick
Smell				
Gills, skin abdominal cavity	Seaweed	No smell of seaweed or any bad smell	Slightly sour	Sour

B. POST MORTEM CHANGES

After death of fish several changes take place in its body. These changes are collectively called 'post mortem changes'. They can be grouped into the following six steps mentioned in the order of their occurrence after death.

1. Hyperaemia
2. Rigor mortis
3. Autolysis
4. Microbial putrefaction or Decomposition
5. Lipid oxidation or Autoxidation
6. Discoloration

1. Hyperaemia:

» The secretion of large quality of slime to body surface after death of fish is called 'hyperaemia'. It occurs as a peculiar reaction of the dying fish to the unfavorable surrounding. The mucus glands of the skin secret large quantity of mucus to the skin surface which forms a thick layer of slime on the fish body. When fish is alive, a thin layer of slime covers its body. It protects the fish body against microbial invasion in three ways as given below.

   1. It has immunological properties.
   2. It contains lysozyme.
   3. It acts as a physical barrier.

» But when fish dies, microbial invasion starts due to the following two reasons.

   1. The immunological properties of the slime are lost.
   2. Microbes attach to the sticky slime

» As slime consists mainly of the glycoprotein, mucin, which is a rich nutrient for microbes, it becomes an excellent substrate for the growth of the microbial flora present on the skin. These putrefy the slime giving offensive smell. They penetrate into the muscle and internal organs through the skin, which becomes still easier if there cuts or bruises on it. They decompose the muscle and internal organs with production of off-odour, off-flavour substances, presence of which distracts the consumers and renders the fish spoiled. Thus the release of slime does not in itself lead to spoilage; rather it acts as a substrate for microbial growth and as a vehicle for penetration of microbes into the fish body, which results in its spoilage.

2. Rigor mortis:

After short lapse of time from the death of fish, its body remains stiffened for a certain period during which the elastic nature of living muscle disappears and muscle becomes hard. Such stiffening of muscle after death is called rigor mortis. Depending upon the state of rigor mortis, fish to be utilized for processing can be in any one of the three states such as (a) Pre-rigor (b) On-rigor (c) Post-rigor.

» **Stages of rigor mortis:**

i. **Stiffening of muscle:** During the stiffening stage, the muscle stiffens gradually resulting in a gradual decrease in the softness from the original softness level. Stiffening continues to a point of maximum stiffness called 'peak'. After this starts the second stage of rigor mortis.

ii. **Softening of muscle:** During this stage, the muscle looses its stiffness gradually and softens gradually. Softening continues till the softness return back to original softness level. Rigor mortis is said to be completely resolved or terminated when the softness of fish tissue reaches the original softness level. However, the gradual softening of the muscle continues unabated even after the termination of rigor mortis.

» **Biochemistry of rigor mortis:**

The most important event occurring after the death of fish, which is responsible for rigor mortis, is the breakdown of its tissue components starting immediately after death. The breakdown of carbohydrates and nucleotides start immediately after death leading to rigor mortis while the breakdown of protein, lipid and nucleic acids takes place afterwards during autolysis, which follows rigor mortis.

The five postmortem hydrolytic processes are given below.

A	Saccharolysis	Breakdown of carbohydrates (glucose & glycogen)	Start immediately after death
B	Nucleotidolysis	Breakdown of nucleotides (ATP)	
C	Proteolysis	Breakdown of protein	Start with onset of autolysis which follows rigor mortis
D	Lipolysis	Breakdown of lipid (fat)	
E	Nucleolysis	Breakdown of nucleic acid (DNA & RNA)	

While the fish is alive, cycles of chemical changes take place continuously in the muscle; these provide energy for the muscle while the fish is swimming, and also produce substances necessary for growth and replacement of worn-out tissue. The compounds that bring about, and control, these changes are known as enzymes. The enzymes in the flesh go on working even after the fish is dead, and some of them act on those substances that normally keep the muscle pliable and lifelike. During life the muscle would contract and become rigid if its two main protein components were allowed to interact and bond together, but the bonding is prevented by the presence of substances that keep the muscle pliable. For so long as the muscles contain a reserve of energy, these substances can be replaced by one set of enzymes as fast as they are destroyed by another; thus the muscles stay pliable for a time after death. But once the energy reserves are used up, the replacement stops and depletion results. The protein components are then able to interact, the muscle attempts to contract, and it eventually becomes hard and rigid. The interaction of the protein components is also influenced by the accumulation of lactic acid, which is produced in the muscle when the energy reserves break down.

» **Changes during rigor mortis:**

The two most important manifested during rigor mortis are

a. Loss of texture of fish as swan by pressure marks left when a post-rigor fish gently squeezed

b. Fall in pH of fish tissue.

Immediately after death anaerobic break down of the glycogen to lactic acid starts in the muscle tissue, which continues during rigor mortis. Production and accumulation lactic acid result in a fall in pH of fish tissue from near neutral (7.05- 7.35) to about 6.2- 6.5. Fall in pH suppresses the development of spoilage microbes and thereby suppresses the spoilage of fish.

» Postmortem muscle structural changes:

Fish muscle is organized as tissue blocks of myomeres attached to myocommata. This structure makes the flesh inherently soft, especially when cooking, because fish connective tissue is soluble at low cooking temperatures. However, there is very little structural change in fish myofibrils postmortem. In fish, I bands are almost never broken. The cytoskeletal proteins of fish are degraded within days postmortem, so it is surprising that fish myofibrils are structurally stable.

» Influence of rigor mortis on quality of fish:

Rigor mortis has great influence on the quality of fresh and processed fish. Rigor mortis has beneficial effects on the keeping quality of fresh fish. It extends the shelf life fresh fish by delaying the activity of spoilage microbes, as the low pH produced during rigor is unfavourable for most of them. In frozen whole fish, it can influence the quality through gaping, toughness of tissue and drip loss. In frozen fillets, it also influences the quality adversely through shrinkage, toughness of tissue and drip loss. It influences the quality of smoked fish in that, when fish is smoked pre-rigor, the treatment in the smoking kiln can cause high temperature rigor effects resulting in much shrunken smoked fillets.

The time taken by a fish to go into, and pass through, rigor depends on the species, its physical condition, the degree of exhaustion before death, its size, the amount of handling during rigor and the temperature. Generally, small fish with low reserves of energy gets exhausted in poor condition, and when kept at a high temperature, will enter and pass through rigor very quickly. On the other hand, large, rested, well-fed fish kept at a low temperature will take a very long time to enter and pass through rigor.

The problems of rigor affect in the fish industry are more acute because there is no control over the nutritional condition or the degree of exhaustion of the fish before their harvest. Rigor creates problems mainly for freezing fish at sea, either as whole fish or as fillets. Rigor can affect the quality of whole fish in three main ways, by causing gaping in wet and frozen fish, and toughness and excessive drip loss on thawing in frozen fish.

A fillet is said to gape when the individual flakes of muscle come apart, giving the fillet a broken and ragged appearance. This happens when the material that binds the flakes together, known as connective tissue, breaks down. There appear to be several causes of gaping, one of which is the rigor process. As muscle goes into rigor, it attempts to contract but, because the skeleton and the connective tissue prevent contraction, tension increases within the muscle. As long as the connective tissue can withstand this increase in tension, the flesh will not gape, but when the tension becomes greater than the inherent strength of the connective tissue, some gaping will occur. The temperature of the whole fish as it goes into rigor can have a marked effect on the amount of gaping; the higher the temperature when it goes into rigor, the greater is the rigor tension and the weaker the connective tissue becomes. Thus the higher the temperature the more the flesh will gape. Rough handling of fish in rigor can also cause gaping, because any attempt to bend a rigid fish will break the muscle or the connective tissue.

The higher the temperature at which a fish goes into rigor, the greater will be the drip loss on thawing and, when the fish is cooked and eaten, it will be tough and stringy. Whole fish frozen pre-rigor tend to have a higher drip loss than similar fish frozen in rigor or post-rigor, but this may be due to what is known as thaw rigor. The safest and most reliable way of avoiding the undesirable effects of rigor is to keep the fish chilled at every stage before freezing. Provided the fish pass through rigor at a low temperature, the effect of rigor on quality will not be serious.

Thaw rigor:

When muscle is frozen pre-rigor and kept for a short time in cold storage, it is still able to contract and go into rigor after thawing. This is known as thaw rigor and, when the thawing is done rapidly at a high temperature, the muscle can then suffer from the defects associated with high temperature rigor.

Thaw rigor is rarely a problem in thawed whole fish because freezing and cold storage has usually sapped the energy reserves sufficiently to weaken the contractions in the muscle; the skeleton restrains the muscle but the stresses are insufficient to break the connective tissue. If damage is seen at all in thawed whole fish, it usually occurs near the tail, where thawing is most rapid.

However, when pre-rigor fillets are thawed, the muscle is free to shrink as soon as the ice within the flesh melts, and the fillets become shrunken and corrugated and lose a large amount of drip. The effects are most severe when the pre-rigor muscle is cooked from the frozen state. The simplest way to avoid thaw rigor is to extend the cold storage time of the stock of pre-rigor fish. Whole fish and fillets frozen pre-rigor can give good products if they are carefully thawed at a low temperature in order to give *rigor mortis* time to pass while the muscle is still frozen.

Frozen whole fish:

	Advantages	Disadvantages
Frozen pre-rigor	• Buffer store not required • No gaping, except possibly from thaw rigor	• Thaw rigor gaping possible • High drip loss may occur • Large processing capacity required to cope with high catching rates
Frozen in rigor	• Uniformly good quality	• Buffer store required • Texture variation possible • Gaping or broken fillets when fish are forcibly straightened or rigor temperature is high
Frozen post-rigor	• Uniformly good quality • Danger of contraction damage avoidable	• Buffer store required • Gaping may occur when held too long or at too high a temperature before freezing

Frozen fillets:

	Advantages	Disadvantages
Frozen pre-rigor	• Buffer store not required • Fillets can be cut by hand or machine	• Large processing capacity required to deal with high catching rates • Fillets shrink when awaiting freezing or after thawing • Rough cut surface • Particularly unsuitable for smoking • May be high drip loss
Frozen in rigor	• Excellent quality possible • No shrinkage	• Buffer store required • Difficult to fillet by machine or by hand • Less yield from hand filleting • Usually unsuitable for smoking • Bent fish yield gaping fillets
Frozen post-rigor	• Uniformly high quality • No shrinkage • Machine or hand filleting	• Large buffer store required for up to 3 days • Usually unsuitable for smoking

3. Autolysis:

Autolysis means "self-digestion". It is the process of break down of fish tissue after death by own body enzyme. It manifested itself mainly through softening of the tissue.

Production of energy in post mortem muscle:

At the point of death, the supply of oxygen to the muscle tissue is interrupted because the blood is no longer pumped by the heart and is not circulated through the gills where, in the living fish, it becomes enriched with oxygen. Since no oxygen is available for normal respiration, the production of energy from ingested nutrients is greatly restricted.

The complex tissue components are broken down or hydrolyzed into their simple building units by the native or endogenous enzymes present in the fish tissue. The five most important tissue components broken down during autolysis are proteins, lipids, nucleic acid, carbohydrates and nucleotides.

Glycogen (stored carbohydrate) or fat is oxidized or "burned" by the tissue enzymes in a series of reactions which ultimately produce carbon dioxide (CO_2), water and the energy-rich organic compound adenosine triphosphate (ATP). This type of respiration takes place in two stages: an anaerobic and an aerobic stage. The latter depends on the continued presence of oxygen (O_2) which is only available from the circulatory system. Most crustaceans are capable of respiring outside the aquatic environment by absorption of atmospheric oxygen for limited periods.

Under anaerobic conditions, ATP may be synthesized by two other important pathways from creatine phosphate or from arginine phosphate. The former source of energy is restricted to vertebrate muscle (teleost fish) while the latter is characteristic of some invertebrates such as the cephalopods (squid and octopus). In either case, ATP production ceases when the creatine or arginine phosphates are depleted. It is interesting to note that octopine is the end-product from the anaerobic metabolism of cephalopods and is not acidic (unlike lactate), thus any changes in post mortem pH in such animals are not related to the production of lactic acid from glycogen.

For most teleost fish, glycolysis is the only possible pathway for the production of energy once the heart stops beating. By this process, lactic and pyruvic acids are produced as end-products. In addition, ATP is produced in glycolysis, but only 2 moles for each mole of glucose oxidized as compared to 36 moles ATP produced for each mole of glucose if the glycolytic end products are oxidized aerobically in the mitochondrion in the living animal. Thus, after death, the anaerobic muscle cannot maintain its normal level of ATP. Thus, post mortem glycolysis results in the accumulation of lactic acid which in turn lowers the pH of the muscle.

The post mortem reduction in the pH of fish muscle has an effect on the physical properties of the muscle. As the pH drops, the net surface charge on the muscle proteins is reduced, causing them to partially denature and lose some of their water-holding capacity. Muscle tissue in the state of rigor mortis loses its moisture when cooked and is particularly unsuitable for further processing which involves heating, since heat denaturation enhances the water loss. Loss of water has a detrimental effect on the texture of fish muscle and there is an inverse relationship between muscle toughness and pH. Unacceptable levels of toughness (and water-loss on cooking) generally occur at lower pH levels.

Types of Enzymes Involved in Autolysis:

With respect to the location in fish body and the mode of regulation of their activities, the enzymes involved in autolysis are of two types.

- Digestive or gut enzymes
- Muscle and internal organ enzymes

i. Autolytic hydrolysis:

The autolytic break down of the important fish tissue components takes place in the following five steps.

(a) Saccharolysis:

Complex carbohydrates —*Carbohydrases*→ Simple sugars

(b) Nucleotidolysis:

Nucleotide (ATP) —*Nucleotidases*→ Phosphates + sugar (D-ribose) + purine (hypoxanthine)

(c) Proteolysis:

Proteins —*Proteases*→ Amino acids

(d) Lipolysis:

Nucleic acid (DNA & RNA) —*Lipases*→ Free fatty acid + glycerol

(e) Nucleolysis:

Nucleic acid (DNA & RNA) —*Nucleases (DNase/RNase)*→ Nucleotides

Out of the above five steps, saccharolysis and nucleotidolysis take place immediately after death. They have been described under rigor mortis. The process of true autolysis, which follows rigor mortis and which manifests itself in softening of the muscle has the following three steps.

a. Proteolysis

b. Lipolysis

c. Nucleolysis

» **Effects of autolysis:** Autolysis has both negative and positive effects.

- Negative effects: Softening of tissue, Penetration of microbes, Growth of microbes, Increase in NPN content, Development of rancidity, Belly bursting, Gaping
- Positive effects: Taste and flavor development, Liquefaction, Maturation of salted fish, Tenderization.

Autolysis and nucleotide catabolism:

The resolution of rigor is a process which results in the subsequent softening (relaxation) of the muscle tissue and is related to the activation of one or more of the naturally-occurring muscle enzymes, digesting away certain components of the rigor mortis complex. The softening of the muscle during resolution of rigor (and eventually spoilage processes) is coincidental with the autolytic changes. Among the changes, one of the first to be recognized was the degradation of ATP-related compounds after death. The degradation of ATP forms adenosine diphosphate (ADP), adenosine monophosphate (AMP), inosine monophosphate (IMP), inosine (Ino) and hypoxanthine (Hx). A formula has been developed for fish freshness based on these autolytic changes:

$$K\% = \frac{[Ino]+[H_x]}{[ATP]+[ADP]+[AMP]+[IMP]+[Ino]+[H_x]} \times 100$$

Where, [ATP], [ADP], [AMP], [IMP], [Ino] and [Hx] represent the relative concentrations of these compounds in fish muscle measured at various times during chilled storage.

The K or "freshness" index gives a relative freshness rating based primarily on the autolytic changes which take place during post mortem storage of the muscle. Thus, higher the K value, lower the freshness level. However, this is not considered reliable as a freshness index for all marine finfish.

Autolytic changes involving proteolytic enzymes:

Many proteases have been isolated from fish muscle and the effects of proteolytic breakdown are often related to extensive softening of the tissue for example belly-bursting in pelagic (fatty fish) species.

a. Cathepsins: These are "acid" proteases usually found packaged in tiny, submicroscopic organelles called lysozomes. In living tissue, lysozomal proteases are believed to be responsible for protein breakdown at sites of injury. Thus cathepsins are for the most part inactive in living tissue but become released into the cell juices upon physical abuse or upon freezing and thawing of post mortem muscle. The proteolytic enzymes autolyze the muscle. In addition to their detrimental effect on texture, catheptic enzymes induce intentional autolytic changes in fermented fish products towards textural changes.

b. Calpains: A second group of intracellular proteases called "calpains" or "calcium activated factor" (CAF) has been associated with fish muscle autolysis and is found in finfish and crustaceans. Tenderness is probably the most important quality characteristic of red meat. Calpains have been found primarily responsible for the post mortem autolysis of meat through digestion of the z- line proteins of the myofibril.

c. Collagenases: The flesh of teleost fish is divided into blocks of muscle cells separated into "flakes" or myotomes by connective tissue called myocommata. Each muscle cell or fibre is surrounded with connective tissue which attaches to the myocommata at the ends of the cells by means of fine collagenous fibrils. During chilled storage, these fibrils deteriorate and collagen gets solubilized due to the action of autolytic collagenase enzymes. It is these enzymes which cause "gaping" or breakdown of the myotome during long-term storage on ice or short term storage at high temperature.

Autolytic changes during frozen storage:

The reduction of trimethylamine oxide (TMAO), an osmoregulatory compound in many marine teleost fish, is usually due to bacterial action but in some species an enzyme is present in the muscle tissue which is able to break down TMAO into dimethylamine (DMA) and formaldehyde (FA):

$$(CH_3)_3\,NO \rightarrow (CH_3)_2NH + HCHO$$

Formaldehyde induces cross- linking of the muscle proteins making the muscle tough and readily loose its water holding capacity. The enzyme responsible for formal dehyde-induced toughening is called TMAO-ase or TMAO demethylase.

Table 4.2: Summary of Autolytic Changes in Chilled Fish

Enzyme(s)	Substrate	Changes Encountered	Prevention/Inhibition
Glycolytic enzymes	Glycogen	Production of lactic acid, pH of tissue drops, loss of water-holding capacity in muscle High temperature rigor may result in gaping	Fish should be allowed to pass through rigor at temperatures as close to 0°C as practically possible Pre-rigor stress must be avoided
Autolytic enzymes, involved in nucleotide breakdown	A T P A D P A M P IMP	Loss of fresh fish flavour, gradual production of bitternes with Hx (later stages)	Same as above Rough handling or crushing accelerates breakdown
Cathepsins	Proteins, peptides	Softening of tissue making processing difficult or impossible	Rough handling during storage and discharge
Chymotrypsin, trypsin, carboxy-peptidases	Proteins, peptides	Autolysis of visceral cavity in pelagics (belly- bursting)	Problem increased with freezing/ thawing or long- term chill storage
Calpain	Myofibrillar proteins	Softening, molt-induced softening in crustaceans	Removal of calcium thus preventing activation
Collagenases	Connective tissue	Gaping" of fillets Softening	Connective tissue degradation related to time and temperature of chilled storage
TMAO demethylase	TMAO	Formaldehyde-induced toughening of frozen gadoid fish	Store fish at a temperature -30°C Physical abuse and freezing/ thawing accelerate formaldehyde-induced toughening

4. Microbial putrefaction:

Microbial putrefaction or microbial decomposition is the process of conversion of fish tissue components by microorganisms into off-odour, off-flavour substances accumulation of which distracts the consumers and thereby results in the spoilage of fish.

Microorganisms are found on all the outer surfaces (skin and gills) and in the intestines of live and newly caught fish. The total number of organisms vary enormously ranging between 10^2 to10^7 cfu (colony forming units)/cm^2 on the skin surface, 10^3 and 10^9 cfu/g both in the gills and the intestines.

The bacterial flora on newly-caught fish depends on the environment in which it is caught rather than on the fish species. Fish caught in very cold, clean waters carry the lower numbers whereas fish caught in warm waters have slightly higher counts. Very high numbers, i.e., 10^7 cfu/cm^2 are found on fish from polluted warm waters.

The bacteria on temperate water fish are all classified according to their growth temperature range as either psychrotrophs or psychrophiles. Psychrotrophs (cold-tolerant) are bacteria capable of growth at 0°C but with optimum around 25°C. Psychrophiles (cold-loving) are bacteria with maximum growth temperature around 20°C and optimum temperature at 15°C. In warmer waters, higher numbers of mesophiles can be isolated.

Species of Microbes associated with spoilage of fish:

It is important to note that generally the endogenous microbes or the natural microbial flora present in fish are spoilage type while the exogenous microbes contaminating the fish during post harvest handling, processing and storage are pathogenic type. The different species of microbes associated with the spoilage of fish through microbial putrefaction are *Pseudomonas, Aeromonas, Achromobacter, Flavobacterium, Vibrio, Moraxella, Arthrobacter,Micrococcus, Corynebacterium, Photobacterium, Bacillus, Shewanella* and *Proteus.*

The microflora on temperate water fish is dominated by psychrotrophic Gram-negative rodshaped bacteria belonging to the genera Pseudomonas, Moraxella, Acinetobacter, Shewanella and Flavobacterium. Members of the Vibrionaceae (*Vibrio* and *Photobacterium*) and the Aeromonadaceae (*Aeromonas* spp.) are also common aquatic bacteria and typical of the fish flora. Gram-positive organisms as *Bacillus, Micrococcus, Clostridium, Lactobacillus* and coryneforms can also be found in varying proportions, but in general, Gram-negative bacteria dominate the microflora.

Aeromonas spp. are typical of freshwater fish, whereas a number of bacteria require sodium for growth and are thus typical of marine waters. These include *Vibrio*, *Photobacterium* and *Shewanella*. However, although *Shewanella putrefaciens* is characterized as sodium-requiring, strains of *S. putrefaciens* can also be isolated from freshwater environments.

Spread of microbes in the body of dead fish:

After the death of fish, microbes penetrate into the muscle and internal organs due to the following reasons:

a. Loss of defense mechanism or immunity of body

b. Loss of microbe-impermeability of the membranous linings of the body

c. Abundance of readily utilizable nutrients for microbial growth and multiplication.

The flesh of healthy live or newly-caught fish is sterile as the immune system of the fish prevents the bacteria from growing in the flesh. When the fish dies, the immune system collapses and bacteria are allowed to proliferate freely. On the skin surface, the bacteria to a large extent colonize the scale pockets. During storage, they invade the flesh by moving between the muscle fibres.

Effects of Microbial Putrefaction:

A clear distinction should be made between the terms spoilage flora and spoilage bacteria since the first describes merely the bacteria present on the fish when it spoils whereas the latter is the specific group that produce the off-odours and off-flavours associated with spoilage.

Comparison of the chemical compounds developing in naturally spoiling fish and sterile fish has shown that most of the volatile compounds are produced by bacteria. These volatile compounds include trimethylamine, volatile sulphur compounds, aldehydes, ketones, esters, hypoxanthine as well as other low molecular weight compounds.

The substrates for the production of volatiles are the carbohydrates (e.g., lactate and ribose), nucleotides (e.g., inosine mono-phosphate and inosine) and other NPN molecules. The amino-acids are particularly important substrates for formation of sulphides.

The TMAO reduction is mainly associated with the genera of bacteria typical of the marine environment (*Alteromonas*, *Photobactetium*, Vibrio and *S. putrefaciens*), but is also carried out by Aeromonas and intestinal bacteria of the Enterobacteriaceae. TMAO reduction has been studied in fermentative, facultative anaerobic bacteria like *E. coli* and *Proteus* spp. as well as in the non-fermentative *S. putrefaciens.*

a. Spoilage of fish:

A number of off-odor, off-flavour substances are produced in fish by microbial putrefaction of its tissue components. Their cumulative accumulation results in the spoilage of fish. Some of these substances include NH_3, H_2S, TMA, indole, skatole etc.

Table 4.3: Typical spoilage compounds during spoilage of fresh fish stored aerobically or packed in ice or at ambient temperature

Specific spoilage organism	Typical spoilage compounds
Shewanella putrefaciens	TMA, H_2S, CH_3SH, $(CH_3)_2S$, Hx
Photobacterium phosphoreum	TMA, Hx
Pseudomonas spp.	Ketones, aldehydes, esters, non-H_2S sulphides
Vibrionaceae	TMA, H_2S
Anaerobic spoilers	NH_3, acetic, butyric and propionic acid

Table 4.4: Substrate and off-odour/off-flavour compounds produced by bacteria during spoilage of fish

Substrate	Compounds produced by bacterial action
TMAO	TMA
Cysteine	H_2S
Methionine	CH_3SH, $(CH_3)_2S$
Carbohydrates and lactate	Acetate, CO_2, H_2O
Inosine, IMP	Hypoxanthine
Amino-acid s (glycine, serine, leucine)	Esters, ketones, aldehydes
Amino-acids, urea	NH_3

b. Increase in PH:

After death, the PH of fish tissue falls from neutral to acidic range due to accumulation of lactic acid .The microbial decomposition of fish tissue results in the production of large quantities of nitrogenous substances, which are basic in nature. The accumulation of these bases increases the PH of fish tissue from acidic to alkaline range (PH-7.1 to 7.2).

c. Increase in NPN substances:

During microbial putrefaction, the proteolytic microbes secret proteolytic enzymes to outside their cell hydrolyze the surrounding tissue protein into amino acids. This results in an increase in NPN of fish tissue.

d. Development of rancidity:

During microbial putrefaction, the lipolytic microbes hydrolyse the tissue fats into glycerol & fatty acids. Accumulation of these free fatty acids leads to the development of hydrolytic rancidity. Some of the microbes, which contain the enzymes lipoxygenase (lipoxidase) catalyse the oxidation of the unsaturated fatty acids of tissue fats by atmospheric oxygen. The end products are peroxides, hydroperoxides, aldehydes, ketones and lower fatty acids. The presence of these oxidation products leads to the development of oxidative rancidity.

e. Food poisoning:

Some of the end products of microbial putrefaction are toxic and can cause food poisoning. They include histamine, phenylethylamine, cadaverine, putrescine, neurine, phenol, cresol, indole, and skatole.

5. Lipid oxidation (auto-oxidation) and hydrolysis:

The two distinct reactions in fish lipids of importance for quality deterioration are:

» oxidation

» hydrolysis

They result in production of a range of substances among which some have unpleasant (rancid) taste and smell. Some may also contribute to texture changes by binding covalently to fish muscle proteins. The various reactions are either nonenzymatic or catalyzed by microbial enzymes or by intracellular or digestive enzymes from the fish themselves. The relative significance of these reactions, therefore, mainly depends on fish species and storage temperature.

Fatty fish are, of course, particularly susceptible to lipid degradation which can create severe quality problems even on storage at subzero temperatures.

Oxidation:

Fish lipids are rich in unsaturated fatty acid. The unsaturation of fatty acids is due to the presence of double bonds between carbon atoms in their hydrocarbon chains. When exposed to atmosphere, the atmospheric oxygen attacks and reacts with the double bonds on the fatty acids. The reaction of the atmospheric oxygen with the unsaturated double bonds of the fatty acids of lipid-molecules is called 'lipid oxidation' or 'autoxidation'.

Autoxidation of lipid takes place in four steps such as:

i) Initiation, ii) Propagation, iii) Decomposition and iv) Termination

The process is initiated as described below by abstraction of a hydrogen atom from the central carbon of the pentadiene structure found in most fatty acid acyl chains containing more than one double bond. Contrary to the native molecule, the lipid radical (L) reacts very quickly with atmospheric oxygen making a peroxy-radical (LOO) which again may abstract a hydrogen from another acyl chain resulting in a lipid hydroperoxide (LOOH) and a new radical L. This propagation continues until one of the radicals is removed by reaction with another radical or with an antioxidant (AH) whose resulting radical (A) is much less reactive. The hydroperoxides produced in relatively large amounts during propagation are tasteless, and it is therefore perhaps not surprising that the widely used "peroxide value" usually correlates rather poorly to sensorial properties.

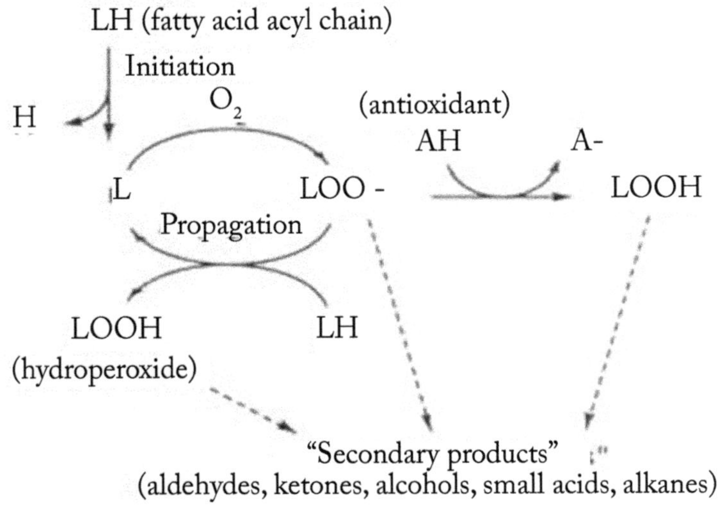

Figure 4.1: Autoxidation of polyunsaturated lipid

The hydroperoxides are readily broken down, catalyzed by heavy metal ions, to secondary autoxidation products of shorter carbon chain-length. These secondary products - mostly aldehydes, ketones, alcohols, small carboxylic acids and alkanes, give rise to a very broad odour spectrum and in some cases to a yellowish discoloration. Several of the aldehydes can be determined as "thiobarbituric acid-reactive substances".

Factor influencing lipid oxidation:

» Composition of lipids

» Temperature

» Degree of exposure of flesh

» Oxygen concentration

» Fat content of fish, its type & deposition

» Location of lipid

» Extraction of lipid

» Presence of pro-oxidants

» Presence of synergists

» Presence of antioxidant

Effects of lipid oxidation:

» Development of rancid flavor

» Flavor reversion

» Discolouration

» Loss of vitamin A

» Loss of essential fatty acids

» Protein denaturation

» Development of toxicity

» Loss of nutritional value of protein

The living cells possess several protection mechanisms directed against lipid oxidation products. The cellular membranes contain the phenolic compound tocopherol (Vitamin E) which is considered the most important natural antioxidant. Tocopherol can donate a hydrogen atom to the radicals L- or LOO- functioning as the molecule AH. Other compounds, for example the carotenoids, may also function as antioxidants. Wood smoke contains phenols which may penetrate the fish surface during smoking and thereby provide some protection against lipid oxidation.

***Hydrolysis*:**

During storage, a considerable amount of free fatty acids (FFA) appears. The phenomenon is more profound in ungutted than in gutted fish probably because of the involvement of digestive enzymes. Triglyceride in the depot fat is cleaved by triglyceride lipase originating from the digestive tract or excreted by certain microorganisms. Cellular lipases may also play a minor role.

In lean fish, production of free fatty acids also occurs, even at low temperatures. The enzymes responsible are believed to be cellular phospholipases in particular phospholipase A2 . The fatty acids bound to phospholipids at glycerol-carbon atom 2 are largely of the polyunsaturated type, and hydrolysis therefore often leads to increased oxidation as well. Furthermore, the fatty acids themselves may cause a "soapy" off-flavour.

6. Discolouration:

- Discolouration is any change in the original colour of fish after death. Post mortem discolouration can be grouped into the following categories.
- Black spot formation in shrimp, Brown discolouration of red meat of fish and Red/ brown discolouration of white fish fillets.

C. HAZARDS ASSOCIATED WITH FISHERY PRODUCTS

Food borne hazards are on the rise throughout the world. A hazard is defined as a biological, chemical and physical agent in food, or a condition of food with the potential to cause harm. An estimate of the probability and severity of the hazard to populations caused by the consumption of food is called risk. The World Health Organization (WHO) has observed that in Asia- Pacific region alone; more than 700,000 people die every year from consuming contaminated food. An estimated 76 million cases of food born illness occur each year in the United State costing between $ 6.5 and $ 34.9 billion in medical care and loss of productivity.

Some of the important hazards associated with fishery products are discussed below.

1. Bacterial pathogen:

A number of pathogenic bacteria have been implicated in seafood- borne diseases. The leading cause of food borne illness during the last few years was salmonellosis followed by Shigellosis, staphylococcus intoxication, and gastroenteritis.

» Pathogenic bacteria indigenous to aquatic environment and naturally present on fish:

Pathogen	Primary habitat	Carried seafood	Quantitative level
Clostridium botulinum	Temperate & arctic aquatic environment Multiplication in aquatic carrion	Smoked fish, white fish, ready to eat products	Mode of action preformed toxin, generally low (<0.1) sporeg^{-1}, but upto 5.3 spores g^{-1}
Vibrio cholera, V. parahaemolyticus V. vulnificus	Ubiquitous in warm (>15 ° C) seawater environment	Shellfish, RTE products, raw oysters clam, crabs	Up to 10^2- 10^3 cfu g^{-1} Up to 10^4-10^6 cfu g^{-1}
Aeromonas spp.	Aquatic environment	Aquacultured fish/ shellfish	Generally low, but up to10^4 cfu in seawater
Salmonella typhi	Environment	Mussel, scallops, oyster	–
Shigella spp.	Environment	bivalves	–
Salmonella spp.	Tropical aquatic environment	Fish, shellfish, bivalves from warm waters, cultured fish	<10^2-10^6 cfu g^{-1}

2. Viruses:

The prominent food born viruses are Hepatitis type A, Norwalk virus, and small round viruses. The symptoms of diseases caused by these viruses include paralysis, meningitis, respiratory illness and myocarditis among others. Consumption of raw molluscan bivalves is a major cause of viral disease as a result of filtration of water by these shellfish.

3. Parasites:

Parasites such as flatworm, round worm, and protozoa infest the gill, viscera, and skin of marine, fresh water as well as farm raised fish and shellfish, and can pose health hazard to consumers. The most common parasites associated with fish that can infect consumers include nematodiases and cestodiases. Fish borne nematodiases can be detected in human as incidental infection, whose natural definitive hosts include marine mammals, birds and pigs. The mode of infection is ingestion of fish containing the infective larvae. Nematodes in fish can be killed by thoroughly cooking the fish, freezing or curing with salt and acetic acids.

4. Insects:

A number of insects such as flesh flies, beetles and mites infest fish, particularly during sun drying. The most destructive pest is the hide beetle, *D. maculates Deg.* A case study of losses in traditionally cured fish showed the infestation of blowfly as the major causes of losses in dried sardine products.

5. Biotoxins:

The large proportion of seafood born disease outbreak is caused by biotoxins and histamine. The main type of fish poisoning include ciguatera, paralytic shellfish poisoning, diarrheic shellfish poisoning, amnesic shellfish poisoning and puffer fish poisoning.

6. Chemical hazards:

Uses of chemical in food production and processing have contributed to increase in chemical hazards. The WHO has identified 18 chemical contaminants in food, which include industrial chemicals, organo- chlorine pesticides, organo- phosphorus and mycotoxins. Incidence of chemical hazards is particularly widespread in aquaculture, since some of the chemicals are also used to sterilize pond soil between production cycles. When used as per good aquaculture practices, none of the water treatment compounds can be considered a hazard. Food additives and processing aids such as nitrosamines are carcinogenic.

D. CONTROL OF HAZARDS

Seafood related health hazards have forced the global regulatory authorities to introduce stringent quality standards in traded fishery products.

Table 4.5: Seafood hazard categories in order of decreasing risks

Category	Description	Example
1	Those consumed raw without cooking	Mollusks, mussel. Clam, oyster
2	Non heat processed raw foods often consumed with additional cooking	Fresh/frozen fish and crustacean
3	Lightly preserved fish products	Salted, marinade, fermented, cold smoked fish
4	Semi preserved fish	Salted, marinade, fermented fish
5	Mildly heat processed fish products	Pre-cooked, breaded fillets
6	Heat processed	Canned, retort-pouch packed items

Table 4.6: Control measure against bacterial pathogens and parasites in seafood

Pathogen	Control measures
Salmonella spp.	Proper sewage disposal, sanitary handling processing, cooking just prior to eating, Prohibit harvesting from polluted waters, maintenance of water quality in aquaculture
Vibrio spp.	Sanitary handling and processing, adequate refrigeration and freezing, cooking just prior to eating
C. botulinum	Correct processing, just prior to eating of food to inactivate toxins
Viral infection	Proper sewage disposal, adequate cooking, prohibit harvesting from polluted areas
S. aureus	Sanitary handling and processing, maintenance of proper hygiene during processing
L. monocytogenes	Check raw materials, Sanitary handling and processing, adequate cooking, prevent cross contamination from raw to processed food
Parasites	Check raw materials, freezing/ handling/ curing

Key Areas for Quality Control for Food Processors:

- Raw material: monitor the quality and reliability of the raw material. It helps to set up a proper traceability of input ingredients and produce quality products
- Sanitation: proper hygiene in prerequisite with all standard operating procedures
- Good manufacturing practice

- Implementation of HACCP
- Analytical testing: microbial, chemical, sensory and shelf life
- Environmental monitoring
- Standard operating procedures (SOP) and document trial through incoming, processing, finish products and outgoing products
- Process verification and validation studies to ensure safe products with extremes. The market storage condition and instruction and instruction to be validated.
- Proper pest control programme to avoid any contamination.
- Employee training and education.

5 Microbiology of Fish and Fish Products

Fish is a main source of animal protein and is an important part of a healthy diet since they contain high quality protein, but typically present a low fat percent when compared to other meats. In addition, most fish contain omega 3-fatty acids and other essential nutrients. Although fish is broadly similar in composition and structure to meat there are a number of distinctive features.

- Protein content in fish varies typically from 16 - 21%. The lipid content, which can be up to 6 - 7%, typically fluctuates between 0.2 - 20%, and is mostly interspersed between the muscle fibres. Fishes are a poor source of carbohydrates, offering less than 0.5%.
- Fish composition can vary significantly within the same species due to feed intake, migratory patterns, and spawning season.
- The lipid fraction is the component showing the greatest variation; it shows a typical season pattern especially in migratory species such as herring or mackerel.
- Fish can be divided into fatty and lean fish; lean fish are those fish that store most of their fat in the liver, while fatty fish have fat cells distributed along their bodies.
- Muscle composition and structure of fish also differ from those found in other meat.
- Fish flesh is dominated by the abundance of white muscle in relatively short segments, giving it its characteristically flaky structure.
- The connective tissue content of fish is also lower than that found in meat, typically 3 and 15% of total weight, respectively.
- Because of their health advantages over red meats, the consumption of fish and seafood has increased.

- Catches can be gathered from seas, rivers and lakes whose water can range from pristine to contaminate. Often contamination is from human and animal sources; thus, fish and seafood can be involved in the transmission of pathogenic microorganisms and toxins.

- Geographical region, season, and, for fish, whether they are pelagic (surface to mid-water) or demersal (bottom) feeders will influence the numbers and types of microorganisms present on freshly caught seafood.

- It was often thought that highest quality fish and seafood products are processed and sold as whole fish and/ or fresh fillets (e.g. Sashimi tuna).

- Products of lesser quality, although acceptable and meeting all safety criteria, are used in chilled and frozen condition.

- Recent developments in improved supply chains, and on-board freezing the products at the point of capture, the frozen products, are now perceived by the consumer as being of a better quality than a fresh product, due to the early cessation of spoilage.

- The phylum *Mollusca*, which includes oysters, clams, mussels, cockles, scallops, squid, octopi, chitons, snails, slugs, whelks, abalones, and tooth shells are a group of soft-bodied animals that usually secrete external protective shells. The phylum consists of five classes: *Amphineura*, *Gastropoda*, *Pelecypoda*, *Scaphopoda*, and *Cephalopoda*. The class having species of economic and public health importance is the *Pelecypoda*. The class *Pelecypoda* includes all of the bivalves such as clams, mussels, oysters, cockles, and scallops. In these forms, the foot is compressed to form a muscular spade for digging and the head is greatly reduced, lying well within the mantle cavity. Pelecypods have little need for locomotion. In most species, the shell, composed of two valves hinged together dorsally, completely encloses the body. Strong muscles, the adductors, can hold the shells tightly shut against enemies.

- Bivalves feed by filtering large volumes of water across their gills to obtain oxygen and food. Plankton and algae form the bulk of their diet from seawater. In the process of filtering water, they also trap bacteria, viruses, chemical contaminants and other impurities on the mucus of the gills. Special tracts of cilia move this mucus toward the mouth, where it is eventually swallowed and digested. The digestion process requires less than 2 hours in actively feeding adult bivalves.

- Many microorganisms ingested by shellfish survive the digestive process. The term shellfish here is limited to oysters, clams, mussels, cockles and scallops.

- Other species of shellfish such as crabs, lobster, and shrimp, while commercially valuable, are not as vulnerable to contamination through pollution of their beds as the bivalves.
- Whatever contaminants are in the water will eventually get into the shellfish, and usually at higher concentrations than found in their habitat.
- The habitat of oysters and mussels are sessile, dwelling on the bottom or attached to structures in the water column. Clams are benthic and live burrowed in the mud, but maintain contact with the water by means of a siphon tube.
- Crustaceans are invertebrate animals and represent a class of the phylum Arthropoda. They possess a segmented body, jointed limbs and a chitinous exoskeleton, and are usually aquatic, e.g. crabs, lobsters, crayfish and shrimp.
- Species eaten from tropical waters include *Penaeus* spp., *Parapeneopsis* spp., and *Trachypenaeus* spp. (deep-water shrimp).
- Shrimps, prawns and scampi are temperate species commonly eaten by man and include members of the families *Nephrops norvegicus* (scampi), *Pandalus borealis* (deep-water shrimp), *Penaeus indicus* and *P. monodon* (inshore shrimp).

A. Initial Micro-flora:

- The subsurface flesh of live, healthy fish is considered sterile and should not present any bacteria or other microorganisms.
- Only after death, the saprophytic flora will start to invade the surface flesh. On the contrary, as with other vertebrates, microorganisms colonise the skin, gills and the gastrointestinal tract of fish.
- The number and diversity of microbes associated with fish depend on the geographical location, the season and the method of harvest.
- In general, the natural fish microflora tends to reflect the microbial communities of the surrounding waters.
- It is difficult to estimate how many microorganisms are typically associated with fish, since they heavily depend on the type of sample analysed and the protocol used for isolation.
- In fact, standard culture-dependent methods can only recover between 1 to 10% of total bacteria present in any given sample.

» Gastrointestinal tracts and gills typically yield high bacteria numbers, although these are influenced by water quality and feed.

» Fish harvested from clean and cold waters will present lower bacterial numbers than fish from eutrophic and/or warm waters. However, potential human pathogens may be present in both scenarios.

» The autochthonous bacterial flora of fish is dominated by Gram-negative genera including: *Acinetobacter*, *Flavobacterium*, *Moraxella*, *Shewanella* and *Pseudomonas*. Members of the families *Vibrionaceae* (*Vibrio* and *Photobacterium*) and the *Aeromonadaceae* (*Aeromonas* spp.) are also common aquatic bacteria, and typical of the fish flora.

» Gram-positive organisms such as *Bacillus*, *Micrococcus*, *Clostridium*, *Lactobacillus* and coryneforms can also be found in varying proportions.

» Human pathogenic bacteria can be part of the initial microflora of fish, posing a concern for seafood borne illnesses. These pathogens can be divided into two groups: organisms naturally present on fish; and those that although not autochthonous to the aquatic environment, are present there as result of contamination (anthropomorphic origin or other) or are introduced to the fish during harvest, processing or storage.

Table 5.1: Indigenous bacterial pathogens typically present in fish

Organism	Temperature range	Estimated minimum infective dose
Clostridium botulinum – non-proteolytic type E	3 - 26°C	00.1 - 1 μg toxin lethal dose
Pathogenic *Vibrio* spp. *Vibrio cholerae* *Vibrio parahaemolyticus* (exception: *V. vulnificus*) *Vibrio vulnificus* other vibrios	10 - 37°C	High for most species, 10^5 - 10^6 cells/g
Aeromonas spp.	5 - 35°C	Unknown
Plesiomonas shigelloides	8 - 37°C	Unknown

Table 5.2: Non-indigenous bacterial pathogens frequently present in fish

Organism	Primary habitat	Minimum Infective Dose
*Listeria monocytogenes** stream water, estuarine strain environments, and mud	Soil, birds, sewage, Variable depending on the	($>10^2$ cells/g)
Staphylococcus aureus	Ubiquitous, human origin	10^5 - 10^6 cells/g. Toxin levels 0.14 - 0.19 μg/kg bodyweight
Salmonella spp.	Intestinal track of terrestrial vertebrates	from $< 10^2$ - $>10^6$
Shigella spp.	Human origin	10^1 - 10^2
Escherichia coli	Fecal contamination	10^1 - 10^3
Yersinia enterocolitica	Ubiquitous in environment	High (10^7 - 10^9 cells/g)

* Some sources considered *L. monocytogenes* as part of the natural aquatic flora

» It is apparent that there is potentially a very diverse range of organisms present on fish.

» However, numbers of pathogenic bacteria in raw fish tend to be low, and risk associated with the consumption of seafood is low.

» In addition, during storage indigenous spoilage bacteria tend to outgrow potential pathogenic bacteria.

» Shelf life depends on the initial microflora on the fish, potential contaminants added during handling and processing, and conditions of storage.

» After capture and during any subsequent processing, handling or preparation stage, other potentially pathogenic and spoilage bacterial contaminants may be introduced from a wide range of sources including the environment, product handlers, and ingredients added to the final product. Contamination will, on the whole, remain on the surface of the fish.

» The edible tissue of some species of fish are more susceptible to gaping than others, while gaping is also dependent on a range of other factors that affect the muscle composition, including seasonal variation, storage, and processing conditions such as dehydration or poor filleting practices.

- The microflora of molluscan shellfish directly reflects the environment from which the shellfish is harvested.
- The initial microflora of molluscan shellfish is comprised of the natural commensal microorganisms and the microorganisms accumulated from the water during feeding. Because molluscan shellfish are filter feeders, they accumulate pathogenic microorganisms from polluted waters. Whatever contaminants are in the water will eventually get into the shellfish.
- The microflora are dependent upon a number of factors such as season, and the environment, which affects the temperature, salinity, pH, nutrient concentration and pollutants of the water column.
- The following pathogenic microorganisms causing illness in humans are characteristically associated with molluscan shellfish: *Vibrio parahaemolyticus, Vibrio cholerae* O1, *Vibrio cholerae* non-O1, *Aeromonas hydrophila* group, *Staphylococcus aureus, Salmonella* spp., *Shigella* spp. *Escherichia coli, Bacillus cereus* and other bacterial and viral agents.
- The microflora of molluscan shellfish at harvest consists predominantly of Gram negative rods while the Gram-positives constitute a minor portion of the flora.
- As with finfish, the level and type of the initial microflora of crustacean shellfish will reflect a combination of factors, which include the environment from which they have been captured or harvested, their feeding and living habits, the geography of the area they are captured from or are grown in, the season, and the temperature and quality of the waters in which they exist.
- Following capture or harvest, the flora will change depending upon the methods of handling and/or the environmental conditions to which they are exposed. The microbial flora of crustaceans from temperate waters differs from that of crustaceans from warmer, tropical waters.
- Spoilage bacteria include *Pseudomonas* spp., *Shewanella putrefaciens* and members of the *Acinetobacter-Moraxella* group, Enterobacteriaceae, *Vibrio* spp., *Pseudomonas* spp., *Alcaligenes*, corynebacteria, *Micrococcus* spp., and *Achromobacter* spp.
- Hence, from the time of capture and processing, bacterial contamination and spoilage are inevitable.
- Microbiological risks associated with crustacean shellfish increase in proportion to

the degree of handling that the product undergoes, particularly after it has been cooked. For example, during the peeling of cooked shrimp, the potential exists for cross contamination from both the processor and from the environment.

» Control measures to prevent these potential hazards occurring include training and supervision of operatives as well as Good Hygienic Practices (GHP). Poor hygienic procedures will increase the probability of contamination with bacteria of public health significance.

» The structure of the circulatory system of crabs is not closed, which means that the haemolymph can be a reservoir of bacteria, particularly for members of the genus *Vibrio*.

» Spoilage begins following death once the immuno-response system of the crustacean shellfish fails to function; hence, for crabs and lobsters, spoilage bacteria are of little importance as, traditionally, only live carcasses are processed, e.g. boiled to kill the crustaceans and cook their meat prior to consumption.

» In comparison, shrimp die immediately after capture, so that spoilage may occur during landing and transportation to the processing plant.

» As with finfish, it is important that shrimp are handled and iced correctly in order to keep them in as fresh and uncontaminated a state as possible.

» Spoilage bacteria associated with tropical species include *Enterobacteriaceae*, *Vibrio* spp., *Pseudomonas* spp., *Alcaligenes*, *Corynebacteria*, *Micrococcus* spp., and *Achromobacter* spp.

» Hence, from the time of capture and processing, bacterial contamination and spoilage are inevitable.

» The rate of spoilage will be a function of time and temperature, whether GHP are in place, and will be dependent upon whether the crustaceans were harvested from cool or warmer waters.

» As with finfish, there is evidence that the shelf life of tropical crustaceans is longer than that of temperate-water species stored correctly in ice - for example, up to 16 days for tropical species compared with 8 - 10 days for temperate water species.

» The psychrotrophic bacteria responsible for causing spoilage are not present on freshly captured crustaceans from tropical waters in such high levels as are found in crustaceans from cooler waters and, thus, the rate of deterioration of the crustacean shellfish by these organisms is reduced, especially if they are stored in ice or iced water.

Non-seafood ingredients:

» The range of ingredients used in the preparation of chilled and frozen prepared seafood products is very diverse, and so are the potential microbial contaminants associated with them. Table 5.3 provides an overview of common raw materials, and associated spoilage and pathogenic microorganisms; however, it is essential that during new product development a full hazard identification and risk analysis be conducted to ensure that the microbial risk for specific ingredients are identified.

Table 5.3: Spoilage and pathogenic micro-flora associated with common raw materials and ingredients used in prepared frozen and chilled products

Raw material	Associated microflora	Pathogenic microflora/toxins
Vegetables (Untreated) including fresh herbs, mushrooms, nuts and fruit	Lactobacilli, coryneforms, spore-formers, coliforms, pseudomonads, moulds, yeasts	*Salmonella*, *Escherichia coli*, *Clostridium perfringens*, *Clostridium botulinum*, *Listeria*, *Shigella*, *Bacillus cereus*, *Aspergillus*, *Penicillium*, *Fusarium*, protozoa and viruses
Carbohydrates		
Potato flake and powder	*Bacillus*, moulds	*B. cereus*
Rice	*Bacillus*	*B. cereus*
Pasta	*Bacillus*, moulds	*Staphylococcus aureus*, *staphylococcal* toxin, *B. cereus*
Flour	*Bacillus*, moulds	*Bacillus licheniformis*, *Bacillus subtilis*
Dried herbs and spices	*Bacillus*, *Clostridium*	*B. cereus*
Dairy products including butter, cream, and milk	Lactobacilli; *Staphylococcus*, anaerobic spore-formers, *Pseudomonas*, yeasts	
Cheese	Lactobacilli, streptococci, Enterobacteriaceae, Moulds	*Salmonella*, *Staph.* toxin, histamine
Water/Ice	Enterobacteriaceae	*Cryptosporidium*, *Vibrio cholerae*, *Aeromonas*, enteric viruses

B. Spoilage:

a. Fresh fish spoilage:

» Food spoilage can be considered as any change that renders the product unacceptable for human consumption. Spoilage of fish starts upon death due to autoxidation (oxidation of unsaturated lipids), reactions caused by activities of the fish's own enzymes, and metabolic activities of microorganisms present in the fish. Over time, loss of the fresh characteristics may be simply measured by comparative visual and smell analysis.

- Loss of freshness and spoilage cannot be separated as processes, but it is related to autolytic degradation and spoilage is more of microbial in origin.
- Degradation of whole fresh fish stored in ice generally follows a set of patterns which is the basis of freshness grading schemes. The eyes turn from convex and clear to concave and opaque, the gills from pink and shiny, with no smell, to brown and slimy with an intense off-odour; the skin turns from iridescent to dull and bleached with bacterial slime; and the flesh turns from bright and elastic to dull and soft.

b. Factors affecting fresh fish spoilage:

i. Temperature:

- The most effective method of reducing the rate of whole fresh fish spoilage is temperature control.
- Fish spoil as a result of the chemical, biochemical and microbiological reactions taking place within and on the fish.
- All chemical reaction rate kinetics (and thereby microbial growth) are temperature-dependent; the lower the temperature of storage, the slower the spoilage processes proceed (within limits).
- Careful temperature control during storage is not only important in terms of quality loss, but also crucial for the assurance of consumer safety.
- The application of ice storage increases the shelf life of fresh fish from 1 to 7 days. This increase has been reported to be moderately dependent on the temperature of the sea from which the fish are taken.
- The reasons for fish spoilage rates may be many, including the ability of the endogenous bacteria to grow at low temperatures and lower endogenous enzyme activity.

ii. Fish-spoilage bacteria:

- The parameters affecting the multiplication of microorganisms in foods have been categorised into two general groups:

- Intrinsic (inherent qualities of the food) and
- Extrinsic (qualities of the food environment).

- The factors are interactive and cannot be completely isolated.
- By altering the gas atmosphere (extrinsic), the pH of the fish muscle may change (intrinsic) and a different microbial population may develop due to the change in atmosphere and pH. The sum of these changes may result in a different spoilage profile.

iii. Recognised specific spoilage organisms (SSOs):

- The degree of spoilage leading to sensory rejection of fish is partly dependent on the perception of the consumer. Not all the bacteria growing on a food will lead to the production of objectionable characteristics; a minority are often associated with the majority of the spoilage.
- The concept of specific spoilage organisms (SSOs) is not new.
- For fresh fish, the bulk of the microbial population on newly caught fish does not cause off-flavours and off-odours.
- The spoilage of a product may be strongly influenced by the conditions under which the product is held.
- Therefore, the characterisation of spoilage of each product must be made before the identification of the responsible agent(s) can proceed.

c. Methods of evaluation of fish spoilage:

- Various methods for whole fish freshness evaluation by a trained panel have been proposed; these are summarised in table 5.4.

Table 5.4: Fish-quality scales

Name of scale	Scoring system
Torry Scale	10 to 0 (≤ 4 reject)
Quality Index Method	0 to 24 (arbitrary reject)
European Grading Scheme	Extra, A, B, C (C = reject)

- Significantly, rejection of whole fresh fish for human consumption may be made without chemical or microbiological evaluation, and for that matter, evaluation of the taste of the fish, is very important.
- When working with fillets, evaluation of taste is unavoidable as the visual pattern of degradation of the eyes and gills is unavailable for analysis.
- Taste panel evaluation of cooked fish may be used to score or grade colour, texture, smell and taste of cooked fish.
- Generally, the cooked analysis of fish passes through four phases of spoilage:
    - Delicate sweet, sea-weedy taste, possibly slightly metallic.
    - Neutral taste, little flavour.
    - Traces of sour, fruity and/or bitter off-flavours; development of sickly sweet, cabbage-like, ammoniacal, sulphurous and/or rancid smells; texture becomes soft and watery or hard and dry.
    - Enhancement of the spoilage characteristic of phase iii, spoiled and putrid.
- The processes of degradation being analysed by the quality scoring methods above are a complex mix of physical, chemical, biochemical and microbiological actions. These processes are strongly influenced by the physical conditions of storage.
- After death, rigor mortis is the first noticeable change in the fish. From being flaccid, the muscles harden as residual adenosine triphosphate (ATP) is reduced and the myosin and actin filaments bind to form actomyosin. After some time, rigor resolves, the muscles relax again and the fish returns to a flaccid state.
- The pH of the muscle will drop after death depending upon the amount of residual glucose or glycogen that is reduced anaerobically to lactic acid with the coproduction of ATP.
- This will generally correlate with the length and severity of rigor. Because fish tend to have relatively little residual glycogen compared with mammals, the pH drop of the muscle is correspondingly less; post-rigor values are typically in the range of pH 5.8 to 6.5.
- After death, the Eh of fish muscle remains relatively high.

- All marine fish have TMAO, which has been ascribed a number of possible functions as:
    - a trimethylamine (TMA) detoxified waste product,
    - an osmoregulator,
    - an anti-freeze,
    - a waste product present due to bioaccumulation.
- TMAO permits a high Eh to remain in the muscle tissue, as little endogenous reduction of TMAO occurs. However, bacterial reduction of TMAO to TMA, an intense odour compound, is significant and may even be responsible for the ultimate sensory rejection of fresh cod and other fish with high initial TMAO content.
- As the fish degrades, the conductance generally increases. Thus, measurements of the falling skin resistance may be made to estimate the time that the fresh fish has been stored in ice, or its remaining shelf life.
- The process of ATP breakdown is used as an indicator of fish freshness. By measuring the concentrations of the six components, a ratio of concentration of the hypoxanthine (Hx) and inosine (HxR) to total concentration (ATP, ADP, AMP, IMP, HxR and Hx) gives a quantity (K-value, %), which increases from 0% towards 100% with time.
- During the initial storage, reduction in ATP and increase in hypoxanthine by endogenous enzymes allow this measurement to be used for determining freshness.
- However, hypoxanthine, which may also be formed by bacteria, is later significantly reduced by bacterial action. So the measurement is effective only during the initial loss of freshness. This measure is highly dependent upon the temperature of storage and may not reflect the rate of loss of quality equally at different temperatures.
- Lipid oxidation and other oxidative changes lead to oxidative rancidity, colour changes, and are especially important in the spoilage of frozen fish, as microbial spoilage is limited by the low storage temperature.
- Lipid oxidation can be a result of enzymic action or a cascade reaction initiated by free radicals produced by aerobic respiration and other forms of metal ion reduction.
- Oxidative rancidity is known to reduce the quality of fatty fish in particular.

- Chemical degradation continues after the initial post mortem phase. However, the importance of microbial action increases with time. Quality indices based upon the products of microbial metabolism do not explain changes in quality until microbial growth produces measurable changes in the fish. Therefore, these measurements are usually used to quantify the amount of spoilage, not to describe freshness.
- Volatile bases are the best-characterised chemical indicators of fresh fish spoilage.
- The variation in post mortem fish pH may likewise influence the amount of bases, Total Volatile Base-Nitrogen (TVB-N, also termed TVN) being liberated to the air and consequently affect the odour characteristics of the fish.
- Comparing results for different fish species, however, does not show correlation between muscle pH and the amount of volatile bases contained within the fish at rejection.
- During spoilage, the majority of volatile bases are produced from the soluble non-protein nitrogen of the fish (free amino acids and other low-molecular-weight nitrogenous compounds), as significant proteolysis is observed only during the latest stages of spoilage and after rejection.
- For some fish species, a correlation can be made between the spoilage of the fresh fish and the production of TVB-N.
- The major spoilage odours and flavours of fresh fish are principally microbial in origin, but rejection of whole fresh fish by sensory methods is based upon non-specific odour detection and physical appearance.

Molluscs:

- Spoilage microorganisms such as *Achromobacter*, *Pseudomonas*, *Flavobacterium*, *Micrococcus*, *Proteus*, *Alcaligenes*, and *Pseudomonas fluorescens* are the responsible agents of spoilage in oysters held at various temperatures.
- Shellfish, particularly shucked meats, are an excellent medium for growth of bacteria.
- Keeping shellfish refrigerated is the most effective method of retarding deterioration and spoilage.
- The initial flora found in shellfish is the major factor of spoilage in bivalves. The environment has an effect on the bacterial population found in shellfish. These

organisms may play a significant role in the spoilage processes.

» The spoilage process in shellfish may be divided into three stages:

  • Increase of acidity,
  • Abundant gas production, and
  • Proteolysis.

» The high incidence of proteolytic bacteria and other types capable of fermenting glucose in the natural flora may be responsible in post-mortem spoilage of shellfish. For example, in oysters, glycogen levels range from 0.47 to 6.8%.

» Glycogen, a carbohydrate, is stored in the bivalve as a source of energy. Glycogen is hydrolysed by the initial flora found in shellfish, thus producing acid and gas.

» This process changes the bacterial profile to microorganisms favouring low pH. These include *lactobacilli*, *streptococci*, and yeasts.

» Lactobacilli seem to be the predominant microflora in shellfish, when stored at 7^0C, whereas bacterium such as *Pseudomonas*, *Achromobacter* and *Flavobacterium* levels decrease during storage.

» In summary, anywhere during the process of shellfish from harvesting to the point of sale, where there is a breakdown of temperature control or prolonged storage of shellfish, spoilage will occur. Temperature abuse will result in high bacterial counts.

Shell fish: *Prawns and shrimps:*

» The main genera of Gram-negative bacteria responsible for the spoilage of crustacean shellfish are the same as those that cause spoilage in finfish; they include *S. putrefaciens* and *Pseudomonas* spp.

» Freshness indicators for prawns and shrimps are for the shells to be crisp and dry and feel sharp and cool to the touch, and have a sweetish smell.

» Heat is produced if the shrimps decompose in a confined space, and a strong ammoniacal odour is produced, indicative of protein degradation and the growth of the principal spoilage bacterium *S. putrefaciens*.

C. Processing and its Effects on the Microflora:

i. Capture, handling and processing:

» Usually finfish are caught by net, hook and line, or traps, with very little control over the condition of the fish at the time of death or the duration of the killing process. This contrasts greatly with the meat industry, in which the health of each animal can be assessed prior to slaughter, and the killing process is designed to minimise stress.

» However, in recent decades, aquaculture practices have been expanding worldwide, offering better control of fish health prior to, and during harvest.

» Physical damage such as loss of scales, bruising and bursting of the gut will increase the number of sites open for bacterial attack and spread.

» In addition, cortisone levels increase during prolonged stress and can alter the quality.

» After capture, the fish may be eviscerated prior to storage, a practice that may have both advantages and disadvantages.

» The action of intestinal enzymes and activity of the gut bacteria on the flesh around the belly cavity may produce discolouration, digestion and off-flavours in un-eviscerated fish.

» In eviscerated fish, however, the cuttings provide areas of exposed flesh that are open to microbial attack.

» If evisceration is carried out at the harvest site, care should be taken in removing all the gut contents and washing the carcass thoroughly prior to refrigerating, icing or freezing.

» As with all foods, careful and sanitary handling during processing is required to reduce the risk of contamination with potential human pathogens, and to limit the loss of quality.

» Good Manufacturing Practice (GMP) and control of the sanitary conditions of the transport and processing environments are essential to limit additional risk of disease caused by fish consumption.

ii. Shell fish Processing and its Effects on the Micro-flora:

- Crabs, lobsters and crayfish are harvested from cages or pots containing bait. They are then taken to market alive in either holding tanks, chilled sea water, or transported directly to processing plants. These animals can survive out of water if the surrounding environment is kept cool and damp. The claws are usually bound together to prevent the animals fighting and causing damage to each other. As the muscles of crustacean shellfish decompose rapidly after death dead animals are rejected.
- In comparison shrimps die quickly after capture. A large proportion of shrimps are produced by aquaculture, or are found in brackish water around the shores, so they are either sold on the local market or are processed in local processing plants.
- Shrimps can be processed whole, with "head on" or "head off" whereby the edible tail is removed from the head, gills and thorax. The latter process can expose the shrimp muscle to incidents of cross-contamination if GHP are not followed; however, by removing the head and thorax a source of enzymes is eliminated that will prevent the occurrence of black spots.
- When the shrimp arrive at the processing plant they are washed and then sorted into size. Shrimp might be further processed with the shell on, or with the shell and central vein removed.
- Scrupulous hygiene is important at this stage to prevent the cross-contamination of the shrimp tails.
- Shrimp will then be glazed and frozen in a raw state, or cooked, glazed and frozen.

D. Microorganisms Associated with Food:

1. Pseudomonas:

- Gram negative, aerobic, rod shape
- The pseudomonads produce proteinase enzymes which catalyse proteolytic reactions and contribute to food spoilage.
- Important in the spoilage of refrigerated, fresh animal products.
- *P. aeruginosa* produces an enterotoxin and can cause gastroenteritis.

» Widely distributed in nature. Psychrotrophic Pseudomonads are found in almost all types of refrigerated and frozen foods.

2. Halobacteriaceae:

» Gram negative, aerobic

» This family contains two genera viz. Halobacterium and Halococcus.

» These are extreme halophiles requiring a high concentration of salt (around 15%) for growth.

» Found in salted foods.

3. Alcaligenes:

» Aerobic

» Optimum temperature for growth is 20 to 37^{0}C.

» Found in soil, water, decaying matter and intestinal tract of animals.

» Involved with the spoilage of protein foods.

4. Escherichia: (*E. coli*)

» E. coli is the most common aerobic organism in the intestinal tract of man and warm blooded animals. Generally the *E. coli* strains that colonize the gastrointestinal tract are harmless commensals, or they play an important role in maintaining intestinal physiology. However, within the specie there are at least 4 types of pathogenic strains:

1. enteropathogenic *E. coli* (EPEC)
2. enterotoxigenic *E. coli* (ETEC)
3. enteroinvasive *E. coli* (EIEC)
4. enterohaemorrhagic *E. coli* / (EHEC) / verocytoxin producing *E. coli* (VTEC)

» When tested with IMVIC (Indole, Methyl red, Voges-Proskauer and Citrate) tests, the organism produces indole from tryptophan, Methyl red +ve, does not form acetyl methyl carbinol and citrate is not utilised (+ + - -).

- Found in soil, water, intestinal tract of animals and animal products.
- Heat sensitive, so present in heat pasteurised/ cooked products.
- Indicator organism of faecal contamination.
- It produces acid and gas from most carbohydrate.
- Cells grow at a temperature from 10^0C to 45.5^0C
- Off-odours and off-flavours produced in foods called "BARNY" by this organism.
- There is no indication that seafood is an important source of *E. coli* infection. Most infections appear to be related to contamination of water or handling of food under unhygienic conditions.

5. Salmonella:

- *Salmonella* are members of the family *Enterobacteriaceae* and they are of Sub genera I, II, III, & IV.
- These mesophilic organisms are distributed geographically all over the world.
- Found in or on soil, water, sewage, animals, humans, processing equipments, feed and various food products.
- Natural habitat is intestinal tract of humans and animals.
- Cause salmonellosis by serotype of salmonella.
- The principal symptoms of salmonellosis (non-typhoid infections) are non-bloody diarrhea, abdominal pain, fever, nausea, vomiting which generally appear 12–36 hours after ingestion. However, symptoms may vary considerably from grave typhoid like illness to asymptomatic infection.
- Contamination of shellfish with Salmonella due to growth in polluted waters has been a problem in many parts of the world. However, it has also demonstrated that Salmonella in aquaculture shrimp products originate from the environment rather than as a result of poor standards of hygiene, sanitation, and poultry manure as feed.
- Most prawns and shrimps are cooked prior to consumption and these products therefore pose minimal health risks to the consumer except by cross contamination in kitchens.

6. Shigella:

» Cause shigellosis, a food borne gastroenteritis.

» The genus Shigella is also a member of the Enterobacteriaceae and consists of 4 distinct species. This genus is specific host-adapted to humans and higher primates, and its presence in the environment is associated with fecal contamination.

» Shigella is the cause of shigellosis, which is an infection of the gut.

» Symptoms vary from asymptomatic infection or mild diarrhea to dysentery, characterized by bloody stools, mucus secretion, dehydration, high fever and severe abdominal cramps.

» The incubation period for shigellosis is 1–7 days and symptoms may persist for 10–14 days or longer.

» The great majority of cases of shigellosis is caused by direct person-to-person transmission of the bacteria via the oral-faecal route. Also waterborne transmission is important, especially where hygiene standards are low.

7. Staphylococci:

» The staphylococci are ubiquitous organisms and can be found in water, air, dust, milk, sewage, floors, surface all articles that come into contact with man and survive very well in the environment. However, the main reservoir and habitat is the animal/human nose, throat and skin.

» Most strains can grow in 7.5% to 15% salt. Found in various food products.

» *S. aureus* produces enterotoxins causing illness.

» The disease caused by *S. aureus* is intoxication. Common symptoms, which may appear within 2–4 hours of consumption of contaminated foods are nausea, vomiting and sometimes diarrhea. Symptoms usually persist for no more than 24 hours, but in severe cases, dehydration can lead to shock and collapse.

» Seafood may be contaminated with *Staphylococcus* via infected food handlers or from the environment. More often the contamination is from an individual with an infection on hands or with a cold or sore throat.

» *S. aureus* is mesophilic with a minimum growth temperature of 10°C, but higher temperatures are required for toxin production (> 15°C).

» *S. aureus* produces a number of enterotoxins, when growing in the food. These toxins are generally very resistant to proteolytic enzymes and heat. There have been no outbreaks reported from foods that have undergone normal canning procedures, but the heat applied in pasteurization and normal household cooking is not sufficient to destroy the toxin.

» Good sanitary conditions and temperature control is necessary to avoid contamination, growth and toxin production - particularly in precooked seafood.

Table 5.5: Growth limiting factors and heat resistance of bacteria originating from the animal/human reservoir

Pathogenic bacteria	Temperature °C			pH	NaCl (%)	a_w	Heat resistance
	minimum	optimum	maximum	minimum	maximum	minimum	
Salmonella	5	37	45–47	4.0	4–5	0.94	D_{60} = 0.2–6.5 min.
Shigella	7–10	37	44–46	5.5	4–5		60°C/5 min.
E. coli	5–7	37	44–48	4.4	6	0.95	D_{60} = 0.1 min. D_{55} = 5 min.
Staphylococcus aureus	7	37	48	4.0	10–15	0.83	D_{60} = 0.43–7.9 min.
Staphylococcus aureus toxin production	15	40–45	46	ca.5.0	10	0.86	High heat stability of toxin

Control of *Enterobacteriaceae*:

» The Enterobacteriaceae, (*Salmonella, Shigella, E. coli*) are all occurring on fish products as a result of contamination from the animal/human reservoir. This contamination has normally been associated with fecal contamination or pollution of natural waters or water environments, where these organisms may survive for a long time (months) or through direct contamination of products during processing.

» Good personal hygiene and health education of food handlers are therefore essential in the control of diseases caused by Enterobacteriaceae. Proper treatment (e.g. chlorination) of water and sanitary disposal of sewage are also essential parts in a control programme.

» Risk of infection with Enterobacteriaceae can be minimized or eliminated by proper cooking before consumption.

8. Klebsiella:

- Found in water, sewage, soil and on fresh produce and frozen foods.
- Members of coliform group of indicator organism.

9. Enterobacter:

- Found in water, sewage, soil and various food products.
- An indicator organism.

10. Vibrio:

- The species *V. costicola* can tolerate 23% salt concentration and is found in cured meats and brine.
- *V. cholera* and *V. parahaemolyticus* are important pathogens, causing gastroenteritis in humans.
- *V. cholera* is found in water and occasionally in food, in intestinal tract of man and animals.
- *V. parahaemolyticus* is found in the ocean, sea foods.
- Most vibrios are of marine origin and they require Na^+ for growth.
- The genus contains a number of species which are pathogenic to man.
- *V. cholerae* occurs in two serotypes, the 01 and the non-01.
- The diseases associated with *Vibrio* sp. are characterized by gastro-enteritic symptoms varying from mild diarrhea to the classical cholera, with profuse watery diarrhea.
- The mechanism of pathogenicity for the vibrios is not entirely clear. Most vibrios produce powerful enterotoxins. A number of other toxins are produced by *V. cholerae*, including a hemolysin, a toxin similar to tetrodotoxin and one similar to shiga-toxin. Pathogenic strains of *V. parahaemolyticus* are known to produce a thermostable direct hemolysin (Vp-TDH), which are responsible for the Kanagawa-reaction.
- A wide variety of foods has been involved in transmission of cholera, including soft drinks, fruits and vegetables, milk, raw, uncooked, or cross-contaminated cooked shellfish.

» Inadequate sanitation and lack of safe water are major causes of cholera epidemics. Therefore cholera can only be reliably prevented by ensuring that all populations have access to adequate excreta disposal systems and safe drinking water.

» Low temperature storage has been proposed as a means of eliminating pathogenic vibrios from food.

11. Aeromonas:

» Main habitat is water.

» Some strains cause disease in fish and some cause enteritis in humans.

» The genus Aeromonas has been classified with the family Vibrionaceae and contains species pathogenic to fish and man.

» In recent years A. hydrophila has received increasing attention as a possible agent of food borne diarrheal disease.

» Aeromonas is ubiquitous in freshwater environments, but may also be isolated from saline and estuarine waters. This organism may be readily isolated from meat, fish and seafood and many other foods. This organism has been identified as the main spoilage organism of raw meat, and fish from warm, tropical waters.

» Species of Aeromonas produce a wide range of toxins such as cytotoxic enterotoxin, hemolysins and a tetrodotoxin-like sodium channel inhibitor. But, the role of these toxins in producing disease in man is not confirmed.

» Aeromonas is very sensitive to acid conditions and to salt and growth is unlikely to be a problem in foods where pH is less than 6.5 and the NaCl content greater than 3.0%.

Table 5.6: Growth limiting factors and heat resistance of *Aeromonas spp.* normally occurring on seafood

Pathogenic bacteria	Temperature (°C)		pH	a_w	NaCl (%)	Heat resistance
	minimum	ptimum	minimum	minimum	maximum	
Aeromonas sp.	0 – 4	20–35	4.0		4 – 5	D_{55} = 0.17 min.

12. Acinetobacter:

» Aerobic, mesophilic

» Found in soil, water and also in man and animals.

» Found in raw and prepared foods.

13. Streptococci:

» Facultative, anaerobic

» Found in air, water, sewage, soil and various food products.

» Heat resistant to 60^0C for 30 mins.

» Potential indicator organism in frozen foods.

» Indicator of faecal contamination.

14. Bacillus:

» Found in soil, water, faecal material, decaying material and in various foods.

» *B. stearothermophillus* and *B. coagulans* cause spoilage in low acid canned foods.

» *B. cereus* is involved in food borne gastroenteritis.

15. Clostridium:

» *Clostridium botulinum* is widely distributed in soil, aquatic sediments and fish.

» Human botulism is a serious but relatively rare disease. The disease is an intoxication caused by a toxin called 'botulin' pre-formed in the food.

» Symptoms may include nausea and vomiting followed by a number of neurological signs and symptoms: visual impairment (blurred or double vision), loss of normal mouth and throat functions, weakness or total paralysis, respiratory failure, which is usually the cause of death.

Table 5.7: Growth limiting factors and heat resistance of *C. botulinum* normally occurring on seafood

Pathogenic bacteria	Temperature (°C)		pH	a_w	NaCl (%)	Heat resistance
	Minimum	Optimum	Minimum	Minimum	Maximum	
C. botulinum						
proteolytic type A,B,F	10	35	4.0 – 4.6	0.94	10	D_{121} of spores = 0.1–0.25 min.
non-proteolytic type B, E, F	3.3	30	5.0	0.97	3–5	$D_{82.2}$ = 0.15–2.0 min. in broth D_{80} = 4.5–10.5 min. in products with high protein and fat content

In contrast it should be noted that fresh and frozen fish has never been shown to cause human botulism. This is probably due to the fact that fresh fish normally spoil before becoming toxic. The ultimate safeguard is very low heat stability of botulinum toxin, which means that normal household cooking will destroy any preformed toxin. Thus the risk is clearly associated with foods that do not require cooking immediately before consumption.

Botulism may be prevented by inactivation of the bacterial spores in heat-sterilized, canned products or by inhibiting growth in all other products. *C. botulinum* is classified into toxin types from A–G, and the types pathogenic to man can conveniently be divided into 2 groups:

1. The proteolytic types A and B, which are also heat-resistant, mesophilic and NaCl-tolerant.

2. The non-proteolytic types E, B and F, which are heat-sensitive, psychrotrophic and NaCl-sensitive. It is primarily the non-proteolytic types which are found on fish and fish products.

Canning processes have generally been designed to destroy a large number of the heat-resistant *C. botulinum* types. Thus the "botulinum cook" has been defined as equivalent to 3 min. at 121°C. This value is also called the F_0 value or the "process value". The F_0 value required for canned fish product is equivalent to 12-decimal reductions of *Clostridium botulinum* spores. Using the highest known D-values (0.25 min at 121°C) the F_0 is therefore equal to 12 x 0.25 = 3. This is the so-called 12 D-concept designed to reduce the bacterial load of one billion spores in each of 1000 cans to one spore in a thousand cans.

16. Lactobacillus:

- In food microbiology, they are useful but also cause spoilage.
- Useful in fermentation in which lactic acid production is desirable.
- Some Lactobacilli form slime and spoil sugar cane. Others cause discolouration in sausage.
- Lactobacillus cause spoilage of vinegar preserved products.

17. Corynobacterium:

- Widely distributed in nature. Found in soil, water, plant and animals.
- These may be human pathogen, plant pathogen or, non-pathogen.

18. Molds:

- Molds are part of larger group of micro organisms called Fungi.
- Fungi have no chlorophyll where as Algae have chlorophyll.
- Important fungi in foods are Penicillium and Aspergillus.
- Molds cause spoilage in foods with low water activity (a_w) or, in fruits with a low pH.
- Penicillium and Aspergillus can grow at lower moisture level (14-18%).
- Molds can be considered as spoilage organism in food products (indicator of spoilage). Although molds associated with food are not pathogenic, some produce mycotoxins. These toxic substances pose a potential health hazard to humans.
- Molds are useful in production of antibiotics.
- Aspergillus and Penicillium together called 'Storage fungi of grains'.
- *Aspergillus niger* grows in sucrose solution and *Aspergillus flavus* produce aflatoxin.
- *Penicillium* is widely distributed in nature and is found in many foods. They are important spoilage organisms. Antibiotics produced by this include Penicillin. Some penicillium cause health hazard and produce mycotoxin causing UTI (Urinary tract infection).

19. Yeast:

Certain strains are important in formation of alcoholic beverages, baking industry, chemicals etc.

a. Saccharomyces:

- Means sugar fungi.
- Vigorous fermentation
- Widespread distribution
- Can cause spoilage in fruits, sugar, syrup, vinegar, fermented foods etc.

b. Candida: cause diarrhoea and malnutrition.

20. Virus:

» Viruses are obligate intracellular parasites. They generally have a limited host range. To survive and replicate, a virus must contact and invade an acceptable host cell. If a host cell is not available, the virus may become inactivated. This inactivation is influenced by temperature, humidity, pH, substrate composition and other factors.

» Viruses have a genetic system with either RNA or DNA. This is surrounded by a protein coat called capsid. The capsid protects the genetic material from nucleases and serves as the 'vehicle of transmission' from one host cell to the other and forms a bond with the host cell during the attachment stage of infection. For a virus to infect a host cell, there must be an attachment site on the host cell. The attached virus injects its nuclear material into the host. Thus the genetic material may attach to the cellular chromosome and the cell becomes lysogenic.

» Bacterial virus (Bacteriophages): Bacteriophage is a type of virus that infects bacteria. All bacteriophages are composed of a nucleic acid molecule that is surrounded by a protein structure. These are the natural enemies of bacteria.

» The phages can have at least 4 relationships with their host cells. These are:

1. Virulent phages cannot combine with the host's chromosomes to establish a lysogenic relationship.

2. Pseudolysogenic phages can establish a carrier state in the host and may act as transducing agents.

3. Defective temperate phages form particles with morphological attributes of a normal phage, but cannot replicate.

4. True temperate phages are able to replicate and lysogenize their host cells.

» The genetic material of temperate phages can become attached to the chromosome of the host cell. The properties of phage are modified and it becomes a 'prophage'. Prophage acts as a bacterial gene. A bacterium that carries a prophage is called'lysogenic'. Agents like UV light, X-rays, organic peroxides etc. can cause the release of phage material from the chromosome with resultant production of phage and lysis of the cell when the phage is released.

» Bacteriophages can be stored by freezing with LN_2, using or not using glycerol (as protective additive).

» Bacteriophages can be enumerated by determining the clearance of susceptible bacterial cultures in broth.

» Function in food: Destroying cultures of lactic acid bacteria during the fermentation of dairy products. Phages can be useful in epidemiological work.

» Prophage can be involved in the production of toxin by *C. Botulinum* and *S. Aureus*.

» A phage may require 40-80 min for replication.

» Heat sensitive phages may be inactivated in about 10 min at 60^0C.

» Heat resistant phage : 25 min at 75^0C or, even 30min at 88^0C.

E. Methods for Control of Microbial Growth:

1. Physical methods:

a. Temperature:

i. High temperature (Heat):

» Dry heat: Gas and Electric Oven. (160 to 180)^{0}C for 1 ½ to 3 hrs

» Moist heat: Sterlisation: Destruction of all forms of life.

- Autoclave: Free flowing Steam under pressure (121.2^{0}C or 15 psi for 15 mins).

* Bacterial spores require temperature above 100⁰C
* Bacterial vegetative cells are killed at (60 – 70)⁰C for 10 mins
* Fungi vegetative cells are killed at (50 – 60)⁰C for 10 mins
* Fungal spores are killed at (70 – 80)⁰C for 10 mins

- Free flowing steam at 100⁰C for 20 mins on 3 consecutive days with intermittent incubation at 37⁰C kills all vegetative cells and spores will be killed during subsequent exposure to 100⁰C. The process is called as **Tyndallization** which takes some more time.
- Boiling water: at 100⁰C for 30 mins result in disinfection where vegetative cells are killed but not the spores.

» **Pasteurization**: Use of low temperature to destroy only pathogenic organisms.

- Low temperature method: 62.8⁰C for 30 mins
- High Temperature Short Time (HTST) method (Flash): 71.7⁰C for 15 secs.
- Ultra High Temperature (UHT) method: Used for the sterilization of low acid foods, **UHT treatment** involves heating the product to over 135 °C. It destroys all microorganisms, making the end product suitable for ambient distribution.

ii. Low temperature (Cold):

» Freezing

» Lyophilization

b. Electromagnetic radiation:

i. X- ray: Ionising radiation. It originates from outside of the atomic nucleus, are representative of ionising forms of radiation.

ii. Gamma ray: Ionising radiation. Originates from unstable atomic nuclei.

Essential cell molecules can be directly affected through loss of their chemical structure and activity brought about by the ionisation.

iii. **Ultra violet light:** Thymine dimerization. Capable of producing a lethal effect in cells exposed to low penetrating wavelengths in the range of 210 nm to 300 nm.

Cellular components capable of absorbing UV light are the nucleic acids, with the DNA as primary site of damage (Pyrimidine – absorb UV wavelength). The major effect of this form of radiation is Thymine dimerization which is the covalent bonding of two adjacent thymine molecules on one nucleic acid strand in the DNA molecule. This dimer formation distorts the configuration of DNA molecule and this distortion interferes with DNA replication and transcription during protein synthesis. Some microorganisms possess enzyme systems for the repair of radiation induced DNA damage. Two different systems may be operational. (a) Dark repair system that functions in the absence of light and (b) Light repair system which is made operational by exposure of irradiated cells to visible light in the wavelength range of 420 nm to 540 nm. Light serves to activate an enzyme that splits the dimer and reverses the damage.

c. **Osmotic pressure:**

   i. Hyper-tonicity: Increased salt and sugar concentration

   ii. Hypo-tonicity : Increased water concentration

d. **Desiccation:** Removal of cellular water

e. **Sonic vibration:** High frequency sound waves

f. **Filtration:** Using microbial filters

Cell wall injury:

» Lysis of cell wall will leave the protoplast (Bacteria having no cell wall is protoplast and having incomplete cell wall is called spheroplast) susceptible to osmotic damage and a hypotonic environment may cause lysis of the vulnerable protoplast. Further, certain agents inhibit cell wall synthesis (periplasmic enzyme).

Cell membrane damage:

» Lysis of membrane which will cause immediate cell death. Also, selective nature of the membrane may be affected without causing its complete destruction. As a result, there may be a loss of essential cellular molecules or, interference with the uptake of nutrients. In both the cases, metabolic processes will be adversely affected.

Alternation of the colloidal state of cytoplasm:

» It will produce irreversible cell damage, as certain agents cause denaturing of cytoplasmic proteins.

Inactivation of enzymes:

» By this cells cannot perform essential life functions. This occurs through the mechanisms of competitive or non-competitive inhibition.

Interference with the structure and function of DNA molecule:

» Some agents have an affinity for DNA and cause breakage or distortion of the molecule, thereby interfering with its replication and role in protein synthesis.

1. Chemical methods:

a. Antiseptic: Chemical substances used on living tissue that inhibit the growth of microorganisms.

b. Disinfectant: Chemical substances that inhibit the growth of vegetative microbial forms on non-living materials.

» Examples of Disinfectants and antiseptics:

- Phenolic compounds: Phenol, cresol, Resorcinol, Thymol
- Alcohols: Ethyl alcohol, Isopropyl alcohol
- Halogens:
  - ◊ Chlorine compounds (Sodium hypochlrite, chloramines)
  - ◊ Iodine compounds (Tincture of iodine, Betadine).

i. Heavy metals:

- ◊ Mercury compounds (Inorganic): Mercury bichloride, mercurial ointments
- ◊ Mercury compounds (Organic): Mercurochrome, Metaphen
- ◊ Silver compounds: Silver nitrate

ii. Surface active agents:

◊ Washing agents: Emulsifiers, soaps, detergents

◊ Cationic agents: Quarternary ammonium compounds, Benzalkonium chloride

◊ Anionic agents: Tincture of green soap, sodium tetradecyl sulphate.

iii. Others:

◊ Acids, alkalis, formaldehyde, ethylene oxide, Beta propiolactone and Basic dyes (Crystal violet).

c. **Chemotherapeutic agent**: Chemical substances that destroy or inhibit the growth of microorganisms in living tissues.

» Antibiotics: Synthesized and secreted by certain bacteria, *Actinomycetis* and fungi.

» Synthetic drugs: Synthesized in laboratories viz. p –amino benzoic acid (PABA), sulfdiazine etc.

6 Wet Fish Handling and Preparation

INTRODUCTION

- After independence, India has made glorious achievements in food production and has witnessed multicolour revolutions in terms of green, yellow, white and blue.
- As compared to food grains, milk eggs and other foods, country's fish production has increased at a higher rate.
- This sector is contributing 1.2% to the national GDP and 5.3% to the agricultural GDP.
- Fish is a highly perishable commodity requiring a cautious handling and careful processing.
- To ensure best quality, appropriate technologies and processes must be followed with clean and hygienic conditions.

A. Keeping Quality of Fish:

- The physical, chemical and bacteriological characteristics of fish are influenced by the species, seasons, catching methods and fishing grounds etc.
- The characteristics of fresh fish are:

  a. Shining, iridescent surface covered with a nearly transparent, uniformly and thinly spread slime.

  b. Protruding eyes with a jet black pupil and transparent cornea

  c. Gills that are bright and free from visible slime

d. Flesh is soft and flabby, tending to retain finger indentations

e. The flesh odour is 'marine' or 'fresh sea weedy' in addition pleasant, margarine like odour in case of fatty fish.

» As the spoilage is developed in fish:

a. Its structure loses bright sheen and colour and slime becomes thicker, turbid and lumpy and lately, slime becomes yellow or brown.

b. Eyes sink and shrink with cloudy and milky pupil and opaque cornea.

c. Gills become bleached, light pink in colour, finally turning to greyish brown covered with a thick slime.

d. Flesh becomes soft and inelastic and may be easily stripped from backbone and extrudes juice under light pressure.

e. Odour changes from 'sweetish' then 'ammoniacal' or 'fishy' and finally putrefaction odour.

B. Causes of Spoilage:

» The fish spoilage may be caused by 3 causes:

a. Bacteria

b. Digestive enzymes

c. Others such as oxidation, etc.

» The flesh of freshly caught fish is sterile while the spoiling micro-flora is present in external slime and digestive tracts.

» Fatty fishes may develop spoilage due to all 3 above mentioned causes.

» If not gutted soon after catch, strong digestive enzymes affect the viscera and belly walls and cause discoloration- 'belly burn' or disruption- 'torn bellies'.

C. Handling Wet Fish at Sea:

» In the fishing vessels, the catch is released from the net on to the deck, gutted immediately, washed, and stowed with ice in boxes or compartments below deck.

- Gutting of round fish like cod, haddock and whiting means slitting the belly from throat to vent, removing the liver and cutting out the guts to leave the belly cavity empty. This operation is traditionally done by hand with a knife.
- The gutted fish are washed to remove traces of blood and debris, and to wash away most of the bacteria present on the skin and in the gills of the fish. Then fishes are transferred to fish room where they are stored in ice.
- White fish, promptly gutted, washed, and stowed in ample ice, will keep in first class condition for 5-6 days, become stale after 10-12 days, and are unlikely to be edible after 15-16 days.
- Boxed stowage with flake ice is usual on smaller fishing vessels.
- Fatty fish are not normally gutted at sea because their small size and the large numbers. The keeping time of fatty fish in ice is much less than for white fish; the attack by bacteria and digestive juices is much more rapid because the fish are ungutted, and the fat absorbs oxygen to produce rancid flavours and odours.

D. Handling Wet Fish on Shore:

- The fish are brought to fish landing centre where they are to be washed properly, iced and then auctioned.
- For large sized fish, gutting followed by washing and icing are to be done prior to auctioning.
- Fish is nowadays mostly carried by road transport from the ports and the biggest companies have their own fleets of insulated and refrigerated vehicles carrying fish either to inland distribution depots or direct to customers, while many smaller merchants at the ports share a long-distance transport pool. Although the better carriers use vehicles with adequate insulation, often supplemented by mechanical cooling units, some fish is still carried under less satisfactory conditions on open lorries.
- Most retailers keep the bulk of their supplies in chill rooms on the premises and only display a selection on ice, often in refrigerated cabinets.

E. Handling of market fish:

- Fish should be purchased as fresh or frozen as possible.

- » Exposure to unsafe temperature should be avoided.
- » Exposure to contaminants should be avoided by keeping fishes in plastic bag.
- » It should be kept in refrigerators at home.
- » Raw fish purchased should not come in contact with cooked fish, other cooked food stuffs and raw fruits and vegetables as raw fish contains bacterial load.
- » It should be stored at proper temperature- if hot, temperature should be 60°C and if to be kept cold, temperature should be 4°C or less.

F. Handling of Fresh Catch:

- » Efforts should be done to keep the fish alive as long as possible- by keeping in metal link basket or live box.
- » The fish should be checked for signs of disease or parasites or for any abnormal growth in the flesh.
- » It should be made certain that the fish is being taken from safe water without any contaminants.
- » If it is difficult to keep fish alive, it should be stored at temperature below 4°C.

G. Why Process Fish and Shellfish?

- » There are a number of reasons for processing fish/shellfish. Five reasons are provided below:

1. To supply food that is safe to eat:

- ◊ To ensure that the fish and shellfish that to be consumed at home or sold at the market are safe to eat, they must be washed, and gutted thoroughly to remove all dirt and slime.
- ◊ Shellfish (as filter feeders which may be found in polluted waters) must be left overnight in water to clean out any internal impurities which is called 'Depuration'.
- ◊ Fresh water shellfish should be kept overnight in fresh water, while marine shellfish must be kept overnight in clean sea water.

2. **To minimise loss/waste of a valuable food commodity:**

◊ In many island countries, seafood may be one of the few food items which provide the protein requirements within a diet.

◊ Care must be taken to ensure that the preparation of fish and shellfish is carried out correctly to ensure that there is no loss or waste of this important source of nutrition.

3. **To meet quality standards and consumer preferences:**

◊ When processing food for the consumer, attention must be paid to hygiene in the handling of the seafood. Thereafter it is assured that the food for consumption is of a high standard.

◊ At the same time, when preparing the seafood, definite preferences on the type of seafood the consumer will eat is to be taken into consideration.

4. **To sell and make a profit by adding value and increasing convenience to the consumer:**

◊ Value can be added by removing a part or parts from the body of the fish/shellfish. For example, removing the scales from a fish, or removing the shell from a shellfish may increase their selling price and will reduce the preparation work of the buyer.

◊ Alternatively, value can be added by adding ingredients in the preparation of seafood.

◊ By marinading to modify flavours, adding breadcrumbs and spices, the seafood can be converted into fish jerky, fish fingers, spicy dried clam, or fish cakes. This will provide the consumer with a variety of seafood dishes.

5. **To extend the shelf life of the foods so that the food can be made available out of season or when it is not possible to catch or purchase fresh food:**

◊ Seafood can be processed to extend its shelf life.

◊ Chilling is a method of processing which is used to preserve the life of fish and shellfish. Keeping the fish at low temperatures slows the rate of growth of bacteria and slows spoilage rates.

◊ In addition, the processing of seafood through salting, drying, or smoking can preserve the food, enabling it to be stored and used when necessary.

H. Definitions Used in Fish Preparation:

1. Fillet:

◊ A strip of flesh cut from the fish parallel to the line of the backbone.

◊ Block fillet or Butterfly fillet: Flesh from both sides of a single fish joined together usually along the backbone.

◊ Single fillet: Flesh from one side of the fish.

◊ Fish can be filleted either by hand, using the knives or, using filleting machines. Fish are often filleted prior to sale fresh as well as prior to freezing, smoking, canning and production of minced products.

◊ Round fish yield two fillets (One from each side) whereas flat fish yield four fillets (two from each side).

◊ Fillets from larger fish can be further portioned into 'supremes' or 'escalopes'. They may be skin-on or skin-off.

2. Gutted/ drawn fish:

◊ Fish from which the guts have been removed.

◊ Gutting is often done at sea prior to storage on ice or before freezing in vertical plate freezers on 'freezer trawler'.

◊ Gutting is usually achieved by cutting the fish along the ventral surface from the vent to the gill opening and removing the intestines. In many cases, the head is removed at the same time as the fish are gutted.

3. Split fish:

◊ Fish are often split during preparation for smoking or drying.

◊ The purpose of this is usually to increase the surface area of the fish so that it can dry more quickly or take up flavours of salt, smoke or other condiments used during processing, more uniformly.

◊ Most of the methods used for splitting fish remove the guts but do not remove bones except perhaps for the head.

4. **Boned fish:**

   ◊ The flesh of the fish from which most of the bones have been removed.

   ◊ For instance, a fillet may contain some small pin bones or ribs but it is often described as boned.

5. **Boneless fish:**

   ◊ Flesh of the fish from which all the bones have been removed.

6. **Semi-dressed fish:**

   ◊ Removing scales, gills and viscera but not head.

7. **Dressed fish:**

   ◊ Dressed fish have had their viscera, head, tail, fins and scales removed.

   ◊ ***Pan dressed fish***: Fish which have been prepared for cooking / frying pan or prepared in a particular way for presentation purpose.

   ◊ ***Bullets***: This term is usually used instead of dressed for sturgeon fish because they have particularly unusual anatomy requiring more cleaning and the end product is a bit different.

8. **Steaks**:

   ◊ Steaks are portions of fish that have been cut across the body rather than along the sides. Often they include a section of backbone.

9. **Cheeks**:

   ◊ Some large fish have portions of meat on their heads, separate from the fillets, that are sufficient quantity to be harvested and sold separately.

10. **Loins**:

   ◊ This is the prime part of a fillet from a large round fish.

   ◊ It is the part of the fillet which is above the spine, sometimes called the top back loin.

   ◊ Loins are typically cut from fish like Tuna, sword fish, marlin.

Gutting, cleaning and cooling:

» The harvested fish should be cleaned and cooled as soon possible due to their strong digestive juices; fish spoil very soon and if not gutted and cleaned promptly may develop off flavour and colour.

» A sharp knife, clean cloth, plastic bags and crushed ice must be kept ready for gutting and cleaning of fish.

» With the help of knife, fish should be cleaned and bled.

» Throat should be cut and gills and entrails are removed.

» Cleaned fish first should be put in plastic bags and then in ice.

» If required steaks may be cut and kept in plastic bags.

» To prevent contamination, hands, working area, cutting boards, knives and other utensils should be properly cleaned with water and soap.

Preparation of fillets:

» Wash the fish in clean cold water to remove blood, microbes and enzymes.

» Filleting is done with a sharp knife with flexible blades.

» The fillets collected should be stored properly at low temperature.

***Process of filleting*:**

» Fish is cut behind the pectoral fin straight down to the backbone and cut is angled towards the top of the head. → Knife is run along one side of the backbone, scrapping the rib bones without cutting them. → The knife is pushed through the flesh near the vent just behind the rib bones and fillet is cut free at the tail. → Flesh is cut carefully away from the ribcage. → The first boneless fillet is removed by cutting through the stomach skin.

I. Requirements for Fish Handling Premises:

» Where possible, all preparation, processing and packaging should be carried out in one building.

- Buildings should be single storied; drainage and handling fish is easier on a single level.
- Floors should be smooth surfaced, but non-slip, well drained and water proof. As far as possible they should be resistant to attack by fish oils, offal and brine. High density concrete containing granite chips can be used, although clay tiles are better.
- The floor surface should be carried up the walls or any permanent fixtures for a height of at least 15 cm.
- Floors should slope into the drainage system to facilitate washing down.
- Drainage channels should be accessible and traps should be installed before discharge into any sewer or soak way. Traps should be easily and frequently cleaned. Drainage channels should be arranged away from through areas; if not, they should be covered with flush fitting removable gratings.
- Walls should be water proof and have a smooth surface. Tiles can be used on walls where the working tables against the wall. All other wall surfaces should be painted, preferably with hard gloss paint.
- Doors should be flush fitting and well painted to give a water proof and washable surface.
- Windows should be of simple construction with a few large sheets of glass, not numerous small sheets.
- All metal work should be painted thoroughly to prevent rusting.
- In tropical countries, efficient fly screening is essential. All doors and openable windows should be screened. Self-closing screen doors should be fitted at all normal points of entry and exit to and from the working areas.
- Ceilings should have, as far as is possible, a continuous unbroken surface, be painted in a light colour and be easily washed. In tropical climates, insulation above the ceiling may be required to reduce over-heating.
- Good ventilation is essential and all ventilation openings must be screened. In very hot and humid climates, an air-conditioning system may be justified.
- All equipments must be easy to clean. All work surfaces should be of stainless steel or aluminium alloy. Cutting boards of composition material are preferable but if not

available cutting boards of hard non-resinous wood should be used.

» Containers for fish and brine should be made from stainless steel, aluminium alloy or a suitable plastic. Wooden fish boxes can be used but they are more difficult to keep clean.

» All electrical installations should be properly earthed with water proof power points. They should be sited well above floor level.

» A piped supply of clean drinking quality water should be available. If suitable piped water is not available a filtration and chlorination plant should be installed to treat the raw water supply.

» A regular cleaning routine must be established for the whole premises to ensure that standards of hygiene are maintained. It is important that adequate supervision is given during cleaning.

» Detergents and sterilants should be used for cleaning. Detergents help to remove dirt, sterilants kill bacteria. Many detergents also have sterilising properties. A detergent must be used to remove dirt, followed by a sterilant to kill any remaining bacteria. If water chlorination equipment is on site, high levels of chlorine in the water can be used to sterilise.

» A piped hot water supply should also be available. Detergents are far more effective when used in hot water.

» High pressure hoses should be used for washing down if at all possible.

» Adequate washing, changing and lavatory facilities should be installed but not in the fish working area. Washing facilities should also be installed in the fish working area.

J. Handling During Processing and Storage of Fish:

» After cleaning and packaging the fish should be stored either by refrigerating or freezing.

Points to remember during storage of fish:

◊ Fish must be cleaned thoroughly.

◊ Refrigerator temperature should be 4°C or low.

◊ Thawed fish should not be refrozen again.

◊ Thawing should be done in refrigerator or cold water that must be changed at every half an hour.

◊ Fish should be frozen as soon as possible after the catch.

◊ The package should be mentioned with content and date.

Handling during processing of fish:

◊ Raw and fresh fish are given different treatments through different processing methods for the purpose of:

- Long storage
- Enhancing sensory characteristics

◊ The common techniques of fish processing includes canning, smoking, salting, drying and fermentation.

1. Refrigerating:

◊ Raw fish can be kept in refrigerator at the temperature of 4°C or lower.

◊ It should be stored in covered containers.

◊ While keeping in refrigerator, fish should be moist but not wet.

◊ For refrigerating fish at home, the fish should be removed from wrapper, rinsed properly in clean cold water, surface dried with paper towel and kept in a plate or pan lined with double layer of paper towels

◊ The pan should be covered tightly with aluminium foil or plastic wrap to avoid leakage.

◊ This covered fish may be kept in the coldest part of the refrigerator.

2. Freezing:

◊ It is done if the fish is not to be used within 2 days of the catch or to be transported.

◊ It must be properly cleaned and handled before freezing.

◊ The temperature of freezing fish should be 0° for less.

◊ Following methods can be used for freezing the fish:

a. Double wrapping:

- Suitable for small whole fish, steaks or fillets
- In this, fish is tightly wrapped individually in plastic wrap and again wrapped tightly with another layer of wrap.
- These wrapped pieces are kept in a sealable freezer bag or may be wrapped tightly in aluminium foil.
- The sealed bags are placed scattered in freezer and when they are frozen, they can be stocked on top of refrigerator/ cold store.

b. Freezing in ice block:

- This method is suitable for steaks and fillets.
- It protects the fish from freezer burn and avoids the exposure from air
- For freezing fish in ice box, it is cut into small pieces and placed in containers. Head space of one inch should be left for expansion.
- Fish is then covered with cold water and kept in freezer to freeze.
- Check whether fish is completely covered with ice or not.
- When freezing is done in a pan, the ice block is taken out by running little cold water at its bottom and ice block is double wrapped and kept in freezer.

c. Glazing:

- This method is suitable for whole fish or large cuts.
- In this method, fish is first frozen by placing them on a baking sheet in a single layer and the frozen fish is then dipped into a container having ice water and allowed to freeze again.
- This process is repeated until an ice coating of at least 1/8" thickness is made on the fish.
- It is then put in an airtight freezer bag or container and kept in freezer.
- At interval, glazing should be checked.

Care to be taken during freezing:

◊ The package should be marked with content and date.

◊ Wrapped package should be sealed tightly.

◊ When fish is frozen, make sure it is fully covered with ice.

◊ Frozen fish brought from market should be put in the freezer at the earliest and in the original package only.

◊ Putting the fish package in freezer bag before keeping in freezer to provide an extra protection.

3. Super-chilling:

◊ This method is appropriate when freezing facilities are not available and fish have to be transported.

◊ If followed properly, freshness can be maintained for six to seven days.

◊ For super chilling, fish are cleaned and wrapped tightly as whole with two layers of plastic wrap or aluminium foil.

◊ A mix of rock salt and crushed ice in 1/20 is prepared.

◊ On this, ice mix layer is put. The process is repeated till fish are covered.

◊ For placing a fish for chilling, a four inch layer of plain crushed ice is put at the cooler bottom and wrapped fish are placed on it in a single layer.

◊ Finally, the cooler is tightly covered.

◊ During transportation, water drain may be left open or cooler should be removed in between and water should be drained off. Accordingly, the ice mix should be put again.

4. Canning:

◊ In canning fish, high heat treatment is given to destroy the pathogens and inactivate the spoilage bacteria and the product is packaged in hermetically sealed containers to protect against recontamination.

◊ Canning provides shelf stable fish and so freezing space is not required.

◊ The shelf life of canned fish is for one to five years.

Handling:

◊ After washing, the raw material should be chilled as soon as possible, and kept chilled until unloading.

◊ The processes and principles involved in preparing fresh fish for canning are, for the most part, similar to those that would be involved in preparing them for marketing as fresh fish.

◊ Preparation of raw materials for canning includes steps such as cleaning, washing, sorting, dressing, brining, precooking, trimming, size cutting, packing in cans and adding ingredients.

◊ If the frozen fish is used for canning, thawing is an important step. Thawing is a process of converting solid ice to water in a frozen fish. Thawing is done either in air or in running water till all ice of the fish melts.

◊ If the fresh fish is used for canning it should be chilled immediately. It is important to keep the fish in chilled condition during all pre-processing operations.

Washing:

◊ Washing of fish has to be done using adequate quantity of chilled water (either by using ice or by refrigeration). Immediately after washing, the fish should be iced in order to avoid microbial and enzymatic spoilage.

Dressing:

◊ It is a process of removing head viscera, gills and fins and the rest including skin and bones (considered as edible) go into cans. Sometimes, heads and small fins are not removed. Some fishes have to be washed, dressed and cut to required size depending on can dimension.

Brining:

◊ It is an operation where the dressed fish is dipped in saturated brine solution for 8-15 minutes depending on the thickness of dressed fish, brine temperature and nature of fish. Brining removes blood, dirt, slime and other undesirable materials attached to fish. Further brining helps in improving the texture of meat by removing moisture.

Precooking:

◊ It is done to remove excess moisture and fat from the fish. This can be achieved by steaming, frying and blanching. By this method the fish becomes firm and also the moisture content comes down. After precooking, fish meat looks bright and more attractive. However, there will be some loss of taste and flavour components.

◊ Precooking by steaming is done in retort for 30-45 minutes which expels moisture and fat and can be removed by draining.

◊ Frying is done using refined vegetable oil at a temperature of 160 to 180°C for one to three minutes. Frying results in reduction in moisture, firming of texture, reduction in size and improvement in taste and flavour.

◊ Blanching is one of the precooking methods where the fish/shrimp is dipped in boiling brine containing 6-8% salt for 4-6 minutes. Use of citric acid in blanching of shrimp is a normal practice to achieve better colouration and texture. Blanching results in contraction and curling of shrimp, reduced moisture content, development of firm texture and attractive colouring. Blanching of shrimp helps in controlling the net weight of shrimp in the processed cans.

Filling of cans:

◊ Placing a definite quantity of well prepared food into clean cans can be considered as packing or filling. The amount of solid material packed into the can is referred as solid pack. The quantity of solid packed depends on the size of the can. The liquid media is filled either by measuring or by weighing. It is important that every can has to be packed uniformly with respect to arrangement of fish inside the can.

Exhausting of filled cans:

◊ Removal of air from the cans after filling the food is an important step where the contact of air with food is avoided so that food could be preserved for long time after heat processing.

Can closing or seaming of cans:

◊ It is essential to close the cans without inclusion of air as soon as exhausting is completed.

Heat processing:

◊ Cans are subjected to heat processing at 15 psi (121.2°C) for 45 mins to 1 h based on type of fish and filling medium. This is mainly to kill the Clostridium botulinun, the mesophilic spore former and a pathogen.

Can cooling:

◊ After heat processing the cans are subjected to chilled water so that all the thermophiles will be killed due to sudden low temperature.

5. **Smoking:**

◊ Smoking is very popular in developed countries for taste.

◊ Smoking is much common in less developed countries where refrigeration and transportation facilities are not very good.

◊ It is done for two purposes:

a. For preservation

b. For texture and flavour

◊ Steps in smoking:

a. Preparation/cleaning of fish - Scales, head, fins, tail, viscera are removed and fish are washed with clean cold water.

b. Curing of fish: It can be done in two ways -.

i. **Wet/ brine curing:**

- Ingredients used for brine curing are 1kg table salt, 1½ cups brown sugar and 4 litres of cold water. Other spices may also be added as per the choice. All these are mixed properly in a non-metallic container.
- Fish is completely immersed in brine. One litre of saturated brine is sufficient for curing half kilos of fish.
- It is then refrigerated or kept in ice bags to keep the fish cool.
- Brining is done for 2 to 6 hours.

ii. Dry curing:

- In this salt is rubbed directly on the fish.
- For this, three parts of brown sugar is mixed with one part of non-iodized salt and fish fillets are placed in a deep non- metallic pan.
- The piece is covered with dry cure mix and kept in refrigerator for 18-24 hours
- After curing, fish are taken out, rinsed lightly, pat dried and placed on paper towel and air dried for one hour.

Drying of fish:

◊ Cured fish are dried before smoking.

◊ Drying forms a pellicle or glaze on the skin, which keeps moisture in and contamination out and allows the smoke to adhere properly to get smoky flavour and uniform colour.

◊ Drying can be done in cool shady place or in smoker racks.

Smoking:

◊ The fish may be cold or hot smoked.

◊ Smoked fish may be kept for many months in freezer but in refrigerator, cold smoked fish may be preserved for several months but hot smoked may be kept only for a week.

◊ Smoking requires a smoke source, a smoking chamber and an air ventilation system.

a. Cold smoking:

◊ The fish is smoked at low temperature forlong time.

◊ For this, fish is placed in smoker having temperature maintained at about 80°F. It should not exceed 90°F else it will become cured and cooked and unsafe for consumption.

◊ The smoking time for small fish may be 24 hours and for large fish it may be 3-4 days of continuous smoking.

b. Hot smoking:

◊ In this, fish is placed in smoker having temperature of not more than 90°F for the first two hours. During this period, fish will get a brown colour.

◊ The temperature is then raised gradually to about 175°F and smoking is done for 4-8 hours.

Cooling and packaging:

◊ After smoking is completed, the fish are cooled and then packaged in suitable packaging material and stored.

6. **Salting of fish:**

◊ Salting of fish is an art as well as a science.

◊ This process is affected by weather, size and species of fish.

◊ For salting, good quality fish should be chosen and proper cleaning should be maintained.

◊ Following operations are made for salting fish:

i. Preparing the fish:

◊ Beheading, gutting, bleeding and cleaning of fish should be done as soon as possible.

◊ Larger fish can be cut into suitable pieces.

ii. Salting:

◊ The bottom of waterproof vat is covered with a thin layer of salt and fish are layered on it with flesh side up.

◊ Fish layer is covered with thin layer of salt leaving no space. In the same manner two or three layers of fishes are placed

◊ Then fish is reversed with skin side up covering them with salt layer. The top layer is always a salt layer.

◊ As due to salt, moisture comes out and brine is formed. The brine is kept saturated by adding more salt.

◊ Because of moisture loss, fish level is fallen and accordingly more fish may be placed along with the salt layer.

◊ The fish may be cured for12-15 days in warm weather and 21 days or more in cold weather.

◊ Properly salted fish flesh is translucent, firm but yields to light pressure.

iii. Washing and drying to remove extra salt:

◊ After salting, fish are washed in unpolluted sea water or fresh brine and put on plain surface, applying pressure on them to remove moisture and make them thinner.

iv. Drying:

◊ Salted fish may be dried in sun or by artificial heat and air generated by fans.

7. Drying and dehydration:

◊ Oldest method of preserving perishable food items like fish and is considered as least costly method.

◊ Two common methods for drying fish are

a. Sun drying:

Advantages:

- Due to high concentration and diminished water content, spoilage due to microbes and enzymes are avoided.
- It is simple and inexpensive method.
- Further, dried fish may be stored at most ambient temperatures.

Demerits:

- No control on over drying conditions.
- Dependency on natural weather
- Greater chances of contamination with dirt and dust and infestation with insects, their eggs and larvae.

b. Dehydration:

- The drying carried out in controlled conditions.
- Dehydrated fish have high quality and longer shelf life than sun dried fish.
- Various driers like cabinet drier, kiln drier, tunnel drier, etc are used.

8. Fermentation:

◊ Fermented fish have been considered as a Southeast-Asian product for long.

◊ Fermented fish products are prepared by salting and fermenting the fish to convert the flesh into simple components.

◊ Fish fermentation lasts for3-9 months and the flesh liquefy or becomes like a paste.

◊ Fermented fish means any fishy product having degradative changes due to enzymatic or microbial activity.

◊ Some common fermented fish products are- Nuoc-man of Vietnam, Nam-pla of Thailand, Sushi of Japan and Patis of Philippines.

◊ The main fermentation procedures are:

a. Fermentation with salting and drying

b. Fermentation and drying without salting

c. Fermented with salting but without drying

K. Transportation of Live Fish:

» Fishes are transported mainly for three purposes:

a. For breeding purposes

b. For stocking in nursery pond

c. For selling purposes

» It plays an important role in the distribution of quality fish fry among a country or even the world.

- Fish Transportation is a process by which fish is transported from one place to another place in a scientific way.

Tolerance to transport:

- Tolerance of fish to transport depends on their ability to resists or adapt to stressful condition. Fishes need tender care and love to remain strong and healthy.
- Tolerance varies with species to species.
- Some species mainly having air breathing organ are more tolerant to transport. E.g. Oreochromis spp., *Clarias batrachus* (Magur), *Heteropneustes fossilis* (Singi), *Anabas testudineus* (Koi) etc.
- Low tolerant species - Silver carp.
- Some medium tolerant species including Bighead carp, Grass carp.
- Timing of transportation is important in regard to temperature management. Late night/ early morning is preferred.

Thermal acclimation:

- At more than 2^0C difference, fish should be tempered.
- Higher temperature difference indicates possible handling problems and leads to thermal shock or death.

Salinity acclimation:

- Salt concentration in vertebrate blood is 10-12g/l.
- Time and efficiency of acclimation to salinity difference depends on:
    a. Species (high tolerant to narrow tolerant)
    b. Salinity difference
    c. Other environmental/ physiological factors
- Fish would try to cope with salinity differences through **Osmosis.**
- Performing osmosis would require energy.

» Energy expenditure will be on the cost of other production traits such as growth.

» If energy requirement is beyond fish's ability, fish will be stressed or die.

pH acclimation:

» pH of vertebrate blood – including fish is about 7.4.

» Some marine species are sensitive to significant pH changes

» *Macrobrachium rosenbergii* will need Acclimation for pH changes especially for significant differences (its favourite level is below 9.0).

» Fish require longer periods for pH acclimation.

Conditioning fish before transportation:

» Conditioning is the pre-requisition of fish transportation. It reduces the mortality rate during transportation.

» The temperature of the conditioning water should not be high – it should be preferably on the lower side of the optimal thermal range of the species.

» Lowering water temperature will lead to:

a. Reduce metabolic wastes (feces, ammonia, CO_2)

b. Reduce DO requirements

c. Increase the saturation level of water to DO

» Fry, fingerlings should not be handled with bare hand- the slime over the fish body should remain intact.

» Fish survived better if they have no food in their intestine.

L. Methods for Packing and Transportation:

» There are two systems for packing of fish fry

I. Open system

II. Close system

i. Open system:

- » Open system comprising with open carrier with or without artificial aeration/oxygenation/water circulation.
- » It is a traditional method.
- » Mainly plastic or metal drums or traditional earthen pot "hundi" are used for fish transporting.
- » In general, two types of hundies are used in transportation: I. 20cm diameter-23L capacity II. 23cm diameter-32L capacity.
- » In this system, dissolved oxygen content is maintained through splashing of water during movement.

Advantages:

- » It is cheap and easier method.
- » Equipment are locally available.
- » It keeps water cool for long time.
- » No need to oxygen supply from outside.
- » More suitable for brood fish.

Disadvantages:

- » Relatively fewer amounts of fish can be transported at a time.
- » Needs close observation.
- » Not suitable for larger distance.

Close system:

- » In this system, adequate oxygen supply is maintained in the water while transporting fish.
- » It is the modern transporting system.
- » Plastic or metal container, Rubber and plastic bags are used in this system for transporting fish.

- One-fourth of the bag is usually filled with water and rest three- fourth is filled with oxygen.
- Woven grass or sometime ice is also used during packing of fry bag to provide insulation from heat.
- After adding oxygen, the bag is sealed shut with a twisted rubber band.

Advantages:

- It is most suitable for commercial fish farm.
- Large number of fish can be transported at a time.
- It can be transported easily from a long distance even from one country to another.
- Mortality rate is very low.
- Oxygen can be supplied easily.

Disadvantages:

- Additional oxygenation is required.
- Need skilled person for packing and handling.
- Packing materials are not locally available.
- It is a cost effective method.

Causes of mortality in transportation:

- Oxygen depletion due to oxidation of organic materials.
- Accumulation of free Carbon-dioxide, resulting from respiration and Ammonia as excretory product.
- Sudden fluctuation in temperature.
- Hyperactivity and stress due to handling.
- Ion-osmotic imbalance due to stress.
- Physical injury due to handling.
- Diseases.

M. Chemicals Used in Fish Transportation:

- Several drugs have been studied for use as anesthetic for transport of fish. E.g. Tricaine methane sulphonate (MS-222), Sodium amytal, Quinaldine, Novocaine etc.
- Drugs and chemicals are either used as tranquilizers and sedatives or as antiseptic.
- Sedatives are mainly used for: I. Reducing metabolic rates II. Reducing excitability of fishes III. Convenience in handling fish.
- Common salt may be used @2g/l to 5g/l to lower the water temperature. However, the salt should not contain iodine.

Conclusion:

- Fish is a highly perishable product and can be handled safely to improve its shelf life.
- Different processing methods like canning, drying, smoking, etc if applied properly to the fish as soon as the fish is caught can enhance the shelf life from months to years depending on the methods used.
- The processed fish can be transported to the distance places without the risk of spoilage and food poisoning.

7 Chilling and Chill Store

INTRODUCTION

- Chilling is an effective way of reducing spoilage in fish if it is done quickly and if the fish are kept chilled and handled carefully and hygienically.
- Immediate chilling of fish ensures high quality products.
- For every 10^{0}C reduction in temperature, the rate of deterioration decreases by a factor of 2 to 3.
- The objective of chilling is to cool the fish as quickly as possible to as low a temperature as possible without freezing.
- Chilling cannot prevent the spoilage together but in general, the colder the fish, the greater the reduction in bacterial and enzyme activity.

I. CHILLING OF FISH

- The important chilling methods of fish and fish products at non-freezing temperature are:

1. Iced storage
2. Chilled seawater (CSW) storage
3. Chilled freshwater (CFW) storage
4. Mechanically refrigerated seawater (RSW) storage
5. Cold air storage.

» The most common means of chilling is by the use of ice. Although ice can preserve fish for some time, it is still a relatively short-term means of preservation when compared to freezing, canning, salting or drying. When used properly it can keep fish fresh so that it is attractive in the market place.

» Icing is the most prevalent method of preserving fish. Ideal icing involves packing crushed ice and fish alternatively in insulated boxes, in the fish to ice ratio of 1:1 (w/w). By this, the temperature of the fish is lowered to near 1 to 2°C in about 2-3 hrs (the melting of the ice needs 80 calories of heat/g and this heat is removed from the fish in contact with ice and hence, the fish get cooled).

» Lowering of temperature due to addition of ice results in:

- Almost all enzyme activity will be arrested,
- About 50-60% of the mesophilic bacteria will be killed and
- Activities and growth of all other bacteria, which are cold-loving (psychrophilic) and cold-tolerant (psychrotrophic) will be slowed down.

» Thus, the spoilage of fish is delayed to a considerable length of time in ice. During iced storage of fish, there is an initial drop of bacterial count due to the death of the cold sensitive mesophiles. The surviving cold tolerant bacteria, however, get adapted to growth in low temperature. Consequently, there is a gradual increase in population, which takes about 6 to 8 days to reach a count of one million per gram or above. By that time, the fish has reached the stage of incipient spoilage.

» During iced storage of fish, the Pseudomonas/ *Alteromonas* group emerges as the predominant group of bacteria at the time of spoilage. This is because most of the psychrotrophic bacteria capable of spoilage belong to these genera. The population of these psychrotrophs is very low in the fresh tropical fish initially. But, they easily adopt to grow at low temperature during iced storage and grow very rapidly and spoil the fish. Further, psychrotrophs have a shorter generation time compared with psychrophiles.

A. Iced Storage:

» In general, two types of plastic fish boxes are used: stack-only and nest/stack boxes.

- Fish handling should be done as quick as possible to prevent quality loss. Pre-chilling is of advantage to compensate lack in handling rate.

- Handling methods, which make it possible to guarantee that the icing procedure is sufficient to chill the fish to 0°C and maintain this temperature until landing.
- The fish hold must be constructed in such a way that safe and easy stacking of the boxes can take place.
- Fish hold insulation should be of high quality. Air temperature in the hold should be +1°C or -3°C.

i. Use of antimicrobial agents in ice and cold liquids:

» By ordinary method of icing, fish can be preserved for 10-12 days.

» By use of antimicrobial agents in ice or cold liquids, the shelf life of fish can be enhanced by 60 to 100%. The following antimicrobial agents were used.

a) Chlorine compounds:

» If liquid chlorine or bleaching powder is mixed with ice or cold liquids, shelf life of fish can be enhanced.

» If the cooling media (ice or chilled liquid) contains less than 30 ppm of chlorine, no damage is done to fish.

» However if the proportion is more than this, the fish gets acrid smell.

» Also due to oxidation, the colour of fish may change.

» By this method fish can be preserved for 18-20 days.

b) Antibiotics:

» By use of antibiotics in ice or cold liquids, shelf life of chilled fish can be increased.

» Among the antibiotics, chlorotetracycline and oxytetracycline are important.

» These are powerful broad spectrum antibiotics and are very effective even at low concentrations.

» Antibiotics were used for chilling fish as follows.

1. By dip treatment in antibiotic solution and then chilling.
2. Incorporating antibiotic in water and then ice is prepared and used for chilling.

3. Incorporating in chilling liquids (CSW or RSW)
4. It has been found that dip treatment of fish in water containing 30 to 100 mg of chlorotetracycline (CTC) for 1-5 min and then icing increases the shelf life by 5-10 days.
5. However at present use of antibiotics in banned as per Codex Alimentarius standards.

ii. Storage methods:

- There are three methods of storing fish in ice on fishing vessels. They are
  a. Bulking
  b. Shelving
  c. Boxing

a. Bulking:

- The fish hold is an insulated chamber located at the bottom side of the boat. It is divided into different compartments using wooden boards supported by an upright beam.
- A layer of ice at least 5 cm thick is spread over the bottom of a compartment followed by a layer of fish.
- Ice is then spread over the fish and around the edges so that the fish are not in direct contact with the sides of the board.
- Further layers of fish and ice are added until a depth of about 45cm ice and fish is achieved, with a layer of 5 cm of ice at the top.
- A horizontal wooden board is now placed over the section which must be supported by the stanchion structures, not by the fish and ice in the lower compartment.
- More fish and ice are added in the same way, again to a depth of 45cm.
- The operation is repeated until the compartment is full.
- Wooden boards and stanchions must be kept clean and out of direct contact with the fish.

b. Shelving:

- » The fish hold is divided into sections as it is for bulking but this time removable shelves spaced at about 23 cm are used for holding the fish.
- » The lowest shelf is covered with a layer of at least 5 cm of ice.
- » Fish are placed in rows on the ice and more ice is used to cover the fish to about 5 cm.
- » Only one layer of fish is to be put on to each shelf.
- » Shelves must be supported by stanchions, not by the fish and ice below.

c. Boxing:

- » Fish boxes come in a variety of sizes and materials. Ideally, a box should:
- » Be strong and robust.
- » Be able to be stacked so that the weight of the top boxes is over the boxes below, not by the fish in the box below.
- » Be able to nest to save on stowage space when empty.
- » Be easily cleaned and, if necessary, sterilized.
- » Allow ice melt-water to flow away outside the box below.
- » Have good thermal insulation.

iii. Icing of Fish in the fish boxes:

- » At the bottom of the box a layer of flake ice of 5cm thick should be placed followed by a layer of fish over the ice. Again over the fish layer another thin layer office will be put and it will continue until the box is almost full. However, the top layer of fish should be covered with at least 5 cm of ice.

iv. Refrigerated fish rooms:

- » Now-a-days a unit cooler with fan is placed in the chill room. It is important to get the maximum cooling effect from the ice that is allowed to melt. For this reason, the temperature in the fish room should not fall below 0°C. In order to prevent melting of ice, it is common practice for refrigeration to be used only before fishing starts when the vessel is carrying ice alone. Under these circumstances, the temperature can be below 0°C.

v. Insulated boxes:

- In small fishing boat, insulated boxes are used to carry ice to sea, and for storing ice and fish.
- The size of the boxes depends on the size of the boat and the amount of fish normally caught in a day's fishing.
- Insulated boxes for small fishing boats can often be made locally at low cost.
- The ambient temperature will govern the amount of insulation required in the box, though about 10-15cm of expanded polystyrene is common.

vi. Calculation of ice requirement:

- When fish is kept in ice in an insulated fish box, the amount of ice required will be calculated as per following:

  a. Heat transferred into the box which melts the ice

  b. Heat transferred from the fish

Heat Transferred into the box:

$$q = A.U.(t_o - t_c) \text{ kcal/day} \qquad (1)$$

Where,

q = heat entering the container (kcal/day)
A = surface area of the container (m^2)
U = overall heat transfer coefficient (kcal/day m^2 °C)
t_o = temperature outside the container (°C)
t_c = temperature inside the container (°C)

- This overall calculation of heat transfer may require to be done in parts, for instance, the lid or base of the container is made of a different material or has a different thickness.
- The heat transfer through the various areas are then added together to give the total

heat transfer.

» The heat entering melts the ice, therefore it follows that:

$$q = Li . mi \text{ (kcal/day)} \qquad (2)$$

Where,

q = heat required to melt ice (kcal/day)
Li = latent heat of fusion of ice (usually taken as 80 kcal/kg)
mi = mass of ice melted (kg/day)

» In order to develop a mathematical expression for ice melting rate during the storage period it is assumed that ice melting inside the containers is only due to heat transferred from the surrounding air. In this steady state approach, quantities (1) and melting (2) should be equal, therefore it follows that:

$$Li . mi = A.U. (t_o - t_c) \qquad (3)$$

» Thus, the ice requirement will be:

$$mi = \frac{A \times U (t_0 - t_1)}{Li} \text{ kg/day} \qquad (4)$$

» The mass of ice needed to cool fish from the initial temperature to the final holding temperature can be calculated from an expression, which equates the heat taken up by the ice with the heat lost by the fish.

» Heat taken up by ice = Heat lost by fish

$$(Mi)(Li) = (Mf)(Cpf)(t_i - t_f) \qquad (5)$$

Where,

Mi = mass of ice which melts (kg)
Li = latent heat of fusion of ice (80 kcal/kg)
Mf = mass of fish (kg)
Cpf = specific heat of fish (kcal/kg°C)
t_i = initial temperature of fish (°C)
t_f = final temperature of fish (°C)

» From equation (5) the ice requirement will therefore be:

$$Mi = \frac{Mf \times Cpf \times (t_i - t_f)}{Li} \text{ kg/day} \qquad (6)$$

» The specific heat of lean fish is 0.8 kcal/kg °C and this value should be used if there is a species mix or if there is a possibility that all the fish are of a lean species.

» The specific heat value, however, may be refined to take account of variations in the oil content of the fish and this refined value may be used if the fish composition is reasonably consistent.

$$Cpf = 0.5\,Xl + 0.3\,Xs + 1.0\,Xw \qquad (7)$$

Where,

Cpf = specific heat of fish (kcal/kg)
Xl = mass fraction of lipids (oil)
Xs = mass fraction of solids
Xw = mass fraction of water

» To illustrate the effect of oil content on the quantity of ice required for chilling the following comparison is made between lean and fatty fish.

Example -1:

100 kg lean fish with 1% lipids, 19% solids and 80% water at an initial temperature of 20°C:

Cpf = (0.5 x 0.01) + (0.3 x 0.19) + (1.0 x 0.8) = 0.862 kcal/kg °C
Mi = {100 x 0.862 x (20-0)}/ 80 = 21.55 kg of ice

Example- 2:

100 kg of fatty fish with 21% lipids, 19% solids and 60% water at an initial temperature of 20°C.

Cpf = (0.5 x 0.21) + (0.3 x 0.19) + (1.0 x 0.6) = 0.762 kcal/kg°C
Mi = {100 x 0.762 x (20-0)}/80 = 19.05 kg of ice

» Since the calculation for fatty fish shows only a small reduction in ice requirement and with most species the oil content is variable, it is advisable to treat all fish as lean fish.

» This is strictly a theoretical calculation and does not take the following factors consideration

1. Ice is also melted by the surrounding air; thus a lot of ice is lost, particularly at high ambient temperatures, unless the fish and ice are protected from the ambient heat, preferably with insulating materials.

2. How the fish are packed in ice.

3. The length of time that the fish needs to be kept chilled once cooled.

4. How quickly the fish are chilled.

vii. Ice used:

» Ice is available in several forms such as blocks, plates, tubes, shells, soft and flakes. Of these, flake ice is the most popular form for industrial use because of its cooling efficiency.

» It is also relatively dry and will not stick together to form clumps when stored. Cooling capacity is more for flake ice due to a large surface area for heat exchange.

» It also causes minimum damage to the flesh.

» To ensure maximum contact of ice with the fish, proper selection of the size of ice particles and good stowage practices are needed.

» The rate of chilling is governed by:

- The size, shape and thickness of fish
- The method of stowage
- Adequate mixing of ice, water and fish (in ice slurries),
- Adequate contact of ice with the fish
- The size of the ice particles

viii. Advantages of icing:

- Reduction in temperature to about 0°C.
- Melting ice keeps fish moist.
- Prevents surface dehydration
- Reduces weight losses.
- Melting water also increases the heat transport between fish and ice surfaces (water conducts heat better than air).
- Melting water drains out the debris attached to fish surface.
- Advantageous physical properties of ice.
  - Ice has a large cooling capacity. The latent heat of fusion of ice is about 80 Kcal/kg. This means that a comparatively small amount of ice will be needed to cool 1 kg of fish.
  - Ice melting is a self-contained temperature control system.
- Convenience.
  - It is a potable cooling method. It can be easily stored, transported and used.
  - Clean, pure water is available in plenty as raw material to produce ice.
  - Ice is a relatively cheap method of preserving fish.
  - Ice is a safe food grade substance.
- Extension of shelf life of fresh fish.
- Short-term storage (up to one month maximum for some species, only a few days for others).
- Storage temperature is 0°C.
- Product resembles fresh fish.
- Relatively low technology.
- Low skills required.
- Portable refrigeration system.

ix. Disadvantages of icing:

1. Icing in the conventional method using crushed ice can bruise the flesh which results in leaching of flavor compounds and water soluble proteins.

2. Prolonged ice storage can cause changes in the texture of the muscle, particularly the reduction in breaking strength and hardness of fillets.

3. Muscle proteases including cathepsin D and cathepsin L, calcium activated proteases (calpains), trypsin, chymotrypsin, alkaline proteases and collaginases are involved in softening of fish tissue during storage.

4. Ice storage has been found to adversely influence protein stability and water holding capacity in salmon and cod fillets.

5. Icing cannot completely arrest the activities of psychrotrophic organisms in fish, which is a quality problem in refrigerated food.

x. Types of ice:

a. Block ice:

» Ice is conventionally manufactured in the block form, which is crushed to smaller pieces for icing. Crushed ice has a large surface area, which will ensure rapid cooling than the large blocks.

» For making the block ice, water in ice cans is placed in tanks of refrigerated sodium or calcium chloride solution. Ice blocks are formed within 12 - 24 hours. The ice blocks are formed in few hours in a rapid block ice plant.

» In one type several tubes are arranged in a tank of water through which a refrigerant is passed. The ice as it builds up fuses with the ice on the adjacent tubes to form a block with a number of hollow cores.

» The block are released from the tubes by a hot gas defrost. In another type the refrigerant is circulated through a jacket around each can of water and also through pipes running through the centre of the can.

» Blocks are removed by gravity after a hot gas defrosts.

b. Flake ice:

- Ice with smooth contours and in the shape of very thin flakes is called flake ice. It has an area of 100 - 1000 mm^2 and a thickness of 2 - 3 mm.
- Flake ice has a very high area per unit mass and can cover large quantity of fish for a given weight when compared to crushed block ice.
- Flake ice is usually made by spraying water onto the surface of a refrigerated drum. The ice sheet formed is scraped off in the form of dry sub-cooled flakes.
- In some other models the scraper and the water spray is rotated inside a stationery drum and the ice is formed on the inner surface.
- The refrigerant temperature is in the range of -20 to -25°C and the flake ice production is automatic, continuous and quick.

c. Plate ice:

- Plate ice is made by spraying water on to the surface of vertical hollow plates through which a refrigerant passes.
- A flat sheet of ice will be formed on the surface of the plate and when it attains the desired thickness it is separated by pumping hot gas through the plates In some cases a water spray .
- Plate ice has the advantages of flake ice, the process is quick and the thickness of ice can be varied as per requirement.

d. Tube ice:

- Ice is made as hollow cylinder of about 50 mm with a wall thickness of 10 to 12 mm.
- The ice is formed on the inner surface of a series of hollow cylinders, the outer surface of which will be surrounded by a refrigerant.
- Ice released by hot gas defrosting will be chopped at about 50mm length.
- Tube ice will have to be crushed to the required size before icing and has all the advantages of the flake ice.

e. Liquid ice or Flow ice:

- Liquid ice or flow ice looks like jelly. It actually consists of amorphous ice micro particles suspended in a non-corrosive, non-toxic solution.
- Liquid ice generator system yields a substance with ten times more cooling capacity than chilled water. Because of this it requires smaller flow rates for the same cooling capacity.
- The liquid ice at about 50% ice concentration is pumped into the fish box. Because of its fluid properties, it buries the fish, making close contact with the entire surface.

f. Soft ice:

- Soft ice is made by freezing a weak brine or seawater in a drum provided with refrigerated walls.
- The crystals of fresh water ice forms slurry in the brine as temperature falls and it is pumped into a storage tank. Ice crystals are skimmed off from the tank to be used as soft ice slurry.
- It has the advantage of making perfect contact with the material.
- Besides it does not cake, or forms air pockets as it can happen with other forms of ice.

B. Chilled Sea Water (CSW) Storage:

- In this case, the fish is surrounded with a mixture of ice and water. When enough ice is added to the system, it will bring the temperature of the water near to 0^0C.
- The water will remove the heat from the fish. The transfer of heat from fish to water and from water to ice will continue until the system is brought to a state of equilibrium.
- The surface contact of the fish and chilling medium will be maximum in CSW than in ice alone.
- Because of the maximum contact between the cooling medium and the fish the cooling rate of the fish in CSW is higher than that of fish in ice.
- CSW can be adopted in fishing vessels as it has many advantages over ice storage.

» CSW system is used for the containerized bulk storage of the catch onboard the vessels. This expedites the handling of the catch.

» The main advantages of the CSW system are listed below:

- Faster cooling rate.
- Maintenance of a uniform temperature.
- Less damage to fish.
- Easy unloading of fish using pumps.

» The recommended ratio of fish to seawater is between 3:1 to 4:1.

» In CSW systems, it is usually necessary to provide pumped circulation of the water to ensure even mixing. A rough calculation of the amount of ice needed to provide adequate cooling in a CSW system can be made quite easily in the following manner:

» **Example:** Suppose 8 tonnes of fish (8,000 kg) are to be cooled from 24 to –1°C and 4:1 ratio of fish to seawater is used, the mixture of seawater and fish would weigh 10 tonnes (10,000 kg) and the amount of heat to be removed would be:

10,000 x 25 x 1 (taking 1 as the specific heat of both fish and water)
= 2,50,000 kcal
But, when ice melts, it absorbs 80 kcal.
Thus, the weight of ice required: 2,50,000/80 =3,125 kg

» Thus, just over 3 tonnes of ice would be required to provide the initial cooling. This would only reduce the temperature of the fish and would not be sufficient to maintain the lower temperature. The amount of additional ice needed would depend on the efficiency of the insulation of the tank or fish room, the outside ambient temperature and the duration of the trip.

» Hence, theoretically about 30% of ice is needed to bring down the temperature from ambient conditions to 0^{0}C.

» However, ice is needed to maintain the temperature as well as to accommodate the heat from the environment.

» Hence, in tropical conditions a 1:1 fish to ice ratio is ideal for ice storage.

- Fish of the same size and species are placed in the same boxes.
- It is always recommended to add about 12 – 20% extra ice to the fish in order to compensate for water loss from melting and bad handling.
- The effectiveness of chilling by temperature exchange depends on the thickness of the layers of fish and the distribution of ice.
- For practical purposes the following rules of thumb can be given to calculate ice requirements:

  ***a. Fish boxes*:**

  - Ice to fish ratio in tropics are 1 kg ice to 1 kg fish, and ice to fish ratio intemperate climate and in insulated van are1 kg ice to 2kg fish.

  ***b. Insulated tanks*:**

  - Water to ice to fish ratio in tropics are1 kg water to 2 kg ice to 6 kg fish and in temperate climate 1 kg water to 1 kg ice to 4 kg fish.

Disadvantage:

- On melting of ice, brine gets diluted
- Carry problems of ice

C. Slurry Ice in Fish Preservation:

- The Deep Chill ice consists of small crystals, either in dry form or pumpable slurry.
- The introduction of Deep Chill ice for direct contact cooling of fish made significant improvements in cooling and handling processes over other forms of ice.
- It has the following beneficial effects:

  a. the cooling process is maintained at a constant level as the ice melts

  b. prevents fish from bruising, because of the soft and flexible nature of ice

  c. easy to mix, handle and transport due to pumpability of ice slurry

  d. possibility of brine drainage to provide almost salt free ice for extended storage of fish

Difference between flake ice and ice slurry:

- » Ice slurry provides more efficient chilling than flake ice.
- » This is because ice slurry effectively eliminates hot spots in the fish container and provides maximum contact of ice surrounding the fish.
  a. It significantly retards bacterial growth and reduces fish tissue degradation.
  b. The temperature drop in the fish body is extremely significant in the early stage of the cooling process.
  c. The final equilibrium temperature of the fish is approximately 1^0C lower, using the same amount of ice slurry compared to flake ice.
  d. The difference in cooling level occurs due to the larger enthalpy change of the ice slurry.
  e. No uptake of salt was detected while using slurry with 0.8% salinity.
  f. The quality of the product, such as the texture, colour and bacteria retardation is greatly improved.

D. Refrigerated Sea Water (RSW):

- » Mechanical refrigeration for cooling the sea water.
- » Temperature of sea water is thermostatically controlled at 0^0C to -1^0C.

Advantage over icing:

- » More efficient and rapid cooling due to thorough and intimate contact.
- » Because of buoyancy nature, fish do not get pressed or crushed.
- » Better control of temperature at -1^0C resulting fish temperature between 1.5^0C to 2.0^0C.
- » Considerable saving of labour.
- » Considerable saving of ice storage space.

Disadvantage:

- » Heavy bacterial load.
- » Mechanical refrigeration is costly.

II. PHYSICAL, CHEMICAL AND MICROBIAL CHANGES DURING CHILLED STORAGE

- The chilling of fish is a process by which temperature of a fish is reduced close to but not below freezing point of water (0°C).
- It delays both biochemical and bacteriological process in fish and consequently prolongs the storage of fish and fish products.
- So long as the chill temperature is maintained, deteriorative changes in fish remain retarded.
- Quality of chilled products depends mainly on the initial quality of raw fish, method and duration of chilling and efficient storage.
- The main merit of the method is that it provides the maximum possibility of preserving the natural nutritional and functional properties of the fish.
- Fish may be chilled by ice or by a homogeneous coolant such as cold air or cold liquid (fresh water, brine or seawater).
- For chilling fish on board of a fishing vessel, ice, CSW, CFW, RSW or cold water is used.
- Chilling is an effective way of reducing spoilage if done quickly and handled carefully & hygienically.
- The objective is to cool fish as quickly as possible, to as low a temperature as possible, without freezing.
- Chilling cannot prevent spoilage, but the colder the fish the greater the reduction in bacterial and enzyme activity.
- Chilling leads to various changes according to the initial condition of the fish. Some of the general factors affecting the chilling process are as follows:

III. FACTORS AFFECTING THE RATE OF SPOILAGE IN CHILLED FISH

1. Temperature:

- It is well known that high temperature increases the rate of fish spoilage and low temperatures slow it down. Therefore, if the temperature of fresh fish is low, then quality is lost slowly.

» The faster a lower temperature is attained during fish chilling, the more effectively the spoilage activity is inhibited.

» Generally, the rate at which fish loses quality when stored in ice is used as the base line when comparisons are made regarding shelf-life at different storage temperatures.

» The effect of temperature reduction on the rate of spoilage and shelf-life of fish is given Table 7.1.

Table 7.1: Effect of temperature reduction on fish spoilage

Reduction in temperature (^{0}C)	Rate of spoilage (%)	Extension of shelf life
0	100	-
5	50	x 2
10	25	x 4
15	12.5	x 8
20	6.25	x 16

2. Physical damage:

» Fish is soft and easily damaged; therefore rough handling and bruising result in contamination of fish flesh with bacteria and allow releases of enzymes, speeding up the rate of spoilage. In addition, careless handling can burst the guts and spread the contents into the fish flesh.

3. Intrinsic factors:

Table 7.2: Intrinsic factors affecting the spoilage rate of chilled fish

Intrinsic factors	Relative spoilage rate of fish stored in ice	
	Slow rate	Fast rate
Shape	Flat fish	Round fish
Size	Large fish	Small fish
Fat content in the flesh	Lean species	Fatty species
Skin characteristics	Thick skin	Thin skin

» Shelf life can be defined as the maximal period of time during which the predetermined attributes of the food are retained.

» The chilled storage life of fish is primarily determined by sensory evaluation.

- The chilled storage life of fish depends on several factors such as composition, microbial contamination and the type of micro-flora present in the flesh.
- The fish spoilage pattern is similar for all species, with four phases of spoilage as mentioned below.
  - ◊ Phase - I (Autolytic changes, caused mainly by enzymes):
    - Fish just caught is very fresh and has a sweet, seaweedy and delicate taste. There is very little deterioration, with slight loss of the characteristic odour and flavor. In some tropical species this period can last for about 1 to 2 days or more after catching.
  - ◊ Phase - II (Autolytic changes, caused mainly by enzymes):
    - There is a significant loss of the natural flavor and odour of fish. The flesh becomes neutral but has no off-flavours, the texture is still pleasant.
  - ◊ Phase - III (Bacteriological changes, caused mainly by bacteria):
    - The fish begins to show signs of spoilage. There are strong off-flavours and stale to unpleasant smells. Texture changes are significant, flesh becoming either soft and watery or tough and dry.
  - ◊ Phase – IV (Bacteriological changes, caused mainly by bacteria):
    - Fish is spoiled and putrid, becoming inedible.

IV. FACTORS AFFECTING THE CHANGES DURING CHILLED STORE

a. **Species (Marine/ freshwater):** Marine fish are more prone to spoilage during chilling due to a high content of TMA which is converted into TMAO by the bacteria. In the case of freshwater fish there is an antimicrobial covering on the skin which is absent in marine fishes. Moreover marine shrimp are more quickly damaged due to high amino acid content leading to softening of the muscle.

b. **Size:** It is found that larger fish get spoiled slower than the smaller fishes. As smaller fishes are tender in nature and their skin not that tough, they easily get bruised and hence accelerate spoilage.

c. **Method of capture and handling:** The quality of chilled fish largely depends on the method of capture and handling. How carefully and properly they are handled will be reflected during the storage life. If the fish is undergoing physical deformities from the beginning itself, then it will get damage sooner.

d. **Fat content:** Fatty fish start spoiling faster than the lean fish. Their abdomen will become soft and sometimes belly bursting occurs. Due to high fat content, their muscles are more soft and susceptible to digestive enzymes.

e. **Type of ice:** If the edge of the ice is sharp and rough then physical damage may occur during chilling. It may puncture the skin and give way to microbes to enter into the flesh. So the storage life will be affected.

i. Physical changes:

» *Color*: During chilled storage, fish gets discoloured. The pigments may get oxidized or affected by some other factors.

» *Texture*: The texture becomes harder due to stiffening of the muscle during rigor mortis process. This happens since the protein gets denatured and actin and myosin fail to slide leading no contraction and relaxation.

» *Eyes, skin, scales*: Due to longer chilled storage the structure of eyes become concave. The scales will get removed and the skin become bare which is undesirable and not economic.

» *Weight loss*: Weight loss occurs due to leaching of amino acids (shrimp), WSP and vitamins from the fish body when the ice melts.

ii. Biochemical changes:

» *Moisture content*: During iced storage moisture contents of both freshwater and marine water species showed a small but erratic, mostly increase fluctuation.

» *TVB content*: Least concern with Freshwater fish. Marine water fish show increase after 3 days or more. It is used for the assessment of degree of freshness.

» *NPN content*: Both Freshwater and Marine water show gradual decrease during ice storage. This shows deterioration of flavor.

» *Alpha amino nitrogen*: AAN content of freshwater murrels and brackish water milkfish in unpacked condition showed a gradual decrease during the period of ice storage.

» *Total nitrogen/ Protein*: The total nitrogen / protein content in most of the cases decreased substantially during iced storage. This decrease may be mainly due to the loss of soluble protein components.

» *Water soluble and sarcoplasmic protein content*: The WSP leaches out when the ice melts and hence decreases.

» *Salt-soluble protein content*: Most of the species of tropical fish shows decrease in salt-soluble/ myofibrillar protein fraction during ice storage period.

» *TMAO*: As TMA is degraded into TMAO by the bacteria, its quantity increases during chilled storage indicating spoilage condition.

» *FFA*: Due to lipid oxidation, the FFA increases after a longer period of ice storage.

» *ATP degradation*: During rigor mortis stage, ATP is degraded and hypoxanthine is produced which is bitter in taste. More amount of hypoxanthine indicates higher degree of spoilage.

iii. Microbial changes:

» During ice storage, proportion of mesophiles decrease and psychrophiles increase.

» *Psychrotrophic* bacteria which include *Pseudomonas spp.* and *Shewanella putrefaciens* dominate the spoilage flora of iced stored fish. Among others are Pseudomonas, Moraxella, Acinetobacter, Vibrio, Flavobactor, Micrococcus are prominent in ice stored fish.

V. TRANSPORTATION OF CHILLED FISH

a. Land transportation of chilled fish is carried out in insulated or mechanically refrigerated vehicles. The refrigerated vehicle used for chilled fish transportation should have a minimum inside temperature of 7^0C.

b. Boxes for land transportation are made of wood, aluminium, high density polyethylene, expanded polystyrene or polyurethane. The ideal fish transportation box should be light weight yet strong enough to withstand the combined weight of ice, fish and stacking and should have good insulating properties. The boxes should be easy to clean. Boxes are usually made of double bottom to collect the melt water. Containers used for air transportation of chilled fish should be water tight.

c. Air shipment of chilled fish requires a light weight and protective container. Modern insulated containers are made of high density polypropylene with polyurethane insulation sandwiched between the inner and outer walls of the double walled container.

d. Instead of ice, pads of nonwoven fabric encapsulating synthetic absorbent powder are used for chilling of air shipped fish. These pads could be soaked in water and deep frozen for use. Special thermal barrier films are used in combination with the pads to protect fish containers from heat.

Microbes related:

» Moulds do not grow at <13 – 15% moisture.

» Yeast do not grow at <40% moisture.

» Moulds and yeast require acidic pH to grow.

» In case of marine fish, 90% micro flora is Gram -ve. Only after landing Gram +ve organisms will contaminate the fish.

» In case of prawns, high rate of spoilage is because of more free amino acids (FAA). Generally 75% bacterial load occurs at the cephalothorax region.

At low temperature, destruction of micro-organisms occurs as:

1. Damage to cell by the mechanised action of ice-crystal formation.
2. Prolonged starvation period.
3. Partial coagulation of protoplasm.
4. Nitrogenous waste will remain in the cell.
5. Cold shock

Freezer burn: Surface gets dried off due to dehydration effect.

Question 1: Fresh water fish have longer shelf life in ice than marine fish. Why?

» Fresh water fish have anti-bacterial substance in muscle layer

» No production of ammonia from TMAO in fresh water fish but, which occurs in marine fish.

Question 2: Non-fatty fish keep longer than fatty fish. Why?

» Higher fat content => Soft and more delicate texture and structure of fish. Hence, non-fatty fishes remain longer in ice than the fatty fishes.

8 Freezing and Cold Storage

HISTORY

Several food preservation techniques were evolved in order to enable man to use food produced at various places from age old times. Such methods of preservation of food are sun drying, salting, keeping food in ice, smoking etc. Fish is a highly perishable commodity as it contains good amount of nutrients and hence there is a greater need to preserve it. Without preservation it is impossible to transport and distribute fish to interior places in good condition.

In olden days, in places which were covered with snow or ice, the flesh of hunted animals and fish were covered with ice thereby fish was frozen and quality is preserved. However, freezing of fish/ foods by artificial methods is a recent development. Towards the end of 19th century compressors were developed and because of this machine, novel methods of fish freezing were developed.

First patent on freezing of fish using cold air was given to V. Benzamine of England in 1842. In the year 1861, similar patent was awarded to E. Piper of U.S.A. who used salt ice mixture to freeze food. Russians were pioneers in freezing fish by using machines. They used compressors on board the fishing vessel to freeze fish and also built a freezing plant (factory) in a place called Astrakhan. In 1898, immersion freezing system was developed by Rouart of France for freezing fish. Nikhoidahl, a scientist from Norway developed spray freezing method with highly chilled brine solution. In 1925, a technician named Cook developed contact plate freezer which was further improved by Karens Birdseye of US in 1929. Similarly methods of freezing and packing it in consumer packs were developed subsequently.

A. Mechanism of Ice Crystal Formation

» During freezing of fish the water present in the fish tissue changes ice. The change of water to ice takes place by a process called 'crystallisation as described below.

i. Crystallisation:

» During freezing, when the temperature of fish is lowered below its freezing point, there occurs a change of state of its tissue water from liquid (liquid water) to solid state (ice). This process is called 'freezing' or 'solidification'. The solidification of tissue water to ice during freezing occurs by a process called 'crystallization'. This is because, the process converts 'liquid water' to a 'crystalline solid' (*i.e.* ice). A crystalline solid is one in which the molecules or atoms are arranged in a definite order. In contrast, in the liquid state of water, the water molecules are present as disordered clusters.

» Thus, crystallization is defined as the formation of a systematically organised crystalline solid phase from a liquid phase. It is necessary to know the principles of crystallisation, if the fish is to be frozen stored and thawed at optimum temperature. Crystallisation, being the passage of molecules of water form the random arrangement of the liquid state to an ordered arrangement in solid state, requires a displacement of molecules which have to move from one location to another. The crystallisation process can be divided into two parts:

1. Nucleation
2. Crystal growth

1. Nucleation:

» Crystallisation begins when conditions are appropriate to bring about the aggregation of groups of water molecules into a tiny ordered particle known as crystal nucleus. The process of establishment of crystal nucleus, is known as 'nucleation"

2. Crystal growth:

» The growth of crystal nuclei is the second step of crystallisation process. Crystal growth is simply the enlargement of the crystal nucleus by the orderly addition of water molecules on it. The product of crystal growth is an 'ice crystal'.

ii. Factors affecting ice crystal size:

» The size of ice crystals formed in the fish tissue during freezing depends upon the following factors.

1. **Number of nuclei:** As the completion of freezing is related directly to the number of nuclei, a few nuclei will cause the formation of a few large ice crystals whereas numerous nuclei will result in numerous crystals of smaller size. Control of crystal size is therefore dependent upon nucleation.

2. **Rate of freezing:** The ice crystals are small and numerous in rapidly frozen fish and large in slow frozen ones.

3. **State of fish:** The impression should not be that all tissue samples frozen at the same rate will contain ice crystals of the same size. Variation is greater between different materials form the samples of the same material and even within the same sample. For example, freezing of a fish at pre-rigor stage may be compared to that at post-rigor stage. When frozen at the same rate the pre-rigor sample contains smaller and more numerous ice crystals than the post-rigor samples. This is because of the great amount of bound water in the pre-rigor sample, which retards the migration of free water and encourages numerous small ice crystals.

4. **Agitation:** Agitation of the system during freezing is known to encourage the formation of small ice crystals, probably due to the creation of more crystal nuclei. Agitation wills increase the heat removal thereby lowering the temperature to the point where nucleation is more likely.

iii. Location of ice crystals and cell damage:

» During freezing of fish, maximum ice crystal formation takes place in the temperature range of 0°C to –5°C (exactly –1°C to –5°C).

» In general, slow freezing produces large ice crystals located exclusively in extra-cellular areas.

» On the other hand, rapid freezing generally results in tiny ice crystals located both extra- and intra-cellularly.

» The extra-cellular crystallisation always occurs prior to intra-cellular crystallisation, because the freezing point of the extra-cellular material (–1.2°C) is higher than that of the intra-cellular materials (–1.6°C).

1. Intracellular ice crystal formation:

» When a fish fillet is immersed in liquid nitrogen, it is frozen within the fraction of a second. When the transverse section of such a frozen fillet is observed under microscope, a large number of small ice crystals are seen. In contrast, when the same fillet is frozen in an air blast freezer operating at –30°C, four big ice crystals are seen.

» These ice crystals develop near the cell membrane and rupture it thereby causing cell damage. This type of cell damage is called **'A-type cell damage'**. This type of cell damage occurs when the time taken to reduce the temperature form 0°C to –5°C is about 20 to 30 minutes.

» If freezing time is still more, *i.e.*, if the time taken to reduce the temperature form 0°C to -5°C is about 100 minutes, then there will be only one big ice crystal is formed intra-cellularly. As only pure water freezes out as ice, the concentration of the intra-cellular solutes in the peripheral unfrozen cytoplasm increases considerably. As a result endosmosis takes place and pure water diffuse into the cell from the extracellular areas.

» In this way the amount of water increases inside the cell and the incremental water also freezes into ice. Therefore, the size of the ice crystal goes on increasing. Moreover, there is an increase in the volume of the intra-cellular water as it is converted to ice.

» When the volume of the ice crystal becomes greater than that the cell can accommodate, the ice crystal presses the cell membrane from inside and finally ruptures it thereby causing cell damage. This type of cell damage is called **'B-type cell damage'**.

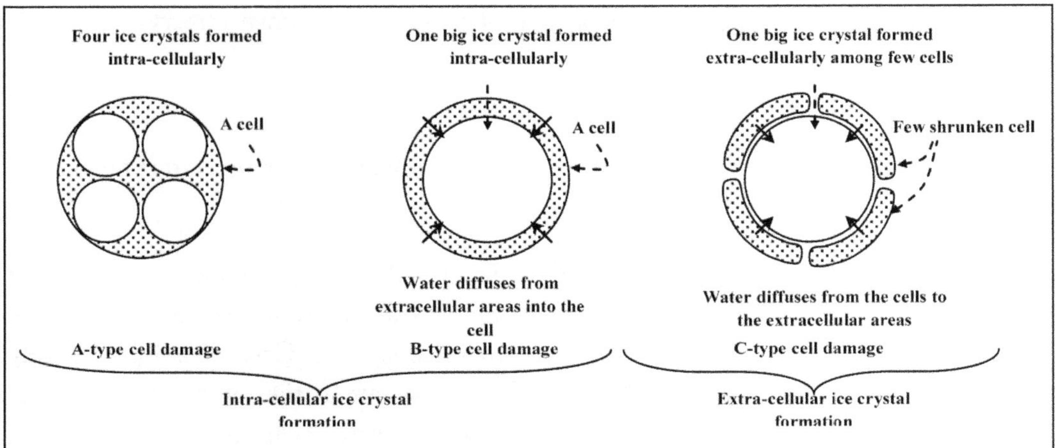

Figure 8.1: Location of ice crystals and cell damage

2. Extracellular ice crystal formation:

» If freezing time is still more, *i.e.*, if the time taken to reduce the temperature of fish form 0°C to –5 °C is about 200 to 500 minutes, one big ice crystal is formed in the extra-cellular space surrounded by several cells.

» As only pure water freezes out as ice, the concentration of solutes in the extra-cellular fluid increases considerably. As a result exosmosis takes place and pure water diffuse out from the cell into the extra-cellular areas.

» The cells get shrunken in size due to loss of water. In this way the amount of water increases outside the cell and the incremental water also freezes into ice. Therefore, the size of the ice crystals goes on increasing.

» Moreover, there is an increase in the volume of the extra-cellular water as it is converted to ice. The big extra-cellular ice crystal presses the cells from the outside and damages them. This type of cell damage is called **'C-type cell damage'**.

» Due to the 'A' 'B' and 'C' type of cell damage, the intra-cellular fluid comes out during thawing.

iv. Recrystallisation:

» Control of crystal size would be quite simple if all crystals once formed remain unchanged in size during subsequent frozen storage and thawing. Such a static situation is rare. Ice crystals generally have a tendency to enlarge during frozen storage and early stages of thawing. This phenomenon is known as recrystallisation.

Probable reasons:

1. Recrystallisation is observed when the samples are exposed to temperature fluctuation (–14°C to –7°C)

2. Migratory crystallisation is defined as the growth of large crystals at the expense of small ice crystals. At constant temperature, this occurrence is the result of free energy difference, between large and small ice crystals. Crystallisation is highly temperature dependent. It occurs at decreasing rate as the temperature is lower below the freezing point. Low and uniform temperature are the obvious ways of minimising recrystallisation during storage of frozen fish. Recrystallisation will occur during thawing. But it can be minimised by rapid thawing.

Freezing Point, Depression of Freezing Point and Eutectic Point

B. Freezing Point

» When a fish is cooled and its temperature is reduced to a definite point there occur ice crystals in it. This definite temperature is called the freezing point of that fish.

» Unlike a pure substance like water, in which the whole substance is frozen completely at its freezing point (say, 0°C for water), in case of fish, the freezing process continues to take place over a long range of temperature. Therefore, another term 'initial freezing point' is used for fish. The initial freezing point of a fish is the temperature at which an infinitesimal small amount of ice appears during its cooling. However, the term freezing point of fish is widely used to mean its initial freezing point.

Table 8.1: Freezing point of foods

Freezing point of pure water is 0°C	
Foods	**Freezing point (°C)**
Common vegetables	– 0.8 to – 2.8
Fruits	– 0.9 to – 2.7
Pork and loaf	–1.7 to – 2.2
Milk and egg	– 0.5
Banana	– 3.4
Groundnut	– 8.3

» Generally most of fresh foods have the freezing point in the range of –1°C to – 2.8°C, except some special cases as banana (–3.4°C) and groundnut (– 8.3°C).

Table 8.2: Freezing point of fishes

Freezing point of pure water is 0°C	
Fishes	**Freezing point (°C)**
Carps	– 0.7
Sardines	– 1.3
Eel	– 1.95
Tuna	– 1.3
Oyster	– 2.0
Crab	– 2.0

- Freezing point of fish varies depending on the species. But in all case it is lower than 0°C, which is the freezing point of water. This is because, the tissue water of fish is not pure; rather it is a solution containing many dissolved solutes which depress the freezing point to below 0°C (Depression of freezing point).
- Among fishes freezing point always differ, but in general marine fishes have lower freezing point than fresh water fishes. Roughly speaking freezing point of surface marine fishes (pelagic fishes) is higher than that of bottom marine fish (demersal fishes). Such differences in freezing point are due to the difference in their content of water and other chemical constituents. However, in general, the freezing point of fish is assumed to be −1°C.
- During freezing, when the fish reaches its freezing point, it does not become so hard as to call it frozen fish. To make the fish hard frozen, its temperature must be lowered much more where comparatively many ice crystals are formed.

Depression of freezing point:

- At standard atmospheric pressure the freezing point of water is 0°C. Water contained in fish is not pure water; rather it is a dilute solution dissolving several simple solutes like salts and sugars present in the fish tissue. Salts and sugars dissolve in water to make true solutions. Therefore, freezing point of fish is influenced by the density of this solution. According to **Raoult's Law** 'the depression of freezing point is proportional to the solute density in the dilute solution'.
- The tissue water of fish not only dissolves simple solutes to make true solutions, it also dissolves many complex tissue components like proteins, fats and carbohydrates forming colloidal solutions. These complex tissue components which exist in colloidal state do not influence the depression of freezing point. So the degree of depression of freezing point depends upon the condition of the solution formed by the tissue water.

Eutectic Point (Cryohydric point):

- The lowest possible freezing point of a true solution is called as eutectic point or cryohydric point.
- At this point maximum solidification takes place.

» When dilute brine is cooled to a temperature lower than 0°C, ice crystals are formed and separated form the solution. These ice crystals consist of pure water. Therefore, the salt concentration of the remaining solution gradually increases with the progress of freezing and the more concentrated salt solution remains in proportion to the more ice crystal formation.

» When the temperature reaches the eutectic point (E), the concentrated solution changes to solid at once together with the water and salt. At this state there are complex ice crystals *i.e* cryohydrates, in which frozen solid contains ice crystals as the mixture.

» The line AE from 0°C to the eutectic point is called **'ice curve'**.

» On the other hand, when the concentrated nearly saturated solution is cooled, it does not produce ice crystals; instead, salt is crystallised and separated from the solution system. This is because; the solubility of the salts becomes lower at lower temperatures.

» When the temperature reaches the definite point, the salt solution changes into solid at once. The solid formed at this point contains salt crystals in it. The temperature of this point coincided with the final point started form dilute solution. This curve is called as **salt curve or saturation curve** (BE) (Figure 8.2).

» Both of the curves (ice curve and salt curve) meet at eutectic point where all liquid changes to solid.

» The eutectic points of some materials are given below.

Material	Eutectic point (°C)
NaCl	– 21 .2
$CaCl_2$	– 55.0
Most foods	– 55.0 to – 60.0

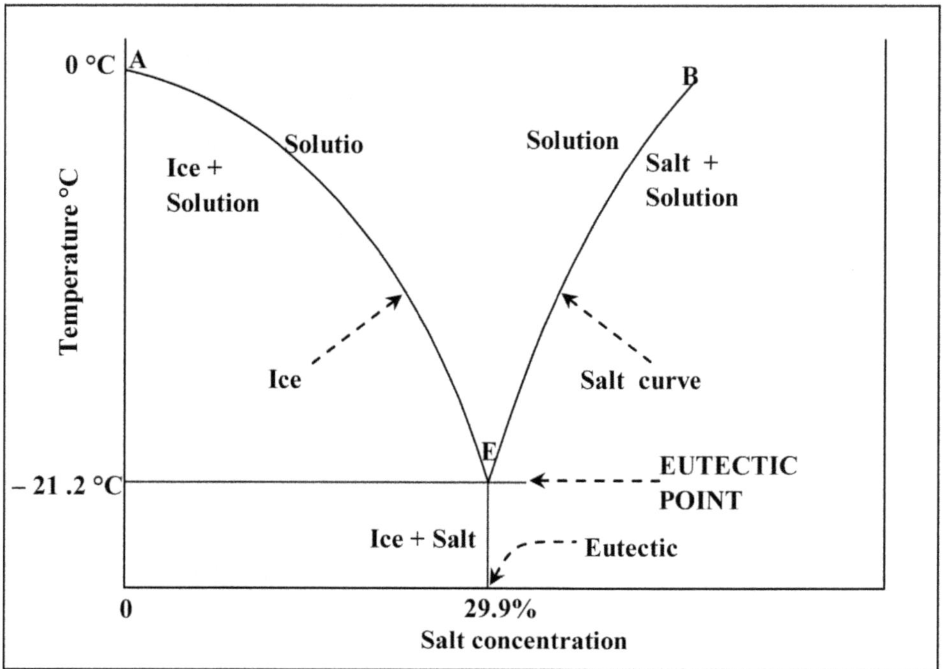

Figure 8.2: Eutectic point of NaCl

C. Freezing Curve

» When a fish is refrigerated, its temperature changes with time. This relationship of temperature and time, when plotted on a graph, a certain curve is obtained. This curve is called 'refrigerating curve' or 'freezing curve'. The freezing curve furnishes some of the useful information needed for the understanding what happens to the fish being frozen, the temperature to which the fish was exposed before, during and after freezing as well as the amount of ice formed.

» Freezing curve is influenced by

1. Freezing method
2. Food varieties
3. Packing method
4. Thermal properties of package materials
5. Thickness of the product

» The freezing of fish takes place in the following four stages. Accordingly, the freezing curve of fish is divided in to four zones each representing a stage of fish freezing as given below.

Stages of freezing	Zones in the freezing curve
Cooling stage	Cooling zone
Super-cooling stage	Supercooling zone
Freezing stage	Freezing zone
Sub-freezing stage	Sub-freezing zone

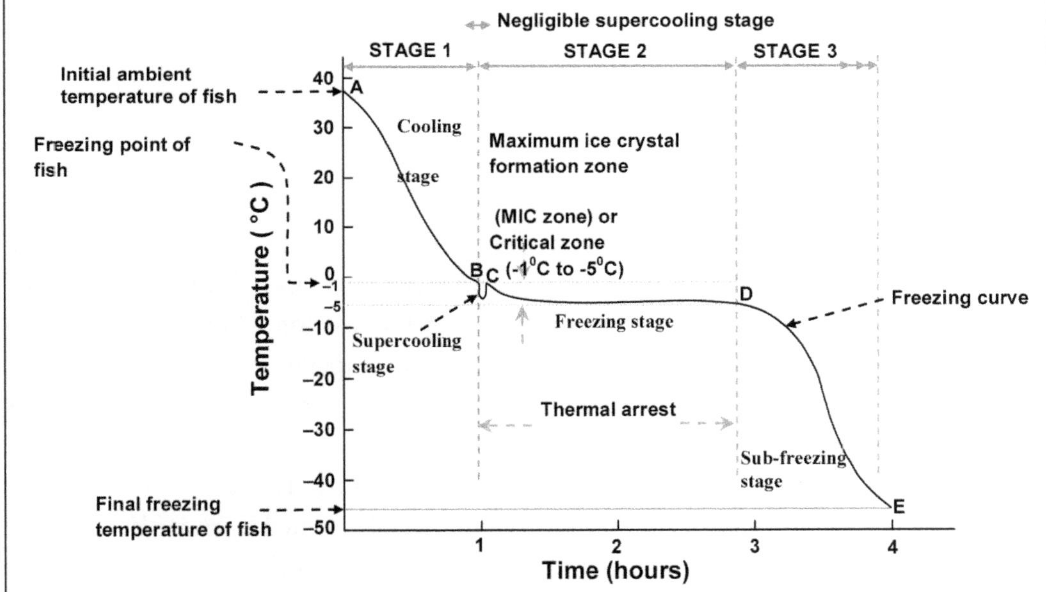

Figure 8.3: Freezing curve of fish

a. Cooling stage:

» When a fish is subjected to low freezing temperature, its temperature suddenly decreases from its 'initial ambient temperature above its freezing point' till its 'freezing point' are reached. This stage of fish freezing is called 'cooling stage'. The portion of the freezing curve (AB) from the 'initial ambient temperature of fish above its freezing point' to its 'freezing point' is called 'cooling zone'. As the freezing point of fish is generally assumed to be −1°C, the cooling zone extends from the initial ambient temperature of fish above its freezing point to − 1°C. It is a slightly sloping curve.

b. Super-cooling stage:

- In this stage, the temperature of the fish goes down slightly below its freezing point without any solidification of the tissue water and again rises back to its freezing point as huge amount of heat is released due to initial crystallisation of the tissue water. The portion of the freezing curve (BC) down from and up to the freezing point is called the '**super-cooling zone**'. It is a small cup-shaped curve.
- As super-cooling is not generally noticeable and not taken into consideration during heat calculations in freezing, it is generally omitted from the freezing curve and very often the freezing curve is represented by the rest three stages as given below.
  a. Cooling stage
  b. Freezing stage
  c. Sub-freezing stage

c. Freezing stage:

- When the temperature of fish reaches its freezing point, freezing of its tissue water starts. As a result, the tissue water starts to change its phase from liquid state of water to solid state of ice by the process of crystallisation. Unlike pure water in which freezing completely takes place at a single temperature of its freezing point (*i.e.*, 0 °C), in case of fish, its tissue water freezes over a range of temperature. The duration during which most of the tissue water of fish freezes is called 'freezing stage'. The portion of the freezing curve (CD) from the 'freezing point of fish' to 'temperature at which maximum ice crystal has been formed from the tissue water of fish' is called 'freezing zone'. As the temperature at which maximum ice crystal has been formed from the tissue water of fish is generally −5 °C, the freezing zone extends from −1 °C to −5 °C.
- During crystallisation, huge amount of heat is removed to cause the change of phase of tissue water of fish from liquid to solid. Almost all the heat removed in this stage causes change in phase (crystallisation) with little change in temperature. As a result, for a small change in temperature, it takes long time. Therefore, this zone of the freezing curve is somewhat horizontal.
- The large proportion of the heat removed from fish in this freezing stage which contributes only to the change of phase through crystallisation is 'latent heat' as it causes no change in temperature. As the removal of this heat causes fusion of the water molecules to form ice crystal, the heat removed is '**latent of fusion**'.

» This zone of the freezing curve also has the following three names.

1. Maximum ice crystal formation zone (MIC zone)

2. Critical zone

3. Thermal arrest period (TAP)

1. Maximum ice crystal formation zone (MIC zone):

» It is named so as maximum amount of ice crystals is formed in this zone i.e. between − 1°C and − 5°C. About more than 75% of the tissue water is frozen to ice by the process of crystallisation in this zone between − 1°C and − 5°C as shown in the following table. The values in the table are obtained by two methods as given below.

◊ **Experimental determination:**

» The experimentally determined percentages of tissue water frozen to ice during freezing of cod with 80.3% moisture are given in the following table (Riedel, 1956).

◊ **Calculation:**

» The percentage of tissue water frozen to ice during freezing of fish, in general, can also be calculated using the following equation proposed by G.B. Chizhov (Zaitsev, 1965).

$$i = \frac{110.5}{1 + \frac{0.31}{\log t}}$$

Where,

i = Percentage of tissue water frozen to ice at the temperature 't'

t = Positive value of temperature at which percentage of tissue water frozen is to be determined(*e.g.*, if temperature is −40°C, take t = 40)

Table 8.3: Percentage of tissue water of fish frozen to ice at different temperatures

Temperatures (°C)	Percentages of tissue water of fish frozen to ice	
	Experimentally determined	Calculated
0	0	#
–1	8.0	#
–2	52.4	54.44
–3	66.5	66.98
–4	73.0	72.94
–5	76.7	76.54
–7	81.1	80.84
–10	84.3	84.35
–18	88.4	88.62
–20	89.0	89.24
–30	90.3	91.33
–40	90.5	92.58
# Maximum amount of tissue water is converted to ice between – 1 °C and –5 °C		

2. Critical zone:

» It is named so, as the time taken to cross this zone has critical influence on the quality of the frozen fish. The time taken to cross this zone depends on freezing rate. Thus, the length of this zone is the index of freezing rate. Therefore, freezing rates are usually defined in terms of time taken to cross this zone (–1°C to –5°C).

» In 'slow freezing' the time taken to cross this zone is long due to which few number of large ice crystals are formed in the fish tissue resulting in greater tissue damage and poor quality of the product.

» Contrastingly, in 'quick freezing' the time taken to cross this zone is short due to which large number of small ice crystals are formed in the fish tissue resulting in less tissue damage and good quality of the product.

» In 'cryogenic freezing' this zone is unnoticeable due to which numerous minute ice crystals are formed in the fish resulting in negligible tissue damage and excellent quality of the product. For example, in sharp freezing, it is long, in contact plate freezing, it is short and in liquid nitrogen freezing it is unnoticeable, thereby producing poor, good and excellent frozen products respectively.

3. Thermal arrest period (TAP):

- As temperature change is arrested (*i.e.*, brought to a negligible level) in this zone due to removal of heat mostly causing change of phase from liquid state to solid state through crystallization, the time period or duration of this zone is also called 'thermal arrest period' or 'TAP'.
- Thus, thermal arrest period is the time taken to reduce the temperature of fish from -1°C to - 5°C. It is important to note that TAP is the 'time period' of the freezing zone while '**MIC zone**' or '**critical zone**' is its 'temperature range' from - 1°C to - 5°C.

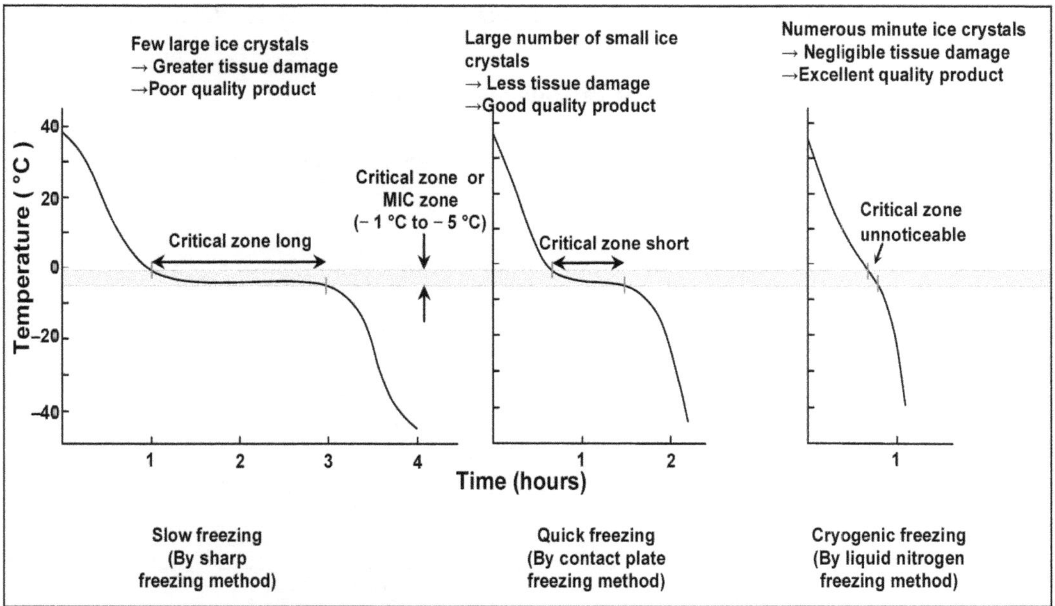

Figure 8.4: Length of critical zone in different freezing methods

d. Sub-freezing stage:

- The right side of the MIC zone is known as 'sub-freezing zone'. It is the third zone of the freezing curve after the MIC zone. Less water is left for freezing. Therefore, temperature reduction is quick and at the same time the remaining water changes to ice gradually.

D. Freezing Rate

- The preserving effect of freezing is due to the reduction in the rate of physical, chemical and biological processes as temperature of fish is lowered.

- It is well known that slow freezing results in the formation of large ice crystals while rapid freezing results in the formation of small ice crystals.
- Therefore, the quality of quick frozen fish may practically be equivalent to that of fresh unfrozen fish.
- Hence, the rate of freezing has long been considered of vital importance for obtaining high quality products.
- Maximum amount of ice formation will take place between -1^{0}C and -5^{0}C. The length of this zone is, therefore, the index of freezing rate. The amount of water frozen at various temperatures has been experimentally determined by Riedal as given below.

Table 8.4: Amount of water frozen at various temperatures

Temperature (^{0}C)	Percentage of water frozen
0	0
- 1	8.0
- 2	52.4
- 3	66.5
- 4	73.0
- 5	76.7
- 7	81.1
- 10	84.3
- 18	88.4
- 20	89.0
- 30	90.3
- 40	90.5

Effect of freezing rate:

Most of the physical and chemical changes take place in the maximum ice crystal formation zone. In order to minimise these changes the time of this critical zone has to be reduced. For this propose quick freezing methods are adopted. Some of the scientists delineated quick freezing as follows:

1. **Moran:** According to Moran, when the temperature change forms +5°C to -5°C takes place within 30 minutes, it is called quick freezing.

2. **Pool:** According to Pool, when the temperature change from -1°C to -5°C takes place within 25 minutes, it is called quick freezing. This temperature zone is called maximum ice crystal formation (MIC) zone.

3. **British Food Ministry:** According to British Food Ministry, when the temperature change from 0°C to -5°C in all parts of the body takes place within two hours it is called quick freezing.

4. **Russian scientists:** According to Russian scientists freezing of fish at a rate not lower than 3.3 cm/hr is called quick freezing. The freezing rate is calculated using the following equation.

$$V_{fr} = \frac{d/2}{Z^{-1}_{-5}}$$

Where,

V_{fr} = Freezing rate (cm/hr)
d = Thickness of fish (cm)
Z^{-1}_{-5} = Time taken to change the temperature at the center of fish from -1 °C to -5 °C (hrs)

Using the above equation, whether a freezing process is slow freezing or quick freezing can be determined as given in the following example.

Example: A 2.5 cm thick fish was frozen in brine at -16°C. The time taken to change the temperature at the center of fish from -1°C to -5°C was 22.5 minutes. Find out whether the fish was subjected to slow freezing or quick freezing.

Solution:

Given, Thickness of fish = d = 2.5 cm
Time taken to change from -1°C to -5°C = = 22.5 minutes = 0.375 hrs
To find out, Freezing rate = V_{fr} = ?

$$V_{fr} = = = = 3.33 \text{ cm/hr}$$

$$V_{fr} = \frac{d/2}{Z^{-1}_{-5}} = \frac{2.5/2}{0.375} = \frac{1.25}{0.375} = 3.33 \text{ cm/hr}$$

It is given that freezing of fish at a rate not lower than 3.3 cm/hr, it is called quick freezing. In this example the fish was frozen at a freezing rate of 3.33 cm/hr.

* The fish was subjected to quick freezing.

5. **FAO:** Instead of classifying the freezing rates into just two groups (slow and quick) as described in the above four methods, FAO has grouped the commercially used freezing rates into four groups as given below.

Types of freezing	Freezing rates (mm/hr)	Methods of freezing
Slow freezing	2	Sharp freezing in deep freezers
Quick freezing	5-30	Air blast freezing, contact plate freezing
Rapid freezing	50-100	Fluidised bed freezing of small products
Ultra-rapid freezing	100-1000	Cryogenic freezing (Liquid nitrogen freezing, liquid carbon dioxide freezing)

E. Freezing Time

» The freezing time of fish is the time taken to cool it from its initial temperature to a given temperature at its thermal centre which remains as the warmest part of the fish body during its freezing.

a. Average of equilibrium temperature of frozen fish:

» After freezing of fish in a freezer, it is subjected to cold storage in a cold store which maintains a very low temperature.

» It is recommended that the average or equilibrium temperature of fish should be reduced during freezing in the freezer to the intended cold storage temperature.

» However, during freezing, the whole body of fish does not attain a uniform low temperature; rather the temperature at the surface becomes very low while that towards to the interior remains high. The temperature is lowest at the surface of the fish while it is highest at its thermal centre.

» Thus, to reduce the temperature of fish to a given average or equilibrium temperature which is equal to the cold storage temperature, the fish should be frozen such that its surface temperature goes down much below the desired cold storage temperature while the temperature at its thermal centre reaches a temperature above the desired temperature.

» For example, if the recommended cold storage temperature is -30^{0}C (as in UK and most of the European countries), the freezing process has to reduce the average or equilibrium temperature of fish to the same temperature of -30^{0}C.

» For this, fish has to be frozen in such conditions that its surface temperature is quickly reduced to near the freezer temperature of say -36⁰C while the temperature at its thermal centre reaches -20⁰C.

» The freezing time in this particular case is the time taken to cool the fish from its initial temperature to −20⁰C at its thermal centre.

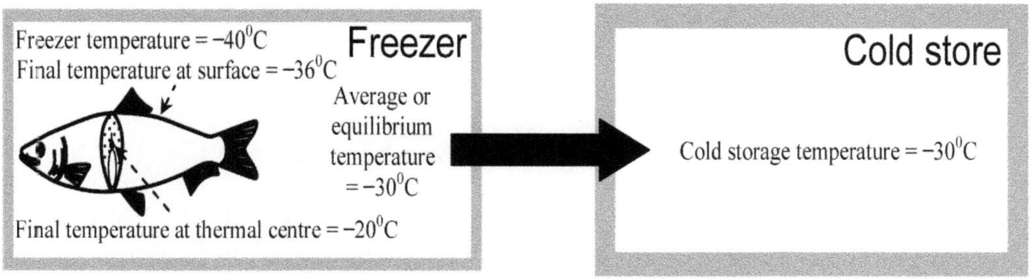

Figure 8.5: Required average or equilibrium temperature of frozen fish for cold storage

b. Factors influencing freezing time:

» The following factors influence the freezing time of fish.

i. **Freezer type:** Different types of freezers operating at the same temperature may have different freezing times for the same fish. For example, a fish freezes faster in an immersion freezer than in an air blast freezer operating at the same temperature. This is due to the higher surface coefficient of heat transfer of the refrigerating liquid medium in an immersion freezer than the cold air medium in an air blast freezer.

ii. **Freezer operating temperature**: When the freezer operating temperature is low, the freezing time is less.

iii. Air speed in Air blast freezer

iv. Product temperature

v. Product thickness

vi. Product shape

vii. Product contact area

viii. Density of the product

ix. Fish species

$$\textbf{Freezing Time } \alpha \ (1 / \Delta T)$$

Where, ΔT = Temperature difference

$$\textbf{Freezing Time } \alpha \ D^2$$

Where, D = Product thickness

F. Changes in Fish During Freezing and Frozen Storage

» The changes which take place during freezing of fish are mainly physical, chemical and biological. Some of these changes favour the preservation of the original composition and properties of fish while others do not.

» In order to preserve the original nutritive and organoleptic quality of fish, it is required to know the character of all these changes so as to control them.

» Favourable factors in the freezing process should be stimulated and unfavourable factors should be suppressed. In order to get maximum reversibility of frozen products, fish has to be frozen under optimum condition.

a. Main characteristics of marine products:

a. Highly perishable

b. High moisture content

c. Some fishes have high fat content

d. High bacterial contamination

e. Soft nature of tissue (as connective tissue/ collagen content is < 3%)

f. Differ in chemical composition a/c to species, size, age, season, feeding habit etc.

g. Marine animals contain lot of USFA i.e. mostly PUFA.

b. Proximate composition of fish:

1. Water:
   - 65% to 85%
   - They participate in chemical reactions.
   - 8 to 10% are only bound water
2. Protein:
   - 12% to 25%
   - Out of this,
     - ◊ WSP – 10-20% (Albumin)
     - ◊ SSP – 70-90% (Globulin)
     - ◊ Insoluble protein (keratin/ Collagen)-Minor quantity
   - Essential Amino Acid: 10 nos (MATTVILPHLy)
   - Fish is rich in Lysine
3. Fat:
   - 0.1 to 20.0%
   - Fish contains mostly USFA/PUFA.
   - Animal fat: More saturated FA – High Melting point and resistant to rancidity
   - Vegetable fat: More USFA but degree is low
   - Fish fat: More USFA but greater degree of unsaturation
4. Ash:
   - 3 to 4%
   - Ca, P, Fe, I are more in fish
5. Vitamins:
   - Vitamin A, D, E, and K are fat soluble

- Vitamin B-complex and C are water soluble

6. Enzymes:
    - Fish contains Cathepsin (Proteolytic enzyme)
    - Digestive enzymes cause belly bursting
    - At more than 75^{0}C, enzyme activity are destroyed
    - When MGP will denature, Ca^{++} will be released. Thus helps in increasing ATPase activity

c. Quality changes in fresh fish:

a. Rigor mortis
b. Autolysis
c. Putrefaction
d. Auto-oxidation

d. Physical changes occurring in fish during freezing:

1. Crystallisation
2. Volume changes
3. Turgor pressure
4. Mechanical and colloidal change

1. Crystallisation:

» It has been described under "Mechanism of ice crystal formation"

2. Volume change:

» Volume change is associated with the change of state of water. At 0^{0}C it expands nearly 9%, when transforming into ice.

» Water alone is responsible for this expansion that occurs during freezing. Any decreases in the amount of water per unit volume decreases the expansion per unit volume.

- » Any food materials capable of binding water strongly decrease the amount of water available for crystallisation and thereby decrease the expansion per unit volume upon freezing.
- » In animal tissues, 8% to 10% of the tissue water is unavailable for ice formation. The increase in volume of fish is more if it has high water content.
- » In case of fish, generally about 7% increase occurs by freezing.
- » When fish is cooled below its cryohydric point, its volume decreases, because below cryohydric point there is contraction of tissue components.
- » In case of plant materials, contraction has been observed because of the presence of air vacuoles or air spaces in the cells. These air spaces accommodate the expanding ice crystals thereby minimising the overall expansion. The expansion of water as it freezes results in local stress which undoubtedly produces mechanical damage in cellular materials.

3. Mechanical damage:

- » When a fish is frozen, there occur several changes in it. A frozen fish can never be the same as it was before freezing.
- » According to the 'mechanical damage theory', the arrangement of muscle fibers, which are the units of fish tissue, are disordered during freezing and they get separated from one another. The cell membrane of the muscle fibers is punctured by the growing ice crystals. As a result, during thawing the cell contents leak out as 'drip'.
- » In case of slow freezing, few large ice crystals are formed. They are located outside the cell and result in more cell damage than quick freezing.
- » In case of animal tissues, the fibres are elastic, for which they can tolerate the mechanical damage to some extent. In contrast, the fibers of plant tissues are poor in elastic nature than that of animal tissues. Besides, it contains more water and yield more ice crystals by freezing. Therefore, plant tissues cannot withstand mechanical damage. If frozen without proper treatment, its texture becomes poor after thawing.

4. Colloidal damage:

- » According to this theory, fish becomes damaged by freezing due to the irreversible changes of its colloidal system which becomes different from its original condition.

The tissue water of fish makes a solution dissolving many tissue components. At the same time, some of the proteins and carbohydrates form a colloidal solution with the fish tissue water.

» When such a colloidal solution is frozen, water molecules adhering to some of its proteins are frozen out as ice crystals and the proteins get dehydrated. As a result, these proteins get denatured and precipitate due to the non-availability of water to keep them in their native state.

» During thawing, complete reversibility is not possible as some water which were bound or adhered to the proteins before freezing do not get bound back to the proteins during thawing.

» Thus, due to the irreversible denaturation of some of its proteins, frozen fish, after thawing, does not entirely return back to its original condition.

» In this way, fish gets damaged in its colloidal system by freezing and thawing.

» Freezing damage of fruits, vegetables, fish and meat be explained by mechanical damage theory while that of milk and such foods without tissue structure may be explained by colloidal damage theory. However, these two theories are not independent of each other and in most of the foods those damages may occur simultaneously.

5. Turgor pressure (Swelling pressure):

» Freezing of fish occurs from the exterior to its interior. As a result, the surface layers of fish form a hard coat of ice in the beginning.

» Therefore, the increase in the volume of fish occurring due to the formation of ice, directs to the centre which is the unfrozen portion.

» Therefore, during freezing there occurs 'turgor pressure' in the interior of the ice coat.

» The turgor pressure can be explained as follows:

◊ When a fish with cylindrical shape is cooled and frozen, there occurs strong outwardly directed pressure on the surface of the whole fish and strong inwardly directed pressure in its interior. This strong inwardly directed pressure in the interior of a body is called turgor pressure.

◊ When the temperature of the frozen fish becomes lower and its diameter becomes bigger due to increase in its volume, the inwardly directed pressure in its interior becomes stronger and fish tissues and organs located in the interior of the fish body get damaged easily.

◊ When fish is frozen with intestine, i.e, in case of whole fish freezing, sometimes the tissues of the intestinal organs are punctured due to the turgor pressure and the digestive enzymes are discharged from the intestine resulting in the deterioration of the meat near the organ. Blood vessels are also often punctured due to the pressure which causes colour change and deterioration around the portion.

Thus, physical and physico-chemical changes during freezing include the followings:

a. Volume change

b. Cellular damage (Type A, B, and C): Mechanical rupture of structural unit through ice-crystal growth.

c. Change in pH, titratable acidity, ionic strength

d. Turgor pressure (Freezing injury): Freezing starts from surface.

   i. Ice crystal at the surface leads to inside

   ii. Pressure on the internal organs develops

   iii. Thus, release of enzymes etc. occurs and cause deterioration.

   iv. Turgor pressure is more in plant tissue than fish as fish muscle (cell membrane) is elastic in nature.

e. Langham-mason effect: Rupture of outer surface layer when interior freeze and expand.

f. Freezer burn: Dehydration during freezing

g. Colloidal damage: Water adhering to protein gets frozen. Protein precipitates causing protein denaturation. Thus reduce in solubility.

h. Disturbance in the structured water in frozen food.

i. Loss of semi-permeability of cell membrane.

j. Loss of lessening of bound water.

e. Quality changes in frozen fishery products:

- Mostly texture of meat products becomes tough due to protein denaturation
- Changes in colour, flavour occurs due to fat oxidation
- Quality changes are observed mainly due to:
    a. Colloidal damage – Protein denaturation
    b. Langham-mason effect
    c. Freezer burn
    d. Freezing injury (inside cell) due to turgor pressure
- Slow freezing results in formation of large ice crystals. This causes the fish tissue to become porous that perforations can be seen after thawing resulting to spongy texture.
- Freezing increases toughness in fish muscle due to destruction of connective tissue resulting the release of fibrils.

f. Effect of low temperature on Enzyme activity (EA):

- Enzyme activity ceases at $<-15^0C$ and $\geq 80^0C$
- Microbial activity stops at -12^0C due to denaturation of enzyme protein.

Rate of enzyme activity depends on –

a. Substrate characteristics viz. Composition, Concentration, pH and presence of inhibitors and activators
b. Rate and extent of freezing

- Lipase – Active even at -30^0C
- When phospholipids break down, FFA gets released and thus, cause rancidity.
- Enzymes resistant to thermal inactivation – a) Peroxidase b) Catalyase
- Enzymes active during frozen storage – Lipase, Lipoxidase, Catalyase, Peroxidase, Protease

» If Bio-colloidal concentration is >1.5% => EA decreases

» EA is maximum at critical temperature (-1⁰C to -5⁰C) i.e. MIC zone because of high substrate concentration.

» When myosin group of proteins break, Ca^{2+} ion will be released and the ATPase activity will increase.

» Freeze induced concentration of salt will enhance EA

» Freeze induced viscocity has no influence on EA

g. Effect of low temperature on Micro-organisms:

» Micro-organisms grow at -8⁰C to +90⁰C

» Enzymes are active between -18⁰C to +80⁰C

» If optimum temperature reduces –

a. Change in duration of lag phase
b. Rate of growth
c. Final cell number
d. Nutritional requirement
e. Enzyme and chemical composition

Group of Micro-organisms	Optimum temperature	Remark
Thermophiles	**55⁰C**	-
Mesophiles	37⁰C	Pathogens
Psychrophiles	12⁰C to 15⁰C	Spoilage organisms
psychrotrophs	25⁰C to 30⁰C	-

» *Vibrio parahemolyticus* cease to grow at 5⁰C to 8⁰C.

» *Clostridium botulinum* can grow and produce toxin up to 3.5⁰C to 5⁰C.

» Salmonella can grow at +15⁰C but toxin production stops at +20⁰C.

» Gm –ve bacteria are more susceptible to cold shock (*E. coli*, Salmonella, Enterobacteria).

h. Effect of freezing on micro-organisms (Freezing injury):

a. Cold shock

b. Concentration of extra- and intra- cellular solutes

c. Dehydration

d. Injury to internal ice formation

e. Mechanical injury

f. Accumulation of waste material – causing toxic action

» Microbial injury is maximum at -2^{0}C to -10^{0}C.

» Microbial activity stops at -12^{0}C.

» Enzymatic activity stops at -18^{0}C.

» Hence, cold storage temperature is recommended for -20^{0}C ± 2^{0}C

» In quick and rapid freezing, salts will be entrapped in ice. This will result in discontinuous distribution of salts and ice crystal. In order to minimise the freezing injury to micro-organisms, cryoprotectants are used during freezing and storage.

» The cryoprotectants include:

a. Polyphosphate

b. Sodium chloride

c. Ascorbic acid – Synargist

d. Citric acid – Chelating agent

e. EDTA – Chelating agent

f. BHA - Antioxidant

i. Protein changes in frozen fish:

» ***Denaturation*** – Change from native state i.e. Native structure will be altered.

» It is mainly due to

a. Fish moisture:

i. Ice crystal formation (Loss of bound water)

ii. Dehydration (Hydrophobic nature)

iii. Increase in salt concentration (Change of ionic strength)

b. Fish lipid

i. Intact lipid

ii. Free Fatty Acid

iii. Oxidised lipids

c. Others

i. Enzymes

ii. Formaldehyde

iii. TMAO etc.

» Fish muscle proteins are hydrophobic (25% to 30%) and hydrophilic (45% to 50%).

» The hydrophilic amino acids on denaturation get converted to hydrophobic resulting reduction in water holding capacity (WHC).

» Fish muscle proteins contains actomyosin, tropomyosin, troponin etc. Out of these, actomyosin is important which is comprised of actin (40%) and myosin (60%).

» Actomyosin starts aggregating as time of frozen storage increases.

» Nearly 80% MGP (Myosin group of proteins) will be denatured i.e.

- Reduced extractability
- Increased precipitation
- Increased ATPase activity
- Reduced water holding capacity

- Reduced fat emulsifying capacity
- Reduced gel forming ability

» As salt concentration (Ca^{2+} and Mg^{2+}) increases, ionic strength becomes high.

» Soluble actin (G-actin) gets converted to Insoluble actin (F- actin). Thereafter, binding to lipid by actin increases.

» NaCl, a pro-oxidant promotes fat oxidation.

» Cross linking of G-Actin with polar and neutral lipid at refrigerated temperature results in formation of Insoluble lipid – protein complex. This reaction is enhanced by heat and ionic strength.

» Free Fatty Acid (FFA): Release of FFA depends on degree of unsaturation. FFA will attack hydrophobically and hydrophillically on proteins and create more hydrophobic regions.

» Fish contains lipids which comprise of 70% Un-saturated fatty acid (USFA) and 30% Poly-unsaturated fatty acid (PUFA).

» PUFA are very susceptible to oxidation and end products obtained are melanaldehyde, hexanol, pentanol, propanol etc. these react with protein molecules and form a complex.

» Lipid also attacks specific amino acid like cystine, lysine, methionine, aspartic acid.

a) Prevention of protein danaturation:

» To pass the MIC zone very fast during freezing

» Prevent fluctuation of storage temperature => prevent re-crystallisation.

» Glazing – to prevent oxidation

» Packaging – to prevent oxidation, dehydration (freezer burn).

» Anti-oxidant – to prevent oxidation of fat.

» Meat picker hole should not be less than 5 mm. If small, more strain.

» Mixing of pro-oxidants viz. metal ions are discouraged.

» Mixing of different species are discouraged.

» Biochemical status of fish – It should be very fresh i.e. before rigor state. If fish on pre-rigor stage are frozen, then during thawing, thaw rigor will occur.

» For fillets:

- Brining/ Firming (7% NaCl for 2 min)
- Addition of 10% Sodium pyruvate
- Addition of Monosodium glutamate (MSG)

» For paste products, 2.5% salt + 5% sugar + 0.2% MSG should be added.

» Protein denaturation can be measured by cell fragility and titratable acidity.

» Protein denaturation of cold stored fish is related to O.D. (Optical density) of the muscle. High O.D. value indicates good quality fish where as low O.D. value indicates myofibrillar aggregation.

» Role of Titratable acidity: Based on the values of 0.1N acid/ alkali required to change the pH from 6.5 to 3.0 or 9.0 respectively, it can be observed that amounts of acids and alkali required to change the pH as referred to above, are gradually reduced as the storage period increases i.e. as the protein denaturation proceeds. During freezing, ice crystal formation takes place resulting increase of ionic strength thus, increase of pH. Thereafter, measuring the quantity of acid/alkali, the degree of protein denaturation of fish muscle can be determined.

j. Lipid deterioration during frozen storage:

» Rancidity generally is the complete or incomplete oxidation or hydrolysis of fats and oils when exposed to air, light, or moisture or by bacterial action, resulting in unpleasant taste and odour.

» The term rancidity refers to 'off' odours and flavours resulting from lipid oxidation or lipolysis (breakdown of oils chemically or by lipase into constituent fatty acids)

» There are four distinct types of deterioration in fish lipids.

1. Hydrolysis
2. Auto-oxidation

3. Polymerisation
4. Flavour reversion

1. Hydrolysis:

» Here, a considerable amount of free fatty acids (FFA) appears.

» The phenomenon is more observed in ungutted fish than in gutted fish as because the digestive enzymes are involved in hydrolysis.

» Hydrolysis of triglycerides involves two steps where temperature plays an important role:

a. Lipase reacts with a primary alcohol group
b. Then it will react with secondary ester group resulting in production of FFA, Glycerol etc.

» In another case, lipase reacts with phospholipids to produce FFA resulting textural change.

2. Auto-oxidation:

» Fish lipids are rich in unsaturated fatty acid viz. USFA and PUFA.

» The unsaturation of fatty acids is due to the presence of double bonds between carbon atoms in their hydrocarbon chains. When exposed to atmosphere, the atmospheric oxygen attacks and reacts with the double bonds on the fatty acids. The reaction of the atmospheric oxygen with the unsaturated double bonds of the fatty acids of lipid-molecules is called 'lipid oxidation' or 'autoxidation'.

» Autoxidation of lipid takes place in four steps such as-

i) Initiation, ii) Propagation, iii)Decomposition, and iv) Termination

i) Initiation:

- Fat molecules are activated by catalyst like heat, light etc. and produces lipid radical and free radicals.

$$RH \rightarrow R + H \text{ (free radical)}$$

ii) Propagation:

- The lipid radical reacts with oxygen molecule to form peroxide radical.

  $R + O_2 \rightarrow ROO$ (Peroxide radical)

- The peroxide radical will then react with another fat molecule and will release hydroxyperoxides.

  $ROO + RH \rightarrow R + ROOH$ (Hydroxyperoxide)

iii) Decomposition:

- The hydroxyperoxides are readily broken down, catalyzed by heavy metal ions, to secondary auto-oxidation products of shorter carbon chain-length viz. aldehydes, ketones, alcohols, small carboxylic acids and alkenes. These give rise to a very bad odour and in some cases to a yellowish discoloration.

  $ROOH \rightarrow RO + OH$

iv) Termination:

- It is the last step where reactions are terminated when radicals react with each other to produce non-radical compounds, either volatile or non-volatile, or, when alkoxyl radical.
- The oxidised products give yellow colouration with bad odour.

$$\left.\begin{array}{l} ROO + ROO \rightarrow \\ ROO + R \rightarrow \\ R + R \rightarrow \end{array}\right\} \text{Non-radical products}$$

» Initiation stage is faster and propagation is slower. Once it is started, propagation step will continue till all fatty acids are absorbed.

3. Polymerisation:

» All commonly used fats and particularly those high in PUFAs tend to form larger molecules (Known broadly as polymers) when heated under extreme conditions of temperature and time.

» Under normal processing and cooking conditions, polymers are formed.

- Although the polymerisation process in not completely understood, it is believed that polymers in fats and oils arise by formation of either carbon-to-carbon bonds or oxygen bridges between molecules.
- When an appreciable amount of polymer is present, there is a marked increase in viscosity.
- Polymerisations are particularly facilitated by high temperature.
- Dimerization and polymerization occurs in three ways:
  a. Radical recombination
  b. Addition of radicals to double bonds, and
  c. Diels-Alder reaction, a reaction between a double bond and a conjugated diene.

4. Flavour reversion:

- Flavour Reversion in fats is defined as the appearance of objectionable flavour from less oxidation than is required to produce true rancidity.
- It may develop during the exposure of the fat to ultraviolet or visible light or by heating.
- It is the breakdown (oxidation) of an essential fatty acid, linolenic acid, found in fish oils and certain vegetable oils, leading to an undesirable flavor change prior to the start of actual rancidity, which revert with extremely slight oxidation.
- It is almost impossible to prepare edible products from such fats and get them to the consumer before they have suffered some loss of palatability.

a. Rate and extent of auto-oxidation of lipid depends on:

i. Degree of un-saturation of fatty acid
ii. Surface exposed to light and air
iii. Pre-freezing holding time
iv. Duration and temperature of frozen storage
v. Oxygen content of micro-environment
vi. Presence of pro-oxidant (metals) and antioxidants (tocopherol)

b. Antioxidants:

i. Synthetic antioxidant: Ascorbic acid/ Butyl Hydroxy Anisole (BHA)/ Butyl Hydroxy Toluene (BHT)/ Monosodium glutamate (MSG)/ Citric acid/ Polyphosphates

ii. Natural Antioxidants: Lecithin/ Tocopherol

c. Antioxidant mechanism:

» It involves 4 steps:

i. Hydrogen donation by antioxidant

ii. Electron donation by antioxidant

iii. Addition of lipid to aromatic ring of antioxidant

iv. Formation of complex between lipid and aromatic ring of antioxidant

d. Preventive measures of lipid deterioration:

i. Lowering the temperature of the product

ii. Prevention of Initiation stage i.e. activators like heat, light are to be reduced.

iii. Exclusion of O_2 from micro-environment

iv. Removal of free radicals by addition of some chemical compounds i.e. auto-oxidation.

G. Freezing Methods

There are several methods of freezing fish. Based on the material with which fish comes in contact during its freezing, these methods can be broadly classified into the following three categories.

a. **Cold air freezing:** Cold air is allowed to come in contact with fish to freeze it

b. **Contact plate freezing:** Cold metal plates are allowed to come in contact with fish to freeze it

c. **Immersion or spray freezing:** Cold liquid is allowed to come in contact with fish by immersion or spray to freeze it

Out of the several methods available for freezing of fish, only a few are used commercially. These are described below.

a) Cold air freezing:

1. Sharp freezing:

It is a method of freezing in which the fish is frozen in a cabinet or tunnel refrigerated to a very low temperature in which the cold air is circulated slowly or not at all. It is the most widely used freezing method for packaged and unpackaged fish. The temperature maintained is -18^{0}C to -29^{0}C. The rate of freezing is very slow as the cold air is circulated slowly or not at all.. The time taken to freeze the fish varies from 3-72 hour. Sharp freezing is not commercially used for the freezing of fish. However, the 'freezer chamber' (not the cooler chamber used for storing vegetables) of 'domestic refrigerators', the 'deep freezers' (both vertical and horizontal) used in the retail markets as well as the 'cold store' used for storing of frozen fish are used for sharp freezing of fish. However, it is important to note that, fish should never be frozen in a cold store, as the fish undergoes severe quality deterioration during the slow freezing process of sharp freezing. Instead, it should be first frozen by a quick freezing method using a suitable freezer and then stored properly in a cold store.

Advantages:

1. The cost of freezing is lower as compared to air blast freezing or contact plate freezing.
2. The sharp freezer have a low maintenance cost.

Disadvantages:

1. Freezing is slower than any other method.
2. Considerable handling of product is required.
3. The loading and unloading of products results in the coil frosting and increases freezing time.

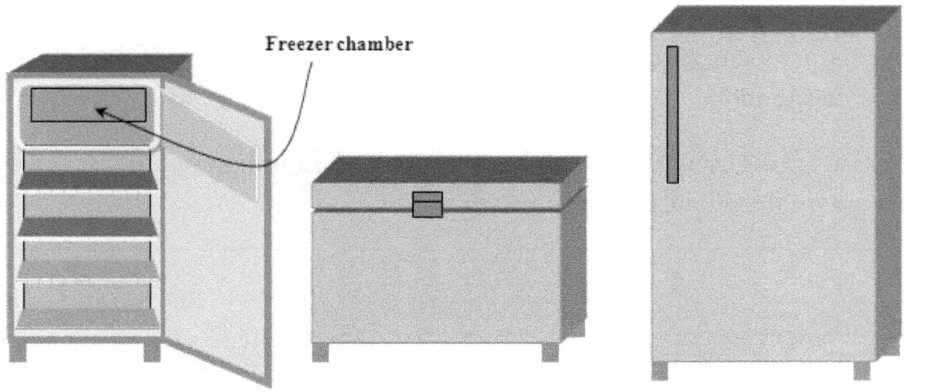

Figure 8.6: Sharp freezers

2. Air blast freezing:

It is a modified method of sharp freezing. Vigorous circulation of cold air on the fish takes place. Freezing proceeds at a moderately rapid rate. In this method, the fish is frozen and placed on trays either loose or in packages. The trays are kept on trolleys and kept in the cooling chamber and cold air is allowed to go through it. In a modified air blast freezer, the air is blown through a refrigerated coil either inside the room or in an adjoining blower room. Inside the tunnel, about -35^{0}C cold air is blasted on the product. The air velocity is 3-9 m/s. The air speed influences the freezing rate. Higher is the speed of air, the quicker is the freezing. However, the freezing rate does not proportionate to the air speed above certain speed limit. It is said that, the most economical and practical speed is about 5 m/s. The fishes are usually frozen as batches and the freezing time for one batch is 4-6 hours. In this case, freezing is quicker and large quantities of fish can be frozen in limited time. Air blast freezing is economical and is capable of accommodating fishes of varieties of sizes and shapes. It can however result in excessive dehydration in unpackaged foods if conditions are not carefully controlled. Undesirable bulging of the packaged fish occurs in case of air blast freezing.

Advantage:

1. Fish of any size and shape can be frozen.
2. 'Freezing time' is less due to which 'freezing rate' is fast and 'quick freezing' takes place.

Disadvantages:

1. The high velocity air takes away a lot of water from the surface of the fish leading to is 'dehydration'.
2. In case of packaged fish, the packages get bulged. In order to minimise it weights are put on the product during freezing.

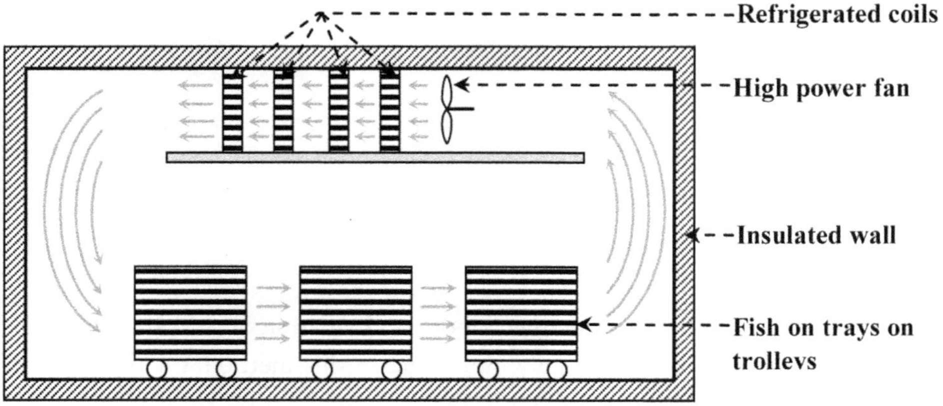

Figure 8.7: Air blast freezer

3. Fluidised bed freezing:

Due to the great demand of individual quick frozen products popularly known as IQF like vegetables, shrimps, fish fillets and fish steaks, fluidised bed freezer is extensively used. Solid particles like small fish and shrimp can be fluidised forming a bed of particles, 1" to 5" deep, on a mesh belt conveyor and then forcing air upward to the belt at a rate sufficient to partially lift or suspend the particles. If the air used for fluidisation is appropriately cooled freezing will be at a rapid rate. An air velocity of at least 375 ft/min is necessary to fluidise suitable particles. The temperature of the air is maintained at about -35°C. The bed depth depends on ease of fluidisation and this in turn depends on size, shape and uniformity of the particles. Fluidised bed freezing is most suitable for relatively small and uniform size products, but many kinds and sizes of unpackaged fish can also be frozen. The freezing time varies according to the condition of freezing. For example the freezing time for different products is given below.

Products	Freezing time (minutes)
Peas	3-4
Large shrimps and fish steaks	12-15
Fish fillets	30

Advantages:

1. Produces IQF products which are very popular.
2. Most efficient in transport of the products.
3. More rapid rate in freezing.
4. Less product dehydration.
5. Less frequent defrosting requirement. For example, dehydration loss of about 1% occurs during fluidised bed freezing of shrimps.

Disadvantages:

1. Large and non-uniform products cannot be fluidised
2. A reasonable air velocity is required.

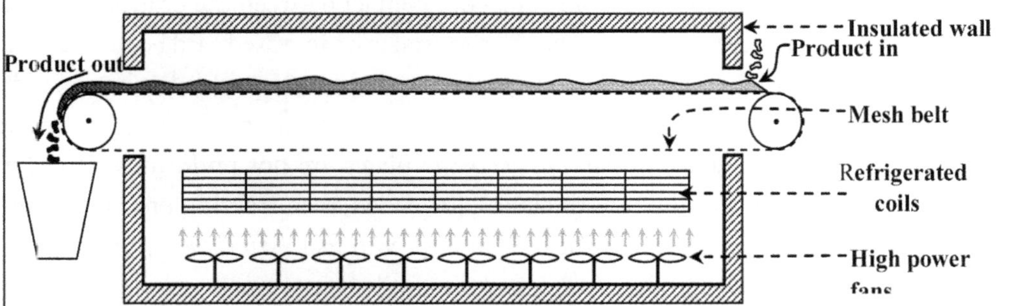

Figure 8.8: Fluidised bed freezer

b) Contact plate freezing:

1. **Vertical contact plate freezing:** In this method, a series of vertical hollow metal plates are cooled by the circulation of a cooled refrigerant through their hollow spaces. Fish to be frozen is put in the spaces among the plates. Subsequently, the plates are moved closer to one another mechanically so that the fish present in the space between any two adjacent plates come in close contact with these two plates from both the sides. Due to pressure from both the sides, the fish present between a pair of plates get compacted and at the end of the freezing process forms a block. The plates are moved away from one another mechanically so as to remove the large blocks of fish from between the plates. These blocks of fish are stored in the cold store.

   Usually vertical contact plate freezers are used at sea 'onboard the fishing vessels' to retain excellent quality of extremely fresh fish just after its harvesting in the sea.

2. **Horizontal plate freezer:** This is an indirect method of freezing. Birdseye and Hall developed it and it was commercially used in 1933. The freezer consists of several hollow metal plates which can be moved up and down by means of a hydraulic press. These plates can be opened to receive the product and closed within any desired pressure. The entire freezing apparatus is enclosed in an insulate cabinet. Fishes packed in aluminium trays are placed between the plates and are cooled to the desired temperature. The temperature is maintained in this case at -35^{0}C to -45^{0}C. After loading, the plates are released by the hydraulic press so that there is a close contact between the plates and the products with slight hydraulic pressure. Poor contact between the products and the plates results in slow freezing and can occur in several ways in waxed card board cartoons. For example, the products may be loosely packed so that an air space is left between the top of the product and the box. This trapped air acts as insulation on the upper surface of the product. If the layers of fish between the plates are not uniform, as it occurs when products of slightly different thickness are loaded, then freezing proceeds fastest at the spots where the plates make the best contact with fish. After loading, the cabinet doors are closed and the products are left there until their temperature reaches about -18^{0}C.

Precautions during operation of horizontal plate freezers:

1. The space between the plates should be adjusted to the height of the freezing trays by use of space bar.
2. The plates should be pre-cooled before loading.

3. Load from bottom and unload from top.
4. Keep the freezer doors closed except when loading and unloading thereby preventing excessive frost building upon the plates.

Advantages:

1. Uniform well-shaped packages with minimum of voids (empty space) are produced.
2. Minimum amount of floor space is required.
3. Packaged fish fillets are frozen quickly and economically.
4. Frequent defrosting of the plates is not required, if the freezer is operated properly.

Disadvantages:

1. Much handling is required in loading and unloading the products.
2. Products with air space in the packages are frozen very slowly.
3. Large storage space is required for freezing trays.

c) Immersion or spray freezing:

1. **Liquid immersion or spray freezing (Brine freezing):** Freezing by immersion of fish in a cold liquid like brine or sometimes by spraying it on the fish is used because of the simplicity of the method and high output. The heat transfer is efficient with moderate circulation, because there is a good contact of the cold liquid with the product. Immersion freezing can accommodate fishes of various sizes and shapes and consumes less power. The aqueous solution of common salt (NaCl) (brine) is the most suitable freezant (liquid) used for immersion or spray freezing of fish. The eutectic point of common salt is -21.2^{0}C, but in practice this medium cannot be used below -18^{0}C. Salt penetration is the major disadvantage of direct brine freezing. The fish takes up salt resulting in reduced keeping quality and in few cases resulting in the development of off-flavour. Salt penetration is serious when the fishes are held in the brine for a very long period and when the brine temperature is not held below -12^{0}C. The fish held in the cold brine after freezing continues to absorb salt. Therefore, the frozen products should be removed immediately after freezing.

The freezants to be used for liquid immersion freezing should have the following properties.

1. Not-toxic
2. Inexpensive
3. Reasonably inert
4. Stable
5. Low freezing point
6. High thermal conductivity
7. Minimum tendency to penetrate the fish.
8. Requires little effort to maintain desired standard for sanitation.

Advantages:

1. Results in rapid freezing.
2. Easily adapted to continuous operation.

Disadvantages:

1. It is difficult to find freezants which satisfy most of the above mentioned requirements.
2. Salt uptake is remarkable if freezing time is very long and freezant temperature is not held below -12^{0}C.

2. **Cryogenic freezing:** Cryogenic freezing refers to very rapid freezing achieved by exposing fish to an extremely cold freezant like liquefied gas like nitrogen or nitrous oxide. The freezing is very quick but the systems are expensive. Initially immersion systems were used for freezing of fish in this method, but now spraying techniques have been developed. The fish is placed on the conveyor belt and is moved into the zone 'A' of the insulated chamber, where it is cooled with moderately cooled gaseous nitrogen (0^{0}C). In zone 'B' liquid nitrogen is sprayed on the fish. Then the fish is passed into zone 'C' where it is allowed to equilibrate to the desired temperature (-18^{0}C or -30^{0}C) before being discharged into the cold store. It should be noted that the final temperature of the fish is usually not different from that obtained during conventional methods of freezing. However, the rate of freezing obtained with cryogenic freezing is much greater than that obtained in air blast or plate freezing. But it is slightly greater than fluidised

bed freezing or liquid immersion freezing. For example, shrimp freezes in 9 minutes in commercial liquid nitrogen freezer and in about 12 minutes in fluidized bed freezer. Freon 12 is used directly in immersion freezing.

Advantages (Liquid nitrogen or LN freezing):

1. Dehydration loss from fish is very less; usually much less than 1 %.
2. Oxygen is excluded during freezing.
3. The individual quick frozen (IQF) fish obtained undergo minimal freezing damage.
4. The appearance of the fish is superior to those frozen by other methods.
5. The equipment is simple and is suitable for continuous flow operation.
6. Adoptable to various production rates and product sizes.
7. Relatively low initial cost.

Disadvantages:

1. High 'operating cost' as the cost of liquid nitrogen is very high.

d) Economics:

Installation cost: Liquid CO_2 (Maximum) followed by Liquid R12, Air Blast freezer, Plate freezer, Liquid Nitrogen freezer.

Energy consumption: Liquid CO_2 (Maximum) followed by Air Blast freezer, Liquid R12, Plate freezer, Liquid Nitrogen freezer.

Labour use: Plate freezer (Maximum) followed by Air Blast freezer, Liquid CO_2 , Liquid Nitrogen freezer, Liquid R12.

Weight loss (Dehydration): Air Blast freezer (Maximum) followed by Liquid Nitrogen freezer, Plate freezer, Liquid R12, Liquid CO_2.

Freezant consumption: Maximum is Liquid nitrogen freezer.

H. Super-Chilling of Fish

» Super-chilling or subfreezing is a method of fish preservation in which, the fish at an ambient temperature is cooled to and maintained at a temperature just below its freezing point (usually to –2.2°C or 28°F) where about half of the tissue water of fish freezes to ice thereby resulting in partial solidification of the fish.

Principle of super-chilling:

» In super-chilling or subfreezing of fish, the temperature of fish body is lowered just below its freezing point (usually to –2.2°C or 28°F) so as to get a longer shelf life than chilled fish.

» The temperature of super chilled fish is slightly lower than that of chilled fish (above – 1°C). This marginally lower temperature can remarkably increase the shelf life of fish as compared to chilled fish, sometimes doubling it.

» For example, if chilled fish has a shelf life of 15 days, super-chilled fish may have a shelf life of 30 days. Thus, super-chilling preserves fish for a considerably longer period of several days more as compared to chilling, though the temperature of super-chilled fish is little lower than that of chilled fish.

Methods of super-chilling:

» Super chilling is usually adopted for preserving fish onboard the fishing vessels, which go on fishing for days together. Super chilling of fish is done by two methods as given below.

a. **Cold air method:** Usually super chilling is done by cooling fish in ice followed by circulating air at about –3°C.

b. **Salt method:** In this method, super chilling is done by adding salt to iced fish, but such method can only be adopted for preserving and transporting bulk quantities of fishes intended for curing or canning where salt uptake and texture changes are of no major consequences.

Precaution in super chilling:

» In super chilled fish, the temperature of fish body must not go down much below –2.2°C, otherwise the loss of quality due to partial freezing of fish at lower temperature than –2.2⁰C would cause more harm than the benefit caused by a longer shelf life. This necessitates a reliable mechanical system to maintain the critical temperature required.

Advantage of super chilling:

» Super chilled fish has longer shelf life (about 32 days) than chilled fish (about 15 days). This allows fish to be stored for more days after harvesting without much quality loss, which in turn enables fishing vessels to go to longer distances and undertake fishing for more days without coming back to shore.

Disadvantages of super chilling:

» At the super chilling temperature below –2.2⁰C 'partial freezing' of fish tissue occurs which has detrimental effect on the quality of fish. This necessitates precise control of fish body temperature through a reliable temperature controller.

» It is difficult to unload fish from the fishing vessel as fish becomes frozen into the crushed ice in which it is cooled.

» It is required to thaw the fish before use.

I. Sanitation in Fish Processing Plants

Every fish processing plant should maintain very high degree of cleanliness and sanitation to prevent microbial infection of fish and subsequent deterioration of its quality and resulting danger of food poisoning by microbial action and there toxins. In case of fish processing operations, particularly of frozen fishery products which are not subjected to sterilisation, perfect cleanliness has to be maintained right from the time of landing of fish though processing, storage, transport and finally marketing.

Degree of cleanliness:

1. **Physical clean:** the object or the surface is clean enough to sight and feel. As this degree of cleanliness can be easily tested by close observation it is not very difficult to obtain.

2. **Chemical clean:** The object or surface should be free from traces of undesirable chemical. Even traces of some chemicals can affect the taste, odour, colour and keeping quality of the fish.

3. **Microbiological clean:** The object or surface should be free from undesirable microorganisms. This kind of cleanliness is tested by determining

    a. Kind and number of microorganisms left on the object cleaned.

    b. Decrease or increase of bacterial load of the fish at the end of the processing line.

    c. Keeping quality of the products.

Cleaning and washing aid:

» For perfect washing and cleaning, selection of suitable wash solution or detergent is very essential, as various factors such as naturalisation, pH, alkaline, buffering capacity etc. of the wash solution influence the effectiveness of cleaning.

Mechanical aids to cleaning equipment:

» Containers and holders, tanks, buckets.

Use of chlorine compounds:

» Chlorine compounds, when correctly used for sanitation of fish processing plant, have proved to be safe and dependable germicides. Addition of chlorine to water used for washing, conveying or cooling of fish or cleaning of equipments has proved to a great value in fish processing. The ability of chlorine even in trace amount in water to destroy the microorganisms has made it possible to produce processed fish with high standards of sanitation.

Terms used in chlorination:

1. **Available chlorine:** All the chlorine compounds are expressed in terms of ppm of available chlorine.

2. **Chlorine doses:** Amount of chlorine added in water is often called as chlorine doses.

3. **Chlorine demand:** It is the difference between the amount of chlorine doses and the amount of residual chlorine remaining at the end of specified contact period. This varies with

    a. Amount of doses

    b. Period of contact

c. Water temperature

d. Types of impurities in water

1. **Available free chlorine:** The portion of total residual chlorine which reacts chemically and biologically as hypochlorous acid is termed as available free chorine

2. **Marginal chlorination:** Addition of small amounts of chorine to water without the types of residue produced is termed as marginal chlorination. This method is used mainly to prevent pathogens in potable water.

3. **Break point chlorination:** The chlorine concentration in water, beyond which the rise in free chlorine concentration is directly proportional to the further addition of chlorine is know as break point chlorination

4. **In plant chlorination:** Addition of chlorine to the entire incoming water supply is known as in plant chlorination.

Use of chlorine compounds:

1. Liquid chlorine

2. Hypochlorites (Calcium or sodium hypochlorite)

Table 8.5: Chlorine dosages

Material to be chlorinated	Minimum	Maximum
Foods plant in general	5-7ppm	10-25 ppm
Washing of boats decks and processing rooms	200 ppm	1000 ppm
Washing of fish (initial washing)	5 ppm	20 ppm
For final products	4 ppm	10 ppm

J. Cold Chain

» A cold chain is a refrigerated food distribution and marketing system connecting the production centre at one end and the consumer at the other end through a network of transportation, storage and marketing activates.

» This distribution system is an ideal method to maintain the freshness of perishable commodities like ice creams, fish, meat, fruits and vegetables for considerable length of time. Fish is most perishable food commodity only next to milk. It spoils within few hours if not preserved properly. About 40% of the spoilage of fish can be minimised by cold chain.

» Cold chain systems are commonly used in advanced countries. In these countries cold chain is linked with various centres by the establishment of cold stores, community locker stores, retail showcases, refrigerator, containers and consumer fridges which are supplied with frozen fish thorough refrigerated trucks, railway wagons and refrigerated ships.

» For organising a similar development in our country, we require a considerable investment of money for machines and equipments. Still it is worth establishing cold chains.

» A skeletal net work of cold chain connecting important fish landing centres in metropolitan cities for the distribution of fish is given below.

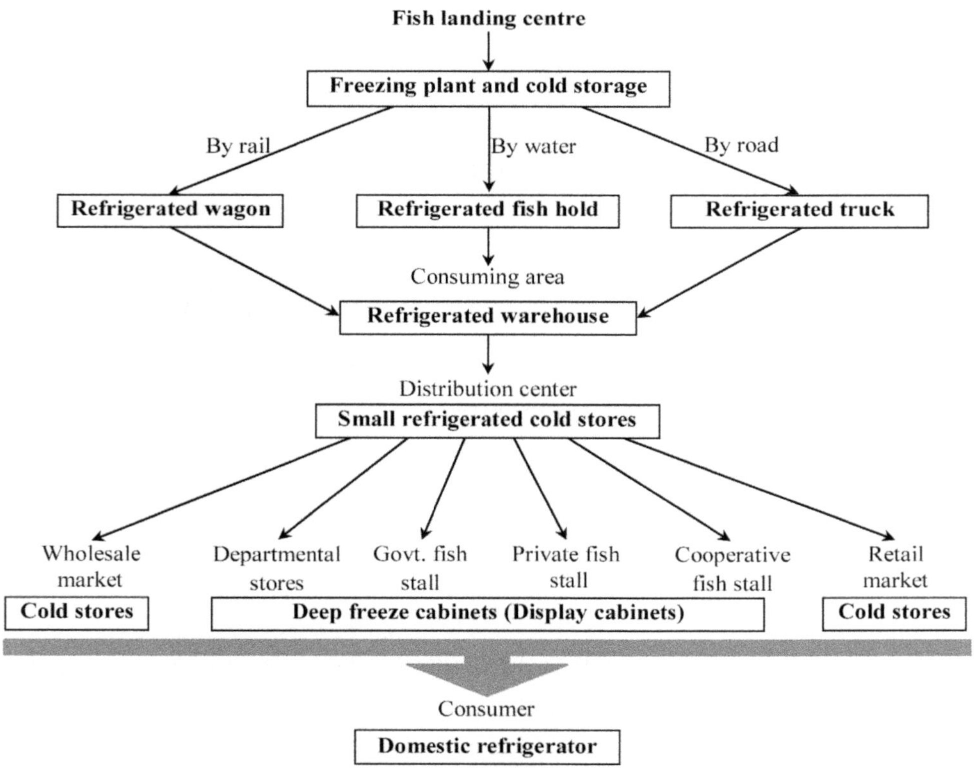

Figure 8.9: A typical cold chain

- One important requirement of cold chain is the setting up of freezing plants especially in a fish landing centre in an organised manner.
- The capacity of such plants should be adequate enough to handle and store the quantity of fish landed in the landing centre.
- These plants should have good facilities to transport the frozen fish in refrigerated trucks, vans etc.
- It is felt that, the transport on road is suitable for limited quantity of fish for short distances whereas ships and rails are convenient for bulk and long distance transportation. Proper marketing of frozen fish can be achieved by a good cold chain.

Merits of cold chain:

1. Frozen fish handled with optimum care can be similar to fresh fish in quality. For long distance transportation frozen fish is better than iced fish. It is reported that 40% of iced fish gets spoiled during transportation. If cold chain is adopted to transport frozen fish, its cost will be almost equal to that of iced fish and even sometimes shaper.
2. Cold chain provides a quick and efficient food distribution system, thus facilitating proper distribution of nutritive food like fish to all parts of the country.
3. It helps to stabilise the price of perishable foods like fish.
4. It is possible to supply fish throughout the year by preserving during harvesting season.
5. Cold chain can give a developmental incentive to other subsidiary industries like refrigeration and processing machineries, packaging, transportation etc.
6. Frozen fish can fetch price equal to fresh fish in regular seasons and better price during off seasons.
7. It provides additional employment.

K. Organoleptic (Sensory) Evaluation for Quality Assessment

i. Quality assessment methods:

- The food manufacturer has to produce a product at a quality which must satisfy (a) the customer and (b) statutory food legislation.

» Much has been written on the analytical techniques developed to establish the nutritional and chemical composition of foods to ensure their compliance with food law.

» In many ways, the quality requirements of the customer are more difficult to satisfy since the customer assesses the quality of a product entirely by subjective means, i.e. by sensory evaluation of a food's appearance, colour, odour, taste and texture, plus the visual appeal of its packaging and presentation.

» Hence, it is the responsibility of the manufacturer to develop methods which can, as accurately as possible, evaluate the sensory properties of a food which the customer finds important.

» This will often mean using ***subjective methods of testing*** such as sensory (taste panel) methods or, in some cases, it may be possible to use objective methods.

» However, it must be remembered that objective measurements of food quality are preferable only if the objective tests can provide a precise measure of the subjective quality being considered.

a. Subjective (Sensory) methods:

» A sensory method is one in which subjective measurements are made by individuals.

» Numerical scoring systems may be developed for such methods, or methods of ranking, or results may be expressed simply in attribute form, e.g. pass/fail, acceptable/defective.

b. Objective (Chemical) methods:

» An objective method uses a calibrated scientific instrument to measure a specific quality parameter usually on a numerical scale.

» This includes proximate composition, Protein quality changes, lipid quality changes etc.

ii. Problems of assessing fish quality:

» Many hundreds of species are sold throughout the world, all with distinct physical, chemical and sensory characteristics.

» Fish, alone among the major items of food, are susceptible to virtually no control before harvesting or slaughtering. In addition, they are one of the most perishable items of food eaten.

» Fish freshness/spoilage may be investigated objectively by assessment based on, for example:

- ***Physical changes***, e.g., measurement of conductivity using the Torrymeter.
- ***Chemical and biochemical tests***, e.g. tests dependent on bacterial action such as estimation of trimethylamine (TMA), tests dependent on autolytic action such as enzymic assays of nucleotide breakdown products, viz. hypoxanthine; tests dependent on fat oxidation, such as peroxide value estimation.
- ***Bacteriological changes***, e.g., counting and identification of various micro organisms.

» However, none of these methods produces results which may be regarded as providing a precise measure of consumer acceptability. Therefore, any system adopted to measure the quality of fish must, at some stage, be related to consumer acceptance. This must involve sensory (taste) panels and consumer research.

» In the following sections, it is planned to consider the problems and methods of establishing a taste panel and assessment of results.

iii. Classification of sensory characteristics:

» Sensory appreciation of food quality may be divided into the following categories:

1. Appearance and colour
2. Odour and taste
3. Texture

» The above three categories cannot be treated in isolation since sensory evaluation is often a combination of several overlapping factors, e.g., 'flavour' includes elements of odour, taste, texture and even the psychological effect of colour.

iv. Taste panels:

Taste panel techniques may be used with an aim:

- To characterise sensory changes in foodstuffs, e.g., changes that occur during spoilage or changes brought about by processing methods such as freezing or dehydration.
- To distinguish (often with the hope of failure) between batches or samples of a particular foodstuff.
- To ascertain whether 'qualify' can be represented by a simple numerical index or whether it is multidimensional.
- To help establish standards, in some defined area, of raw or unprocessed food products.
- To grade products according to some agreed quality classification system.
- To help establish a usable relation between sensory data and consumer acceptability.

v. Methods to achieve the above aims roughly divide into two categories:

a. Product rating
    - Grading on some agreed numerical or descriptive/hedonic scale is carried out by a small panel of well-trained judges.

b. Difference testing
    - Usually done by a panel of sensitive judges, without any great expertise or elaborate training.

vi. Panel selection and training:

- Prospective taste panel members must be in sound health, self-motivated and of even temperament. They should first be screened for primary taste sensitivity and reliability.
- Members should be consistently able to detect and recognise the following levels of the four primary taste components:
    1. **Sweetness**- 1.5 per cent sucrose
    2. **Saltiness**- 0.15 per cent sodium chloride
    3. **Sourness**- 0.06 per cent citric acid
    4. **Bitterness**- 0.005 per cent quinine sulphate
- Many individuals will be able to detect levels considerably lower than these values.

- It is also important to remember that initial detection of a component (say differentiation from water) is likely to occur at a lower concentration than recognition of the primary taste concerned.
- If evaluation of odour is a major part of the sensory test then prospective judges must also be able to demonstrate a good 'odour memory'.
- On a general level, it is wise to find out how well the candidates will perform on an odour recognition test carried out with stoppered bottles of some of the commoner chemical smells such as acetic acid, ammonia, amyl acetate, aniseed, benzaldehyde, linseed oil, menthol, peppermint, vanillin etc.
- Finally, prospective members should be able to display discriminatory skill with different qualities of the foodstuff under test.
- Discriminatory skill need not be general, e.g. a good wine taster may not be a good judge of teas or fish.
- Having chosen members for the 'taste panel', a preliminary training period should be carried out, designed to acquaint the tasters with the quality factors involved in the fish product under test.
- The group should be asked to evaluate a selection of fish of the same type, at different stages in deterioration, from fresh to wholly unacceptable.

vii. Tasting room:

- Control of environmental factors is universally recognised as being of value in sensory work with foods. Thus, a special 'taste panel' room where as many variables as possible can be controlled is highly recommended.
- For maximum privacy and concentration, the use of screened tasting booths is recommended.
- Although natural or white fluorescent lighting should be used where possible, a coloured light facility should be made available to disguise any minor colour differences between samples that may occur on occasion and which would otherwise distract the taster.
- All possible sources of extraneous odour should be excluded, e.g. samples should be prepared in a separate room to eliminate cooking odours.
- Beware also of odours from detergents, floor polishes and other cleaning compounds.

viii. Preparation of samples:

» The cooking method selected should not add any extraneous flavour to the food, e.g. frying is a poor method, but all sample flavour should be retained where possible, e.g. for fish, steaming "casserole) is recommended.

» All samples should be presented at the same temperature.

» In 'difference testing', the difficulty of presenting a 'standard reference sample' is acknowledged since: (a) any standard will deteriorate with time and (b) any standard will be quickly used.

» Many workers use a standard control taken from a frozen stock of acceptable material.

ix. Bias:

» In order to neutralise errors associated with the order of presentation of samples, the order in which items are presented should be balanced between the possible alternatives, e.g. in a triangular test:

» AAB ABA BAA ABB BAB BBA

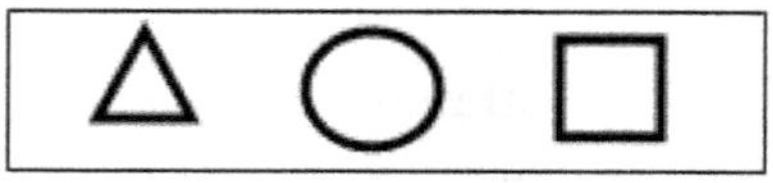

Or, **Red, Blue, Green** } is better than labeling with { A B C / Or, 1 2 3

» Samples should also be labelled in such a way as to eliminate any subconscious bias in the taster.

x. Fatigue:

» The maximum number of samples which can be reasonably assessed in one session will vary depending on the nature of the product.

» In general, fewer strongly tasting foods may be tested in one session than more bland products.

» In order to avoid transfer ('carry-over') of flavour from one sample to the next, a warm water rinse between samples is recommended. (Some authors also recommend the use of white bread, unsalted biscuits, apple, lemon juice etc.).

» Some commonly used test designs

xi. Methods of product rating:

- The sensory panel should consist of 4 to 6 individuals, who have undergone a period of intensive training both with the product under test and with the procedures being used.

xii. Ranking:

- Ranking tests require that judges arrange a series of two or more samples in an ascending or descending order of intensity of a specific characteristic. Samples may be ranked in order of degree of acceptability, or in order of general quality, or by specific attributes of colour, texture or flavour intensity.

xiii. Grading:

- The panel is asked to grade samples according to some agreed numerical or descriptive hedonic scale.

Numerical scoring systems:

- Freshness, or degree of spoilage, may be assessed in a raw fish from:

    i. the general appearance of the fish including that of the eyes, surface slime, and texture of the flesh and

    ii. the odour of the gills and belly cavity.

- In addition, marks may also be awarded for cooked fish odour, texture and flavour.
- Various numerical scoring systems have been devised to cover some or all of the above sensory parameters.
- The most intensive investigation of this type of sensory evaluation has been that of Shewan and co-workers, who have devised detailed descriptive schemes in which numerical scores are given to:

    i. Raw fish - appearance, odour and texture;

    ii. Cooked fish - flavour, odour and texture.

- Such schemes have been devised for cod, haddock, whiting and redfish, and similar schemes have been developed for various other species.

» Generally speaking, the full Shewan's scheme is probably too complex for regular commercial use although abbreviated versions may prove useful for routine quality grading.

xiv. Hedonic scales:

» For example,

1. Excellent, Good, Average, Poor, Bad
2. Like definitely, Like slightly, Neither like nor dislike, Dislike slightly, Dislike definitely

» For a scoring or grading system to be effective certain prerequisites must be met:

1. With numerical scoring systems, quality factors must be properly 'weighted' to reflect their relative importance.
2. The scale should be such that a difference in score reflects a reproducible variation in the quality factor being scored, i.e. scale not too large with too many elements.
3. Agreement is necessary between judges as to quality relating to specific scores given.
4. Whenever possible, physical and chemical analysis of the commodity should be carried out to supplement sensory evaluation.

xv. Methods of difference testing:

» The panel should consist of 10 to 20 individuals with, ideally, 3 or 4 replications per judge per 'difference test'.

» The panel do not require any intensive training as long as members are reasonably motivated and have passed the sensitivity requirements.

» The three most commonly used test designs are the pair comparison, duo-trio test and the triangular test.

» ***Pair comparison*** involves simultaneous presentation of one coded sample each of material A and B with the question: -

'Which is the..................er of the two?' or 'Which is the regular sample?'

» In the ***Duo-trio test***, three samples, two of A with one of B, or one of A with two of B, are presented. One of the duplicates is coded, say, S and the other samples are coded,say, 1 and 2 and the question asked is 'either 1 or 2 is identical with S; which is it?'

» In the ***Triangular test***, three coded samples are presented, of which two are identical and one is different but, this time, the question is 'which one is the odd sample?'

» Statistical calculations can be carried out to establish whether the number of 'correct' answers obtained is sufficient to demonstrate a significant difference in the flavour of the 'odd' sample.

Conclusions:

» The introduction of sensory panel evaluation of fish quality requires research, good organisation and proper training.

» An established visual and organoleptic scale is probably one of the easiest and least expensive ways of evaluating fish spoilage. Its main disadvantage is that it is often open to discussion and disagreement.

» Each country should develop its own standards under its own conditions. Consumer acceptability must always be the criterion on which to base methods. If fish products are for export, retraining of taste panel personnel and review of quality methods may be necessary to meet new market requirements.

L. Storage Life of All Frozen Fish Products Depends on

a. Quality and freshness of raw material

b. Fish handling

c. Processing and freezing techniques

d. Kind and quality of packaging

e. Storage temperature and humidity

» **Pre-cooling temperature:** +5^{0}C

» **Effect of washing the raw material using chilled potable water:**

a. Cleaning dirty matter/ slime etc.

b. Reduction of bacterial count

c. Pre-cooling of fish

» Potable water of 200ppm chlorination should be used for washing and water for final washing should not contain less than 10 ppm.

» Protective treatment:

Protective treatment during freezing is given to –

a) ***Prevent drip loss*** – Phosphate treatment (12% Sodium tri-polyphosphate) + 8.6% Sodium di-hydrogen phosphate) or, 2% Citric acid treatment

b) ***Oxidation of fat*** –

i. *Anti-oxidant* - 0.1% to 0.5% Ascorbic acid/ 0.02% BHA or, BHT or Propyl gallate/ Tocopherol with no limit)

ii. *Synergist* – Ascorbic acid

iii. *Chelating agent* – Citric acid

c) ***Prevent belly bursting for small fish*** – Cold blanching with 15% NaCl for 30min.followed by freezing.

d) ***Improve flavour and appearance*** – Brining / Firming (15% brine solution for 5min) + Sodium benzoate (0.15 to 0.20%) or, Benzoic acid/ acetic acid (0.15%)

» **Antioxidant:** It prevents oxidation of fat.

a. Ascorbic acid – 0.1 to 0.5%

b. BHA/ BHT/Propyl gallate – 0.02%

c. Tocopherol (Natural antioxidant) – No limit

Method of application:

a. Dipping in solution before freezing

b. Use of antioxidant solution in glaze solution

Mechanism:

a. Hydrogen donation by antioxidant

b. Electron donation by antioxidant

c. Addition of lipid to aromatic ring of antioxidant

d. Formation of complex between lipid and aromatic ring of antioxidant

» **Synergist:** Cooperative action of discrete agencies such that total effect is greater than the sum of the effects of each used alone. Example: Ascorbic acid

» **Chelating agent:** Form a complex between metal ion and chelating agent and thus reduce the action of metal ion on food. Example: Citric acid

» **Phosphate treatment:**

a. To prevent drip loss

b. To improve flavour

c. To preserve colour

d. To prevent microbial growth

e. To provide antioxidant effect

1. Example: 12% Sodium tri-polyphosphate + 8.6% Sodium di-hydrogen phosphate or, 2% citric acid

2. Method of application: Dip treatment with phosphate solution for 2 min.

» **Glazing:**

a. To provide a continuous film or coating that will adhere to theproduct and retard both moisture loss and rate of oxidation.

b. It is done by dipping frozen block to the glaze solution or by spraying glaze solution over the frozen block.

c. Antioxidants/ antibiotics may also be incorporated tothe glaze solution

» **Packaging:**

a. To prevent from desiccation / dehydration
b. To prevent from weight loss
c. To prevent fat oxidation

1. Types of packaging:
   a. Product wrap (Plastic – Flexible/ Semi-rigid/ Rigid)
   b. Inner case / Carton: Polyethylene sheet + waxed carton
   c. Outer case / Master carton: Corrugated board
2. Quality of packaging materials:
   a. High water vapour impermeability
   b. High resistance to low temperature
   c. Water repellent properties
   d. Satisfactory gas barrier
   e. Satisfactory aroma tightness
   f. Light protection
   g. Odourless and no chemical action with the product
   h. High resistance to tearing and breaking
   i. Reasonable price
3. Lamination:
   a. Plastic film/ paper/ metallic foil
   b. Two or more substrates cemented together by heat/ adhesive
4. Polyethylene (PE) is a useful package because:
   a. Low cost

b. Excellent clarity

c. Resistance to tearing

d. Easy sealable

e. Relatively low water vapour transmission rate

M. Cold Storage

» Desirable cold storage temperature with RH 90 ~ 95% for

Lean fish = -20^{0}C

Fatty fish = -30^{0}C

» Advantages of -30^{0}C over -20^{0}C temperature:

a. Dehydration – Low temperature and high RH will prevent dehydration

b. Scope of handling, transportation and temperature rise – -30^{0}C to -20^{0}C temperature is acceptable but -20^{0}C to -10^{0}C is not acceptable.

c. Rate of deterioration: Much slow at -30^{0}C

d. Operating cost – Under specific condition, total costof operation at -30^{0}C is only 4% higher than while operating at -20^{0}C.

» Classification of cold storage:

a. Based on temperature:

i. Super A : ≤ -20^{0}C - Fishery products

ii. A : -20^{0}C to -10^{0}C - Butter, Ice-cream

iii. B : -10^{0}C to 0^{0}C - Dry fish, cured fish, smoked fish

iv. C : ≥ 0^{0}C - Fresh fish

b. Based on location:

i. Storage on location

ii. Storage in transit

c. Based on construction:

   i. Jacketed cold store

   ii. Gridded cold store

   iii. Finned cold store

   iv. Unit cooled cold store

» Gross Registered Space (GRS): Area from wall to wall and ceiling to floor.

» Gross Usable Space (GUS): Portion of GRS actually available for use.

GUS = 60% of GRS

» Net Occupied Space (NOS): Portion of GUS which is actually used.

» Packing density: It is expressed in terms of weight in kg/m^3 of area.

» Storage capacity of Cold store depends on:

i. Packing density

ii. Kind of product

iii. Method of storage

iv. Kind of container

v. Size of container

vi. Net wt (kg)/container

» Space utilisation factor (SUF) of cold store: It is the ratio of useful area to the whole building area. This depends on size of cold storage.

SUF = GUS/ GRS

Cold store size	**Space Utilisation Factor (SUF)**
Less than 500 t	0.7
500 – 1000 t	0.75 to 0.8
1000 – 3000 t	0.8to0.85
More than 3000 t	0.85 to 0.9

- Cold storage is affected on room size: If small room, it requires more space for refrigeration. Because:

  i. Greater % of total change of air while opening and closing the door (Air change load).

  ii. Greater average % of wall surface surrounded by high temperature.

- Most important design factors are:

  i. Low temperature

  ii. Uniform temperature

  iii. Steady temperature

  iv. Good air distribution

  v. Minimum rate of air circulation

  vi. Minimum heat ingress

- The design of cold store affects the product storage life through its ability or inability to maintain a low, uniform and steady temperature with a minimum of air circulation.

- The advantages and disadvantages of some different cold store designs are virtually important as regards their optimal operating efficiency. They are:

  i. Insulation: The walls must be adequately insulated and vapour sealed to prevent the passage of heat and frost heave.

  ii. Prevention of heat ingress: Good air lock design is essential.

  iii. The proposed method of storage and handling

  iv. The amount of refrigeration required

i. Insulation:

- A good insulating material is one which, when placed between two bodies of different temperatures, will slow the rate of heat exchange.

- A measure for heat conductivity (or efficiency of insulation) is called the 'k' value. This is a measure of the amount of heat in kilocalories passing every hour through

1 m^2 of material 1 m thick when there is a temperature difference between the two surfaces of the material of 1^{0}C. The more efficient the material, the lower the k-value.

Material	k-value
Wood	0.14 – 0.16
Brick	0.70 – 0.80
Cork	0.40
Glass fibre	0.03
Polyurethane	0.02 – 0.03
Expanded polystyrene	0.02 – 0.03
Air	0.02

» The use of good thermal insulation is a pre-requisite in cold store design and the type of insulation and its deployment are vital to efficient operation.

» The choice of insulator not only depends on its low thermal conductivity but also on the followings:

- Water and water vapour resistance
- Ability to withstand rot and vermin
- Strength for the walls and floors.
- Toxicity
- Flammability
- Price.

1. Cork:

» Expanded cork was one of the original insulators used.

» It has good thermal conductivity characteristics but can allow the passage of water vapour.

» It is strong and does not burn well but is expensive due to scarcity of the trees from which it is obtained.

2. **Glass fibre:**

   » Glass fibre mats have better insulator properties than cork.

   » Fireproof

   » But, it settles and has no structural strength.

3. **Expanded and extruded polystyrene:**

   » These are widely used insulators and are produced in panel form for structural strength.

   » The compressibility is poor.

   » They are destroyed at fairly low temperature viz. 75^0C.

4. **Polyurethane foam:**

   » Thermal conductivity is low initially but gradually increases with time.

   » It can be foamed into a cavity or on to a surface.

   » The vapour resistance is good.

   » It burns giving off toxic substances.

5. **Granular materials**:

   » Saw dust, crumb polystyrene, granular cork, vermiculite etc are often cheap and readily available.

   » These tend to settle and allow water vapour to pass through.

b. Vapour barriers:

» Air at ambient temperature is able to hold more moisture than cold air i.e. the partial pressure of the water vapour outside a cold store is generally higher than that inside. This difference in partial pressure creates a gradient down which the water vapour will pass from outside to inside through the insulation of the cold store walls.

» As the water vapour cools, it condenses, and at 0^0C, solidifies into ice. This will continue until the ice has built up so much that it affects the insulation properties of the wall and also weakens the structure.

» To prevent this from happening, a vapour barrier outside the insulator wall is essential. This must be a gas seal to prevent the ingress of any water vapour anywhere on the exterior surface of the building.

» Vapour barriers can be provided by coating the walls with bitumen or cladding them with sheet metal sealed at the edges.

c. Frost heave:

» This occurs as a result of inadequate floor insulation and is caused by a build up of ice under the floor. As the quantity of ice builds up, it can cause the floor to lift and crack, and may also affect the foundations and walls. There are two methods of preventing frost heave:

» Sub-floor heating under the insulation, usually by hot glycol circulated through pipes.

» Sub-floor ventilation. The building is elevated on to support piles. This can be advantageous for loading of lorries etc.

ii. Air ingress:

Air from the outside brings with it both warmth and moisture. This moisture quickly builds up on the cooler and a small rise in temperature will enhance water loss from the stored products. There are a number of ways of preventing air ingress:

» Air locks: These are unpopular since they are regarded as a hindrance.

» Air curtain with door: This type is often abused and the door is often left open for too long.

» Plastic curtains: These can be suspended from an inner and outer opening and form a very workable air lock.

» Hatches: These are small openings usually combined with mobile conveyors and mounted on the wall to prevent loss of cold air.

iii. Product handling and storage:

» During the design stages, consideration must be given to the proposed methods of handling and storage of the product.

» For example, if the cold store has been designed with sub-floor ventilation to prevent frost heave, loading bays can be constructed outside the store which will accommodate different lorry designs by lifting up or down hydraulically. These bays must have sufficient manoeuvring space for sorting out pallets.

» In tropical countries, it is advisable to have a refrigerated loading and sorting area. This acts as a buffer zone to the warm moist air and should be enclosed and refrigerated to about 5 to 10°C. The approximate area should be 25% of the store area.

» The layout of the store will be determined by the type of product stored as well as the stowage methods. Passage ways should be clearly defined and loading bays should be charted by the operators.

iv. Refrigeration requirements:

» The calculation of exact requirements for refrigeration is very complex and this is dependent on a great number of indefinable variables.

Example:

Store specification:
Capacity - 1000m^3, Insulation thickness – 0.25 m of cork, Surface area – 771.5 m^2, Maximum ambient temperature: +35°C, Store temperature: -30°C

Refrigeration load calculation:

1	Insulation heat leak through walls, ceiling and floor:	
	$$\text{Heat leak} = \frac{\text{(Surface area x Temperature drop)}}{\text{Thickness of insulation}} \times \text{K (Thermal conductivity)}$$ $$= \frac{(771.5 \times 65)}{0.25} \times 0.037 = 7.476 \text{ kcal/h}$$	7.476 kcal/h
2	Air changes:	
	Average of 2.7 air changes in 24 hours Heat gain (35°C and 60% RH air = 40 kcal/m^3 Store volume = 1000 m^3 $$\text{Therefore, air change heat gain } \frac{1000 \times 2.7 \times 40}{24} = 4500 \text{ kcal/h}$$	4500 kcal/h

3	Product load:	
	5.5 kcal/kg for fish load at average temperature of -20⁰C Fish loaded = 35000 kg/day Product load = $\frac{3500 \times 5.5}{24}$ = 8020 kcal/h	8020 kcal/h
4	Men working:	
	1man working at -30⁰C gives off 378kcal/h Therefore, 2 men working at -30⁰C gives off = 2x378 kcal/h = 756 kcal/h	756 kcal/h
5	Lights (left on during working day) 1000 watts	860 kcal/h
6	Fan load 3 x 1/3 hp	923 kcal/h
7	Defrost heat 1 defrost of 8400 watts for 1 h (Recovered over 6 hours) = 1209 kcal/h	1209 kcal/h
	Total calculated refrigeration load=	23744 kcal/h
	Allowances 23744 x 4/3 = 31658 kcal/h	31658 kcal/h

» **Before ordering for a cold store, certain specifications need to be considered:**

i. Required storage capacity

ii. Operating temperature

iii. Ambient temperature in that area

iv. Maximum heat of product when loaded

v. Rate of stowage per day

vi. Plan of proposed site

vii. Insulation proposed (thickness)

viii. Method of cooling proposed

ix. Type and size of door

x. Requirement of air curtain

xi. Stowage methods proposed

xii. Expected gross heat load

xiii. Installation expertise required

xiv. Condition of electric supply

xv. Ratings of equipments

» **Calculation of heat load/cooling load:**

a. Wall heat gain load

b. Air change load

c. Product load

d. Miscellaneous load (which includes electric bulb, people, fan/ motor)

» **Various factors for proper warehousing:**

a. Temperature: -20^0C or, -30^0C and centre temperature should be -18^0C. It is desirable to have ante room (temperature of $+5^0$C)

b. Humidity: High RH => leads to mould growth. Less RH => leads to freezer burn (dehydration)

c. Storing and handling methods: Stacking should not be done to walls. It is desired to store in cartons.

d. Odour control

» **Maintaining clean air in cold store:**

a. Air filter

b. Electrostatic cleaning

c. Activated charcoal absorption

d. Ozone generator and

e. UV lamp

» **Disinfection of cold store:**

a. Rodent control –

i. by gasing (CO_2)

ii. by chemicals (Barium carbonate, Zinc phosphide)

b. Bacterial and mould control –

i. White washing and painting

ii. Treatment with 2% sodium phenyl phenate

iii. Treatment of Antiseptol

» **Disinfection of air of cold room:**

i. Fine spray of 3% Lactic acid

ii. Fine spray of Sodium hypochlorite with 0.6 to 0.8% active chlorine

iii. Fine spray of 2% Sodium phenyl phenate solution

iv. Irradiation lamp

v. Criteria for cold store design:

» The efficient use of a cold store or chill store not only depends on the choice of materials used for construction but also on the actual functional design and operation of the building.

» The following points should be borne in mind:

a. The drainage for chill stores is most important. With wet floors, there is a serious risk of slipping and a suitable high-friction surface should be used.

b. When deciding on size of the store, allowance should be made for the passage of people, trolleys etc.

c. In order to maintain a steady temperature, air locks should be used with air curtains if possible.

d. Location of the store should be such as to minimise heat gain, i.e. inside a building.

e. The floor must be strong as well as insulated: the passage of fork lift trucks can easily disrupt a weak flooring material. Freezing of the ground beneath the store when adequate floor insulation is used could result in frostheave where the subsoil freezes, expands and disrupts the foundations.

f. The amount of refrigeration required depends on a number of factors which must be taken into account at the design stage:

- Ambient temperature and insulation type
- Quantity and temperature of products entering the store
- Number of doors an door openings in a normal working day
- Size of store

» Once the store is operational, maximum efficiency can be achieved by following these rules:

a. Products entering the store should be either pre-frozen to the cold storage temperature or pre-chilled.

b. Products in boxes or pallets should be clearly marked and dated , and strict product rotation practice should be observed.

c. Free flow of air around the products is essential to maximise the refrigeration efficiency.

d. Monitoring of store temperatures should be routine and full records should be kept.

e. Frozen products must be protected by glazing or packaging.

f. If two doors are in operation, only one should be used at a time.

g. There must always be ice surrounding the fish in a chill store.

vi. Design and construction of fish working premises:

1. Site:

i. The cold storage plant should be erected as close as possible to the source of supply of raw material to avoid delay in handling. A site where the fishing vessels can berth is best, provided that other requirements can be met. These take the following factors into account:

ii. Some industries producing offensive odours should be sited away from the cold store plant and also the town area.

iii. Many operations produce quantities ofliquid effluent containing protein residues; the need to dispose of these without causing pollution problems should be borne in mind.

iv. Areas where other industries produce smoke, dust or odours which may taint fish should be avoided.

v. The cost of erecting buildings depends on the nature ofthe ground; areas whereexpensive piling may be needed should generally be avoided. Such land may be cheap, however, and the total cost of land purchase and building on alternative sites should be considered.

vi. If the product is to be taken away byland, all-weather roads are needed. Shipping by rail can be considered. If perishable high value products such as live lobsters are to be shipped, proximity to an airport is an obvious advantage.

vii. Many operations need large quantities of potable water. This may be taken from the local authority supply or from a private bore hole or dam. Clean sanitary sea water can be used for some operations e.g. for washing down.

viii. Premises should not be sited near refuge dumps, these always harbour vermin such as flies and rats.

ix. Some operations have a high power demand. It is always cheapest to use the local authority power supply if this is reliableand can be broughton to the site at a reasonable cost; bringing power on to a site can be expensive. A diesel generator is noisy and makes for unpleasant working conditions.

x. Good telephonic communication is important.

xi. The neare tothe market theplant can be sited thebetter.

xii. Many operations require large quantities of ice. If the plant can be sited near to a reliable source of cheap ice this is an advantage. If not, ice making machinery will be needed. The possibility of generating a surplus for sale to offset costs can be considered.

xiii. Supplies of packaging materials will be needed;if the suppliers are far away, costs will be high.

2. Buildings:

i. Buildings must be big enough to avoid crowding of people and equipment.

ii. All buildings should be at least 3 m high internally. They must be well constructed and kept in good repair.

iii. All areas where the product will be handled must be kept separate from any part such as peeling area, pre-processing area, living quarters or offices or engineering facilities.

iv. Construction must prevent entry of insects, birds and vermin such as rats and mice.

v. Single storey buildings are preferable to multi-storey buildings; this avoids having a load bearing second floor, and makes the movement of raw material and products easier. It is also difficult to arrange adequate drainage above ground floor level.

vi. Non-porous material should be used for all constructions; wood is not suitable for walls or floors.

3. Floors:

i. All floors should be hardwearing, non-porous material.

ii. They should be washable and should slope evenly to the drains. 5 cm thick clay tiles are preferred. Granolithic cement is also suitable; ordinary cement soon wears, is porous and subject to attack by acids and fish oils.

iii. Whatever surface is used in the fish handling areas, it must be such that it can be scrubbed and hosed down. The floor material should be carried up the walls for 15 cm and the joints should be rounded (coved) where this is practicable.

iv. Using trolleys with rubber tyre wheels prevents the damage which occurs when heavy boxes or tubs are dragged across the floor.

v. The floors should fall to the drains. A 1in 100 slope is suitable. Slopes of more than 1 in 40 are dangerous.

vi. If any ribbing is used, the ribs should run towards the drains, not be parallel with them. All joints must be smooth and close fitting.

4. **Drains:**

    i. Floor drains should be wide and deep and have an even fall so that water cannot remain in them.

    ii. They should be covered with removable metal gratings which should be level and flush with the floor.

    iii. Outlets should have a catch basin outside the process area made of water proof concrete.

    iv. Drainage pipe should have a minimum internal diameter of 10 cm.

    v. When a drain goes through a wall, a vermin proof seal is essential.

    vi. All drains should be designed so that cleaning with a rod is easy.

5. **Walls**:

    i. Walls in the process area must be water proof, smooth surfaces and washable.

    ii. Ceramic tiles are best but these are expensive.

    iii. Corrosion-resistant metal such as stainless steel or aluminium is also good.

    iv. Such protection should be carried up the wall above shoulder height. Where concrete is used, it should be painted with high gloss, light coloured, non-toxic paint.

6. **Ceilings:**

    i. These should be in a continuous, smooth, unbroken surface, painted with white, high gloss, non-toxic paint.

    ii. Flush light fittings and ventilation fittings are desirable.

7. **Doors:**

    i. These should be wide and self-closing.

    ii. They must be smooth surfaced for easy cleaning.

    iii. Metal plates should be fitted flush with the floor surfaces to prevent rat damage and entry.

    iv. A metal or washable well painted light coloured surface is needed.

v. Wherever fly screening is needed, either a metal mesh screen or air curtain should be fitted.

8. Windows:

i. Natural light is both better and cheaper than artificial light, so large simple windows should be fitted.

ii. Glass is not essential and is a potential hazard to the product. If glass is used, it should be in a few large panes, not a large number of small ones.

iii. All windows should be at least 1 m above the floor.

iv. Windowsills, if any, should be of concrete and sloped downwards at an angle of 45 degrees.

v. Window frames should be of aluminium or well painted galvanised steel; where wooden frames must be used, these must be well painted.

vi. If glass windows which open for ventilation are used, fly screens which can be removed for cleaning should be fitted. Fly screen mesh can often be used instead of glass windows.

9. Ventilation:

i. Air conditioning is not really needed and is expensive to install and operate in commercial premises but good ventilation is essential.

ii. Air conditioning is not a substitute for using plenty of ice.

iii. Exhaust fans are very useful; these must be fitted so as to exclude insects when not running-a correctly weighted flap is adequate.

iv. Ventilation ducts should be flush fitted and all ventilation openings screened with fly mesh on frames which can be removed for cleaning (except where fans are fitted).

10. Lighting:

i. Fluorescent lights provide the best alternative to natural day light.

ii. In the general working area transparent roof glass should be provided one in 4 square meter area to provide light.

11. Toilet facilities:

i. Rooms should have walls of smooth, washable light coloured surfaces.

ii. Floors should be non-porous, washable and easily cleaned.

iii. Rooms should be well lit and must be kept clean at all times.

iv. They must not open directly on to the processing area.

v. Flush toilets should be provided at the following ratios.

vi. One additional toilet for every additional 30 employees.

vii. A hand washing facility is best provided in the factory entrance so that it is used every time any one enters. It should have hot running water and soap.

viii. Single use towels or paper towels should be provided.

ix. Where paper towels are used, a bin must be provided and the contents burned at intervals well away from the process area.

Number of employees	Number of toilets
1 to 9	1
10 to 24	2
25 to 49	3
50 to 100	5

12. Water supply:

i. This is one of the most critical factors in making products which will be safe to eat and will meet the required microbiological standards.

ii. Water used for lavatories may contaminate the product with faecal bacteria.

iii. Dock water is always filthy, so cannot be used in the processing plant.

iv. Local authority water supplies may need both filtration and chlorination to suitable levels for use in the plant.

v. The water supply should be checked for microbial contamination at frequent intervals.

vi. Both hot and cold water supplies should be provided for use in process lines and in wash basins.

vii. A high pressure system should be available for washing down the premises and equipments at a pressure of 400psi (28 kg/cm^2).

viii. All water used on the process line or for cleaning down should be chlorinated. It will reduce both the number of micro-organisms and prevent or reduce odours.

ix. A maximum of 5 ppm of free chlorine should be present in water which comes into contact with the product.

x. ppm should be used for cleaning down the premises where a two to three stage cleaning process is employed.

xi. Chlorination maybe effected by: (a) Injection of chlorine gas, (b) Injection of sodium hypochlorite liquid, or (c) Adding calcium hypochlorite as a powder.

13. Equipment:

i. All equipment must have smooth, easily cleaned surfaces.

ii. Tables should be of metal, preferably stainless steel.

iii. Toxic metals viz. copper must not be used.

iv. Plastic cutting boards should be used.

v. Plastic, plastic reinforced with glass fibre (GRP), aluminium, and stainless steel are all suitable materials for containers such as fish boxes and tubs; freezer trays should be of metal.

vi. Sinks and other facilities for cleaning equipment should be in a separate room from the process area.

vii. Equipment should be washed with detergent in preferably hot water, scrubbed with a plastic handled nylon brush and then rinsed in water chlorinated to 50 ppm.

viii. Containers should be left upside down to dry.

14. Clothing:

i. Protective clothing such as caps, gowns or coats, gloves and boots should be provided.

ii. Cloth items should be washed, preferably in boiling water.

iii. Plastic and rubber items should be washed inside and out and then soaked in water chlorinated to 50 ppm and then drained.

iv. They should be inspected before reissue.

v. The laundry should be outside the main process area.

vii. Monitoring of Storage life of fish products:

» Storage life is the effective storage time up to which product is acceptable. It depends on:

a. Time-temperature tolerance (T-T-T factor) (Monitored by quality changes)

b. Product (Raw material) quality

c. Product processing characteristics } PPP factor

d. Packaging

» Storage life is of two types:

a. Practical storage life (PSL) –

i. Period up to which product is marketable

ii. Determined by consumer acceptance study

b. High quality storage life (HQSL) –

i. Decided in the laboratory by scientists

ii. Product is compared with control

» HQSL is shorter than PSL.

» Quality changes are cumulative and irreversible.

» Monitoring of storage life is done by two methods i.e. (1) Numerical method, and (2) Graphical method.

» Temperature history helps to find out the quality loss in storage life.

» **Numerical method:**

$$\Delta C = Z_1 \frac{\Delta C}{\zeta_1} + Z_2 \frac{\Delta C}{\zeta_2} + \ldots\ldots\ldots + Z_n \frac{\Delta C}{\zeta_n}$$

Where,

ΔC = Maximum allowable quality loss according to sensory scene
ζ = Maximum storage life at a given temperature according to TTT curve
Δc = Daily quality loss at a given temperature
Z = Effective storage life at a given temperature

» **Graphical method:** It is for summing up quality changes during irregular temperature history.

N. Thawing of Frozen Fish

» Melting of frozen products is called thawing (melting or defrosting).It is done to make the frozen fish warm and return it to the fresh condition. It is the reverse change of freezing.

» The purpose of freezing fish is to reproduce fresh fish as near as before freezing. Therefore, thawing is as important as freezing.

» During thawing, fish is subjected to all possible sources of damage like physical, chemical and biological. Faster thawing reduces the opportunity of damage to the fish.

» During thawing, the temperature of the fish rises very rapidly to the melting point and remains near it for sometime just like during freezing.

» Near the freezing point, the undesirable reactions occur at maximum rates. These reactions include recrystallisation, microbial growth and other physical and chemical changes.

» Therefore, thawing should be done at a very fast rate so that the temperature of the fish remains near its freezing point for a very short time.

Methods of thawing

» There are several methods of thawing of frozen fish. Based on the material with which fish comes in contact during its thawing, these methods can be broadly classified into the following categories.

1. Air thawing
2. Water thawing
3. Vacuum thawing
4. Electrical thawing
    a. Electrical resistance thawing
    b. Dielectric thawing
    c. Microwave thawing
5. Hybrid thawing

» Out of the several methods available for thawing of frozen fish, only a few are used commercially. These are described below.

1. Air thawing:

» In this method, the frozen fish is kept overnight in air at room temperature. To reduce thawing time, air blast thawing is done in which 95% to 100% relative humidity is maintained to prevent dehydration of the fish by the blasting air.

Advantages:

a. Loss of flavour and soluble nutrients is very less, unlike in water thawing.

b. Cost of thawing is less as compared to other methods

c. Less use of water, unlike in water thawing

d. Less contamination of fish, unlike in water thawing

Disadvantage:

a. Thawing time is long.

2. Water thawing:

» Here, the frozen fish is kept immersed in running water at room temperature.

Advantages:

a. Thawing time less as compared to air thawing.

Disadvantage:

a. Flavour and soluble nutrients, like water soluble proteins, are lost considerably as several soluble constituents present in the fish get washed away in the water in which fish is thawed. . s of is very less, unlike in water thawing. Loss of flavour can be minimised by addition of common salt (NaCl) to the thawing water before its use for thawing.

b. Huge quantity of water is required as fish is thawed in running water.

c. If the thawing water is contaminated, it contaminates the fish which subsequently reduce the quality of the fish after thawing. Such contamination may also lead to food poisoning.

3. Vacuum thawing:

» In this method, fish is thawed in a vacuum chamber in which steam is allowed to condense on the frozen fish releasing the latent heat of vapourisation on it. This huge amount of heat thaws the fish.

4. Electrical thawing:

» The maximum speed of thawing in air or water is limited by the rate of conduction of heat through the fish. At present, the only practical methods are those depending upon electrical heating. The methods of thawing which use electric energy are of the following three categories.

a. Electrical resistance thawing:

» When an electric current flows along a wire or any other material then that material gets warm. It is difficult to produce heat in materials that do not conduct electricity easily, but good conductors, whether in the form of wires or slabs, are easily heated by the passage of electricity.

» If the electrical properties of the material are uniform, then the flow of electricity will be uniform and the heat will be produced uniformly.

» This principle is used in the method of thawing of fish recently developed by the Torry Research Station in UK. This method is called electrical resistance thawing.

- In this method, the voltage from the transformer depends on the type of block to be thawed. The part of the circuit containing electrodes is fully isolated from the main supply. The current is allowed to flow for sufficient time to supply enough heat to complete the thawing. The total time varies from 20 minutes to 1 hour according to the type of block. It is 2 to 3 times faster than air or water thawing.

- Because of low initial cost and running cost, it is a very economical method of thawing. But irregular blocks cannot be used because good electrical contact and uniform flow of current cannot be obtained.

b. Dielectric thawing:

- In this method, blocks of frozen fish are carried on a conveyer belt. It possesses one or more pairs of electrodes; one above and one below the conveyor. To each pair of electrode is applied an alternate voltage of high frequency of current of about 500 thousand volts.

- The blocks pass through without touching the electrodes. The fish absorbs the electrical power which is turned into heat. The heat is produced almost uniformly throughout each block and in principle there is no limit of thickness of block. The time required for thawing in this case is less than those required in case of electrical resistance thawing, air thawing or water thawing. It ranges from about 10 minutes per block of whole fish to about 40 to 50 minutes depending on the size of the blocks. One advantage of this method of thawing is that the fish can be removed from a thawer at a uniform temperature; if necessary slightly below the point at which fish is completely thawed.

c. Microwave thawing:

- Here, the frozen fish is thawed by means of radiation. This form of radiation is known as microwave. It is produced at a very high frequency of about 3000 million cycles/sec. It is possible to force heat into fish by means of multiple beams of microwaves at a rate of 10 times that possible by dielectric heating. Thin pieces of fish have been thawed experimentally in a mater of seconds. The main drawback, apart from high cost, is that the strength of the beam is reduced as it goes to the deep of the fish flesh so that the outer layers are warmer than the inner ones. This method does not apper to have any advantage over those already available.

5. Hybrid thawing:

» In this method, a combination of the above methods is used, *e.g.*, a combination of water and dielectric thawing or a combination of electrical resistance and air/water thawing.

9 Canning of Fish

» Canning is a method of preservation in which the product remains under sterile condition and can be preserved for at least a year at room temperature.

» Father of canning is Nicholas Appert (1809).

I. Canning

» It is a method of preservation in which selected foods are prepared as per schedule, packed in suitable containers, capable of being sealed air tight and processed by heat so as to destroy spoilage organisms and cooled promptly to avoid overcooking.

» Canning of fish is a method of preservation where, it does not maintain the original state of fish but, the canned fish is regarded as a different and desirable product.

Principle:

» During canning, the contents of the can are made sterile by heat processing which alters the characters and flavour of the material and both enzymes and bacteria including pathogens and spoilage organisms are permanently inactivated preventing re-infection of the product, thus preserving for a very long time.

Advantages of canning:

a. Simple method

b. Applicable to all kinds of foods

c. Adaptable for any scale of operation

d. Most of operations by hand or machine

e. Ready-to-eat product

f. Safe and nutritious

g. Long shelf life – At least 1 year for acid food and 2 years for Non-acid foods

Disadvantage of canning:

a. Can cost is more

II. Cans Used for Canning the Fish

a. Tin can:

» Generally tin cans are used for canning. Tin cans are steel sheets coated with thin layer of tin on both sides by electrolytic plating.

» Canned sea foods are sulphur staining foods and they produce sulphur ions during processing. They react with tin coating to form tin sulphide or if steel base is exposed, black iron sulphide are formed. Though this is not detrimental to health but only bad appearance makes the product not acceptable. Hence, sulphur resistant lacquers such as vinyls (PVC), phenolics, oleoresinous lacquer and epoxide lacquers are provided on both the sides over the tin plate.

» **External lacquering:**

◊ It is provided to protect the can from rusting.

- Vinyl lacquer
- Phenolic lacquer
- Oleoresin lacquer
- Epoxide lacquer

◊ Desirable qualities:

- Non-toxic
- Non-peeling
- Heat resistant

- Not imparting any colour or flavour to food
- Adhere strongly to can
- Should not react dissolve to food

b. Aluminium can:

Advantages:

- Absence of sulphur blackening
- No metal dissolving to food
- Easy opening of can
- Light weight

c. Glass can:

Advantages:

- Pure
- Easy to clean
- Corrosion free
- Leak proof
- Transparent

Disadvantage: Breakage

Table 9.1: Can type and size

Can type	Diameter	Height	Remark
A-1	301	409	301 = 3 1/16 inch
A-2	307	409	409 = 4 9/16 inch
A- 2 ½	401	411	411 = 4 11/16 inch
A-10	603	700	700 = 7 inch
8-oz	301	206	206 = 2 6/16 inch

III. General Considerations in Establishing a Cannery

» Availability of raw material

» Site for cannery, type of building and division of space

» Duration of canning season

» Transport facilities for raw material and finished product

» Labour availability

» Water supply and disposal of cannery waste

Personal hygiene:

» Proper sanitary facilities

» Adequate headwear

» Clean uniform

Methods of safeguarding food quality:

» Destruction of pathogens and spoilage organisms

» Inhibition of spoilage organisms

Prevention of entry of thermophiles by:

» Heating the equipments either above or below the temperature range at which multiplication of organism takes place.

» Frequent cleaning in potable water

IV. Different Processing Stages of Canning of Fish

a. Raw material

b. Preparation of fish

c. Precooking

d. Packing

e. Filling

f. Exhausting

g. Sealing – closing the can by double seaming

h. Heat processing – Retorting

i. Cooling

j. Labelling and Boxing

A. Raw material

» Factors to be considered before purchase of food materials:

» Nutritive value

» Appetite appeal

» Amount of waste

» Cost

» Variety of choice

» Easy availability

» Availability in convenient amount and package

Fish as raw material:

» Fatty fish: Small fish like sardine, mackerel, herring, sprats etc. are gutted and washed at sea.

» Belly bursting is observed in such fishes as gut enzymes digest the belly wall and soften flesh.

» Belly bursting can be prevented by:

◊ Chilling the fish immediately after catch. This can be done by two methods.

- By mixing with ice.
- By immersing the fish in chilled sea water (CSW) (fish : CSW = 4 : 1) where fish can be preserved up to 5-6 days. In this some salt will be absorbed by the fish and flesh becomes little tough, thus preventing belly bursting.

◊ Cold blanching by dip treatment at 15% brine for 30 mins and then freezing

» In case of fishes with soft flesh like pilchards descaling is difficult. Descaling can be only after chilling the fish immediately after catch.

» Careful handling of small fish is very important as broken, damaged or squashed fish are not suitable for canning.

» Big size fishes like seer, tuna, cod, mullet etc can be preserved either in ice for 5-6 days or frozen stored for a long period.

» Fish in poor condition should not be kept in frozen condition as during thawing fish will be spoiled and thus will be unsuitable for canning.

» Frozen herring, mackerel should be stored at -30^{0}C. At this temperature they can be kept in good condition for at least6 months.

» During freezing, glazing must be done to the frozen block to avoid fat oxidation so as to reduce spoilage during canning.

Classification of foods based on pH:

» The nature of the food, primarily the pH of food, is the most significant determinant of how severely the food will be processed in the body. So, based on pH, all foods can be divided into four categories as given in the table.

Table 9.2: Classification of food based on pH

Group	Type of food	pH	Food items
Group I	Low acid	≥ 5.0	Meat, Marine products, milk and certain vegetables
Group II	Medium acid	5.0 - 4.5	Meat and vegetable mix
Group III	Acid	4.5 – 3.7	Tomato, pineapple and other fruits
Group IV	High Acid	≤ 3.7	Grape fruit, citrus pickles

pH 4.5 : Line of demarcation because of *Clostridium botulinum* which can't grow at pH < 4.5

pH 3.7 : Line of demarcation because of *Bacillus coagulans* (*Bacillus thermoacidurans*) which can't grow at pH < 3.7

Table 9.3: pH values of various fish and sea foods

Food	pH		Food	pH
Fresh fish	6.6 – 6.8		Herring	6.1 – 6.4
Crab	7.0		Clams and Oysters	6.3 – 6.5
Tuna fish	5.2 – 6.1		Sturgeon	5.5 – 6.0
Salmon	6.1 – 6.3		Shrimp	6.8 – 7.0
Fresh water fish	6.9 – 7.3		Whitefish	5.5

Microorganisms of significance:

- The microorganism, *Clostridium botulinum* is most important from the point of canning.
- It is an obligate anaerobic mesophilic organism.
- It can produce spore and can grow at a pH >4.5.
- It produces a lethal toxin, the botulin and cause botulism in human body.
- Since, canning creates an anaerobic condition, this organism is taken utmost care.

a) Based on free oxygen requirement:

- Obligate aerobic (Only in presence of oxygen)
- Obligate anaerobic (Only in absence of oxygen)
- Facultative anaerobic (Both in oxygen or not)
- Micro-aerophiles

b) Based on growth temperature:

- Thermophiles
- Mesophiles
- Psychrophiles
- Psychrotroph

Organisms of pH > 4.5 => Low and medium acid foods:

1. Obligate aerobic:
   - » *Bacillus subtilis* and *Bacillus cereus* – They are not significant in canned foods.
2. Obligate anaerobic:
   a. Thermophiles:
   - » *Clostridium thermosaccharolyticum* – Attack carbohydrate and sugar which breaks to CO_2 and Hydrogen gas resulting Hard swell.
   - » *Clostridium nigrificans* – Attacks protein – produces Fe_2S resulting Sulphide stinker.

   b. Mesophiles:
   - » *Clostridium botulinum* type E – Cause Botulism and toxin produced is Botulin
   - » *Clostridium histolyticum* and *Clostridium putreficans* – Not so important for canned foods
3. Facultative anaerobic:
   a. Thermophiles:
   - » *Bacillus stearothermophillus* – Cause Flat sour
   - » *Bacillus coagulans* – Highly heat resistant and acid producing organism

   b. Mesophiles:
   - » *Bacillus circulans* and *Bacillus cereus* – Not so important for canned foods
4. Non-spore former:
   - » *Streptococcus fuschans* and *Achromobacter sps.*
5. Yeast:
   - » *Torula lactis condensi*
6. Mould:
   - » Aspergillus repens

Organisms of pH < 4.5 => Acid and High acid foods:

» These are less heat resistant organisms.

1. Sporeformers:

   a. Thermophiles:

   » *Bacillus coagulans*,

   b. Mesophiles:

   » *Clostridium butyricum* (Tomato cans),

   » *Clostridium pasteurianum* (Carbohydrates and sugar)

2. Non-spore formers:

   Lactobacillus brevis – Spoilage of tomato ketch up

   Leuconostoc pleurofructi – Spoilage of fruit juice

3. Yeast: *Byssochlamys fulva* - Spoilage of fruit juice

4. Mould: *Penicilliumm* and *Aspergillus* - Spoilage of fruit juice, pickle

B. Preparation of fish

» Fish – Descaling – Beheading – Degutting – Washing in potable water – Size cutting – Brining.

» For frozen fish, the frozen block must be thawed and soon it should be used for canning.

» For shrimps, beheading, removing exoskeleton and deveining are necessary. Deveining is taking off the alimentary canal in shrimps.

Table 9.4: Size cutting of fish for canning

S. no.	Type of can	Size cutting	Standard weight
1	No. 1 Tall can	10.8 cm	454 g
2	8 – oz can	5.5 cm	210g
3	Quarter Dingley can	9.5 cm	121g
4	1 lb Jam can	8.5 cm	340g

Brining:

» Fish are then dipped in saturated salt solution with 1% acetic acid.

» Brining can be done by either batch or continuous processing.

» Continuous brining is preferred as it reduces the amount of manual labour, gives greater control of the process, and results in a uniform product.

» Brine used should be recirculated and filtered.

» Brine should be made with pure salt and should not contain any appreciable quantity of magnesium chloride, a common contaminant of unrefined salt.

» The magnesium chloride in salt increases the risk of struvite formation.

» The struvite is magnesium ammonium phosphate which forms crystals confusing with glass pieces generally observed in canned shell fish like crab.

Brining time:

» Larger fish require longer than small fish

» Fatty fish require longer than lean fish.

» Fillets require less time than the whole fish

» Brining time is between 15 to 45 mins for batch brining and less for continuous brining.

Advantages of brining:

» It cleans the fish.

» It gives salty taste by absorbing salt. A salt content of 1 – 2% is normally acceptable.

» It gives little dehydration effect.

» Texture becomes stiff.

» It imparts desired flavour to the fish.

» Brining also brightens the appearance of the fish, removing any remaining slime.

C. Precooking

» Any cooking prior to sterilisation is termed as precooking.

» There are four methods of precooking. They are:

  a. Steam cooking (gives excellent product)

  b. Water boiling (Hot blanching – Heating in 8% salt solution for 8 mins)

  c. Heating in hot air oven

  d. Frying fish in oil (It gives very good taste)

» The cook loss during precooking depends on

  - Nature of fish
  - Time and temperature
  - Freshness
  - Method of cooking
  - Position of fish during cooking

» For 8 oz can, precooking is done at100^{0}C for 30 min.

» For No. 1 Tall can, precooking is done at100^{0}C for 45 min.

» For fatty fish, the precooking loss will be about 20% where as for lean fish it is about 30 – 33%.

Advantage:

» When fish is heated, it removes water, the amount depends on species.

» It improves the texture.

» To coagulate protein so that no loss during subsequent washing.

» To reduce the exudates in final product. If exudates are more in the can, it gives bad appearance for canned fatty fish. When oil mixes with exudates, it causes rancidity. Hence, exudates should be less than 10%.

D. Packing

» The empty cans should be washed and dried before packing.

» The type of the can is governed by the size of the fish and quantity of fish being packed.

» The net weight is around 70% of the standard weight of the product to be packed.

S. no.	Type of can	Standard weight	Net weight (60 ~ 70% of Std Weight)
1	No. 1 Tall can	454 g	318g
2	8 – oz can	210g	147g
3	Quarter Dingley can	121g	85g
4	1 lb Jam can	340g	238g

» During packing, little space is left above the food level in a can for creating vacuum during processing called 'Head space'.

» Gross head space = Countersink + Net head space.

» Counter sink is closed during seaming process.

» If head space is too small, it indicates the container is overfilled.

- The can remained as under-sterilisation
- There is danger of can ends of becoming permanently distorted by expansion of contents during processing and,
- There is swelling up of can ends due to production of hydrogen gas.

» If head space is too large, it indicates the container is under weight.

E. Filling

» Filling media are brine, edible oil or sauce.

» Since fish is a delicate food item and easily can be damaged and fragmented by mechanical handling process, the can must be filled with brine, edible oil or sauce soon after packing the fish in cans and before it is sealed.

» Filling media will prevent the fish pieces to be in direct contact with the can wall. Thus, preventing charring of fish during heat processing.

» Filling media helps air bubbles to come to the surface or head space.

» Uniform heating to the fish portions because of convection process of heating of filling media.

F. Exhausting

» Removal of air and any other gas from the head space is termed as 'Exhausting'.

» Small cans when filled can be closed immediately and heat processed. But, large cans need to be exhausted before being closed in order to produce a partial vacuum in the head space that will remain after the can is heat processed and cooled.

» When can is open, the head space is filled with a mixture of air and water vapour. After getting sealed when heat processed, the water vapour converted to gas and simultaneously to liquid water on cooling. Thus headspace pressure will dropdown and a partial vacuum is created in the head space.

» Vacuum in cans (Pressure difference between atmospheric pressure and can pressure) is necessary:

- To keep can ends flat or concave
- To reduce strains on can during processing
- To reduce the chance of damage to the seam
- To help in identifying the blown cans
- To reduce the oxygen content

» Measurement of vacuum:

- Piercing type vacuum gauge
- Flip vacuum test
- Measuring Newton's rings

» Three principal method of exhausting:

- Heating the food before filling
- Heating the filled can with or without the lid and then closing
- Closing the can in a chamber in which a high vacuum is maintained

» When closing temperature increases, final vacuum of the can also increases.

» After a can is packed, the lid is lightly clinched to the body in such a way as to allow free passage of gas, and the can and the contents are heated by passage through an exhaust box.

» Exhaust boxes vary in design but consist basically of a long, well insulated box, heated by steam to a temperature of about 95^{0}C. The can passes through a conveyer and is subjected to jets of steam which play on the sides so that air in the can is displaced by vapour.

» The whole process takes about 20 to 30 mins.

» **Clinching**: First seaming operation done loosely so that lid holds on to the can body but it is not air tight.

» Advantage of clinching:

- To get better vacuum
- Longer shelf life of the product
- Lid will not be displaced while moving in exhaust box
- No spilling of food
- Food contents will be uniform – Head space will be uniform and can vacuum will be uniform

» Final vacuum in can depends on:

- Time and temperature of exhausting
- Temperature of food
- Fill and head space of can
- Kind, nature and quality of fish
- Freshness of product

» Delay between exhausting and seaming => Air will enter

» Delay between seaming and retorting => Food spoilage occurs

» Bulging of cans due to:

- Microbial decomposition of contents
- Hydrogen produced by corrosion of the container/ can.

- Over-filling
- Effects of high temperature in altitudes

» An alternative to use of exhaust box is vacuum seaming. Vacuum seamers are difficult to maintain and slow to operate.

G. Sealing the can

» The sealing of can is generally done using double seaming machine.

» **Double seam**: it is a method in which the flange of the can body and the curl of the lid are rolled, hooked and pressed together in two successive operations, so as to form an air-tight joint.

Objectives of closing air tight:

- Perfect seam to be maintained for vacuum of can
- To prevent contamination with the outside atmosphere
- To prevent loss of contents by evaporation or by any other processes
- In preventing oxidative and other chemical changes

» Different units of double seamer:

- 1st operation seaming roll
- 2nd operation seaming roll
- Base plate
- Seaming chuck

» Seaming operations:

- 1st seaming operation: Forming and engaging the hooks
- 2nd seaming operation: Tightening the seam and giving it the final shape.

» Proper care and maintenance of seaming equipment is vital, and its performance should be checked at frequent intervals throughout the working day.

» The can, with lid placed or clinched on top, stands on a base plate which is raised to that a chuck fits into the countersink portion of the lid, holding both lid and can in position.

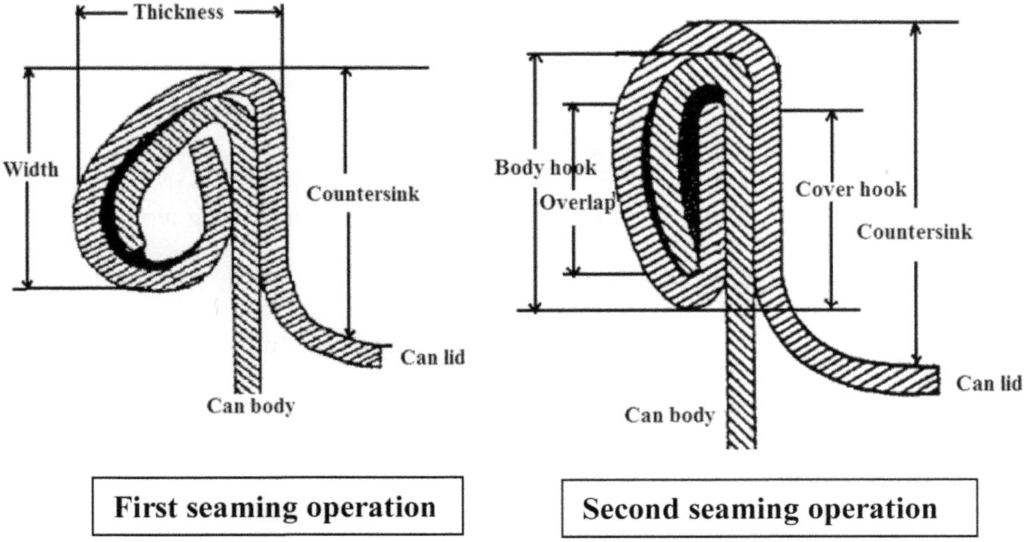

Figure 9.1: Seaming operations

- » The first operation roll is brought to bear on the unformed seam and forms a cover hook interlocking loosely with a body hook.
- » The second operation roll flattens the seam and presses the layers of metal against each other. While closing round cans, the can and lid revolve, while the rolls remain stationary.
- » After seaming, the cans are washed with detergent added warm water to remove any greasy substance adhering the outside of the can.
- » The body hook and cover hook overlap should be more than 50% for a good seam.

H. Heat processing/ Thermal processing

- » It is the operation in which canned foods (closed containers) are subjected to right degree of heat treatment to such an extent that all the pathogens are destroyed and most of the spoilage organisms are either destroyed or inactivated without destroying the desirable qualities of the food so that the product remains unspoiled under ordinary condition of transport and storage.

Objectives:

- » Safety of the consumer by destroying pathogens

- Safety to food by destroying spoilage organisms
- Adequate cooking

The process is divided into 3 distinct sections:

a. Come-up-time- To reach from normal temperature to processing temperature

b. Holding at processing temperature

c. Cooling – To bring down the processing temperature to normal atmospheric temperature

- The whole principle of thermal processing is based on 2 aspects:

a. Physical aspect

b. Microbiological aspect

- The effectiveness of a thermal process is the extent to which the bacterial population is reduced.

Thermal process determination:

1. Determination of heat resistance of most significant spoilage microorganism in the food.
2. Determination of rate of heat penetration in the particular canned food.
3. Correlating the bacterial heat resistance data with heat penetration data and arriving at a calculated thermal process.
4. Verification of the theoretical process under practical (commercial) conditions.

Source of microorganism:

- Raw material
- Container
- Water used
- Plant and equipments
- Personnel

Botulinum cook: Cooking sufficient to destroy botulinum spore. This is 2.5 mins at 121^{0}C.

Processing temperature and time:

» Processing time decreases with increase in processing temperature and in general a short process at a high temperature is to be preferred. Fish packs are usually processed at temperature of 121.2^{0}C (or 15lb pressure) for a period of 75 to 90 mins based on different packs.

» Almost all fish packs consist of a mixture of solids with a small quantity of free flowing liquid, so that heat transfer inside the can is restricted. Because of this the fish close to the walls of the pack tend to become over processed, resulting in discolouration and loss of flavour and texture.

» Heat penetration within the pack can be accelerated and the processing time minimised by choosing a can with at least one of the dimensions small, by increasing the proportion of sauce to fish, and by increasing convection by agitating the can during processing.

Heating:

5. **Retorting:**

   Retort

   (a) Horizontal (b) Vertical

   (a) Continuous type and (b) batch type

6. **HTST method:**

   » High temperature short time – A few seconds to 6 minute

   » Temperature is more than 121^{0}C

   » Flame sterilisation: (1000 -1500)^{0}C – 'In can' process

   » Flash sterilisation: Generally done for liquids

7. **UHT method:** Ultra high temperature

Dole and Martin (USA) developed an Aseptic Canning Unit:

- Sterilise the product
- Sterilise the can and lid separately
- Filling aseptically
- Closing filled cans aseptically

Measures of heat resistance of microorganisms:

1. **Thermal death point:**
   - Minimum temperature beyond which when heated fora short time, an organism looses its capability to grow and reproduce.
2. **Lethal temperature:**
   - Any temperature beyond this thermal death point.
3. **Thermal death time (TDT):**
   - Minimum time in mins required to destroy a given population of bacteria or spores by heating ataspecified lethal temperature.
   - TDT value for *Clostridium botulinum* is

Temperature in ^{0}C	Time in minutes
100 (212^{0}F)	330
120 (240^{0}F)	4
121.2 (250^{0}F)	2.5

TDT Curve:

- When TDT plotted on the log scale and temperature of heating on linear scale, a straight line of best fit is known as thermal destruction curve or thermal death time (TDT) curve.
- Drawn on Semi-log Paper
- X-axis: Temperature of heating(^{0}C) in Ordinary scale
- Y-axis: Death time (Min) in log scale

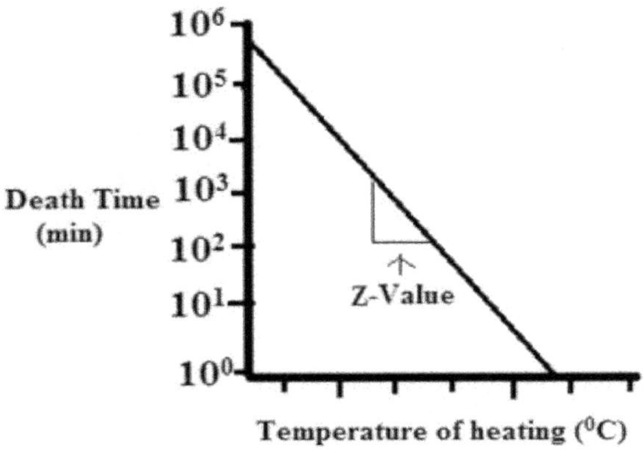

F-Value:

» Time in minutes required to destroy a particular organism at a particular temperature read from TDT curve.

Z-Value:

» Number of ^{0}F to be increased to reduce the time by 1 log is called Z- value.

» The Z value of 18^{0}F is taken as standard.

» The Z value is the increase or decrease in temperature required to reduce or increase the Decimal reduction time by 1 decimal. It is a measure of the change in death rate with a change in temperature.

» The number of ^{0}Frequired for a TDT curve to traverse 1 log cycle.

» This is the temperature increase required to reduce the TDT by a factor of 10.

» The Z-value gives an indication of the relative impact of different temperature on a microorganism, with smaller values indicating greater sensitivity to increasing heat.

4. Decimal Reduction time (D-Value):

» Minimum time in minutes required to destroy a given bacterial population by 90% at a specified temperature.

- D_{250} = 0.1 to 0.2 min
- D-Value < TDT value

- Decimal reduction times (D-values) of bacteria important in low acid canned foods.

Organism	D value (min. at 121.2 °C)
B. stearothermophilus	4.0 - 5.0
C. thermosaccharlyticum	3.0 - 4.0
D. nigrificans	2.0 - 3.0
C. botulinum (A & B)	0.1 - 0.23
C. sporogenes (P.A. 3679)	0.1 - 1.5
B. coagulans	0.01 - 0.07

- Microorganisms show logarithmic rate of destruction

$$D_t = \frac{n}{\log a - \log b} = \frac{n}{\log (a/b)}$$

Where,

n = Duration of heating
a = Initial number of spores
b = Final number of spores

Death rate curve / survival curve:

» At slightly elevated temperatures most microbes will grow and multiply quickly. At relatively high temperatures, microbes can be destroyed. However, there is a lot of variation within any one population of microbes of the same species – most will be killed relatively quickly, others can survive much longer. If a population of microbes is held at a constant high temperature, the number of surviving spores or cells plotted against time (on a logarithmic scale) will look like the following graph – which is referred to as the 'death rate curve'.

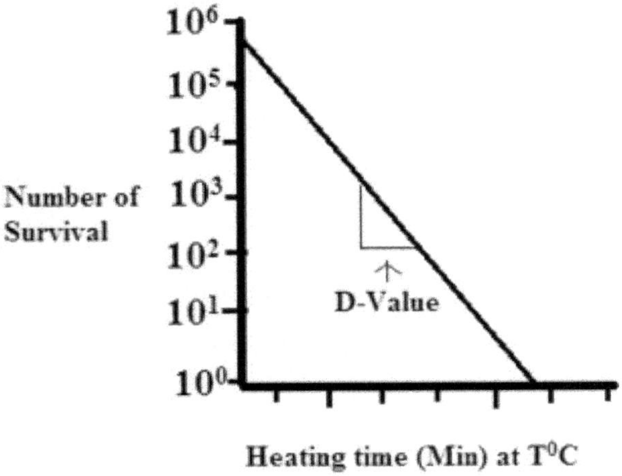

» It is drawn on semi-log paper.

» X-axis: Heating time (min) at T⁰C – in ordinary scale

» Y-axis: Number of survival – In log scale

» The time period for each "log reduction" is referred to as the decimal reduction time or D value.

» The decimal reduction time is dependent on the temperature, the type of microorganisms, and the composition of the medium containing the microorganism. Thus, after an organism is reduced by 1 D, only 10% of the original organisms remain. The population number has been reduced by one decimal place in the counting scheme. When referring to D value it is proper to give the temperature as a subscript to the D.

» In fish canning, canned fish are subjected to a time/temperature process that will reduce the probability of the survival by the most heat resistant *Clostriduium botulinum* spores by 12 logs or 12 D at 250⁰F. The temperature used in the calculation of most commercial 12 D processes is 250⁰F and the D-value for this organism at 250⁰F is 0.21 min.

$$F_0 = D_{250F} (\log a - \log b)$$
$$= (0.21 \text{ min}) (\log 10^1 - \log 10^{-11})$$
$$= 0.21 \times 12$$
$$= 2.52 \text{ min.}$$

Factors influencing Heat resistance of microorganisms:

- Nature, kind, species, strain of bacteria
- Form of bacteria (Vegetative/ spore former)
- Initial number (Spore concentration)
- pH of heating medium
- Presence of salt, starch, sugar, spices in the heating medium
- Presence of chemical preservatives, antibiotics etc in the food
- Presence of fat
- Age of spore
- Sporulation condition of medium

Methods of determination of heat resistance of microorganism:

- TDT single tube technique
- TDT multiple tube technique
- Heating in a flask
- TDT tank method
- TDT can method
- Thermo resistometer method
- Capillary tube method

Heat penetration in canned foods:

- Heat transfers through conduction, convection and radiation. But in canned foods, heat transfer occurs partly through conduction and partly through convection for solid and liquids respectively present in can container. No radiation.
- **Cold spot**: Point which receives heat last or, point of maximum heat lag.
- Cold spot of a can is determined by a thermocouple with copper constantan elements designed by Eocklund of USA.
- For all conduction heating, cold spot will be at the geometric centre of can.
- For all convectional heating, cold spot will be slightly below the geometric centre depending on can size.

» Retorting process has 3 main parts (a) Come up time (CUT), (b) Process time (PT) and (c) Cooling time (CT).

Heating cooling curve/ Heat penetration curve:

» Heating curve is a straight line.

» Slope of heating curve and cooling curve is same.

» For products which change from sol to gel during heating gives broken heating curve.

» When the straight line is extended backward, initial temperature is obtained which is known as 'Pseudo initial temperature'.

Graph in ordinary graph paper

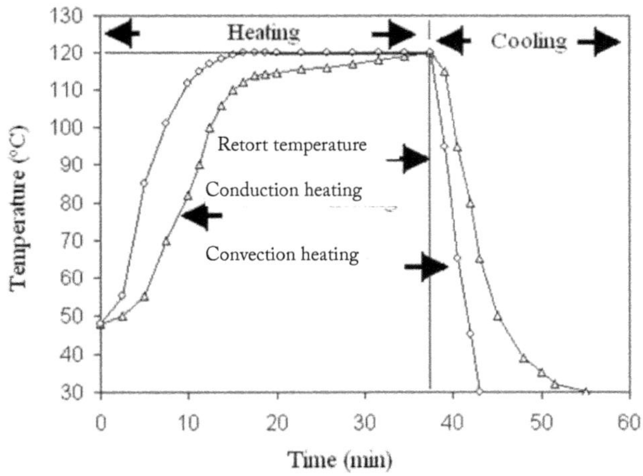

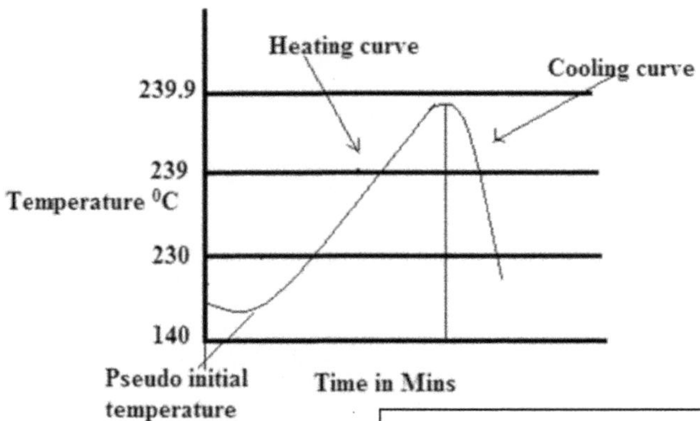

Graph in semi-log graph paper

Figure 9.2: Heating cooling curve

Factors influencing rate of heat penetration:

- » Temperature gradient (RT-CT)
- » Container characteristics
  - Material: Thermal conductivity and specific heat
  - Wall thickness
  - Size and shape
- » Content characteristics
  - Specific heat and thermal conductivity
  - Solid to liquid ratio
  - Consistency
- » Fill and head space
- » Retort characteristics
- » Loading the can
- » Agitation during retorting
  - To and fro or reciprocating motion
  - Axial or rolling
  - End over end

Process calculation methods:

1. General method / Graphical method (Bigelow)
2. Formula method (Ball and Olson)
3. Nomogram method (Olson and Stevens)

Advantages of General method:

- » Applicable to all conditions, all foods, all can sizes
- » Takes care of CUT, PT and CT
- » Applicable even when semilog heating curve is not a straight line

Disadvantage:

» Laborious

» Process values obtained will be for these conditions only which are in heat penetration test.

General method:

» Lethal rate = 1/TDT

» From Heating Cooling curve, temperature of the can at a particular time is found out.

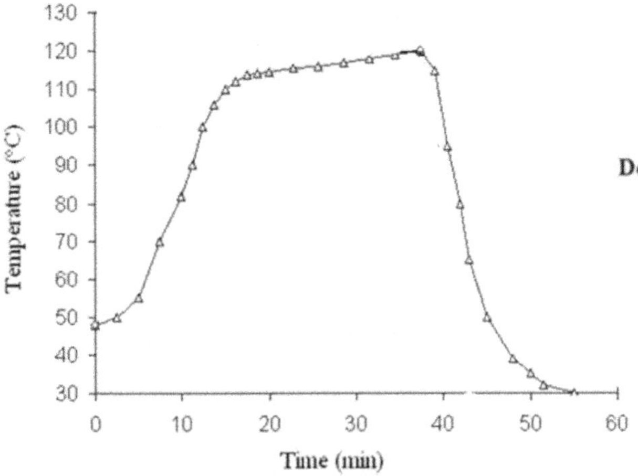

Heating –Cooling Curve

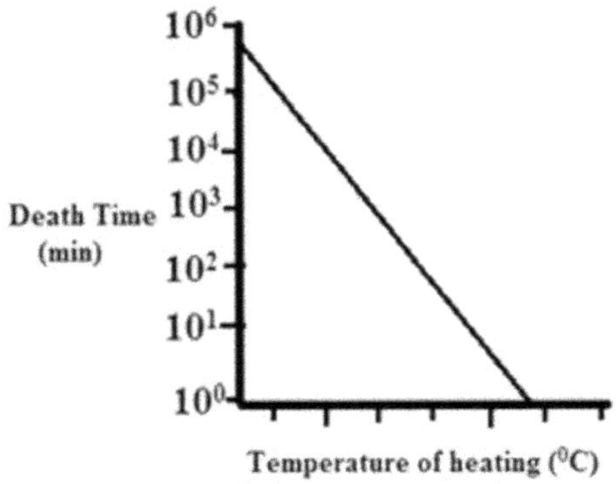

TDT Curve

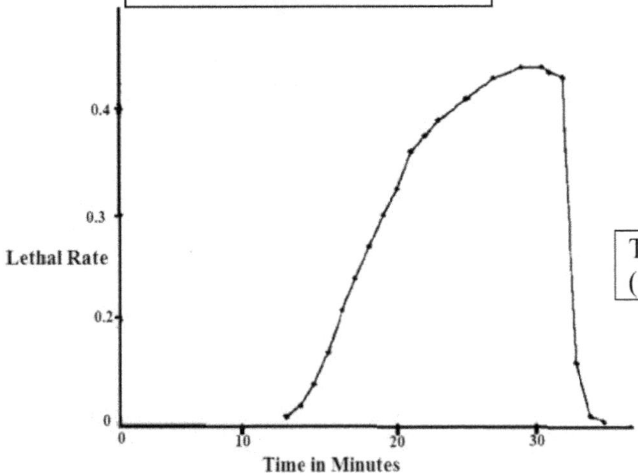

Lethal Rate Curve

Lethal rate curve

Time (mins)	Can Temp.	TDT	Lethal Rate

Figure 9.3: Determination of Lethal rate curve

- From TDT curve, time (TDT value) with respect to that temperature of the can is to be obtained.
- Then find out Lethal rate LR = 1/TDT.
- Finally LR curve are plotted against time of retorting and LR curve is drawn on ordinary graph paper. The area under this curve represents total lethality.
- If total lethality = 1 (Adequate heating)

  < 1 (Inadequate process)

  > 1 (More than adequate process)

Time (mins)	Can Temperature	TDT	Lethal Rate

- At a constant time interval LR values are to be taken.

Calculation:

- LR = ½ of 1st and last lethal rate value, add to all values and multiply by constant time interval.

Formula method:

- Adaptable to any problem
- Can be used when heating curve is broken.
- Written documentation of each step in calculation is required
- Involves many calculations
- Time consuming
- Laborious

Nomogram method:

- Easy and quick method
- **Disadvantage**: Not applicable to broken heating curve

- It can't be used when $Z \neq 180F$
- This method requires a Nomogram set of scales arranged in a particular way.

I. Cooling

- The cans after removing from the retort should be cooled immediately in chilled chlorinated water.
- Cooling in chilled water immediately after retorting gives a lethal shock to the thermophiles which might have been survived during processing leading to lethality to the thermophiles.
- Cooling must be done up to a temperature that they retain just sufficient heat to dry off the surface moisture and thus prevents the external rusting.
- Efficiency of any cooling system may be estimated by measuring:
    - Temperature of inflowing water
    - Temperature of outflowing water
    - Time of cooling
- **Cooling media**: Air/ water/ iced water. Chilled water is most effective.
- **Cooling method**: (a) Cooling canal, (b) Cooling tanks and (c) Spray cooling.
- **Cooling water:** It must be potable water.
    - TPC < 100/ml
    - Coliform < 1/100 ml
    - Pathogen: Nil
    - Inplant chlorination: 4 – 10 ppm free residual chlorine
    - Can cooling discharge (outlet) water chlorination: 0.5 – 1.0 ppm free residual chlorine

J. Coding and labelling

- Coding is mainly provided to identify the fish can type viz sardine in brine or oil or in sauce, mackerel natural, prawn in dry pack etc.

» The lebel is made giving date of manufacture, place of manufacture, batch number etc.

» The coding in tin cans is made by three methods.

- Embossed code
- Printed code
- Electrolytic method of coding

V. Examination of Finished Can

1. Swells/ Bulging/ Buckling/ Peaking: (Mainly by microorganisms)
    - **Flipper**: One end flips out and can be pushed back on light pressure.
    - **Springer**: one end flips out. If pushed, the other end will flip out.
    - **Soft swell**: both ends flip out. On pressure, ends will be pushed back but on releasing pressure again bulges out.
    - **Hard swell**: Both ends flip out and no pushing back. Mainly by *Bacillus thermosaccharolyticum*.
    - **Bursting**: Both ends flips out. On pressure bursting will occur.
2. **Flat sour:**
    - Ends of the can are flat but food item gets sour (acidic).
    - It is mainly by thermophiles viz *Bacillus thermosaccharolyticum*
    - It may occur due to underprocessing.
3. **Leakers:**
    - Leakage of cans occurs due to physical damage such as
        - ➢ Faulty seaming
        - ➢ Faulty labelling
        - ➢ Corrosion
        - ➢ Mechanical damage
        - ➢ Nail holes

- ➢ Bursting

4. **Panelling:**
    - High vacuum in cans
    - Sudden increase of retort pressure
    - This can be prevented by increasing retort pressure slowly.
5. **Dents:**
    - Rough handling of cans after seaming, during retorting or later.
    - It can be prevented by handling the cans with care.
6. **Rusting:**

    Due to:
    - Over chlorination of cooling water
    - Over cooling
    - Rusted crates damaging can surface

    Prevention:
    - Control cooling and chlorination
    - Avoid damage to can
    - Store in cool, dry roomswith minimum temperature change
7. **Surface blackening:**
    - Due to: Sulphur in food acting on iron of tin container
    - Prevention: By using S-R lacquered cans and storing in cool temperature
8. **Curd formation:**

    Reason:
    - Proteins in fish meat precipitation because of unfresh material

- Improper brining and precooking

Prevention:

- Use of fresh fish
- Brining and precooking sufficiently
- Use of dip treatment in 5% tartaric acid for1 min.

9. Struvite formation: (Struvite – $Mg NH_4 PO_4, 6H_2O$)

- Seen in canned tuna, crab and shrimp
- Mainly due to use of stale material
- Slow cooling of processed can

Prevention:

- By using soft water, fresh fish and quick cooling
- Use of sodium hexametaphosphate/ lecithin/ Sodium sulphide

10. Honey comb and green meat:

- Here proteins get denatured and honey comb like structure will be formed
- Seen in canned tuna.
- Due to damaged fish material and stale fish
- Prevention: Use of fresh fish

11. Blue meat:

- Seen in canned crab.
- It is due to haemocyanin pigment present in the blood
- It occurs in case of damaged and stale crabs

Prevention:

- By bleeding

- Use of very fresh or live crabs

12. Stack burn/ retort burn or dry top burn:

- If temperature is maintained for more time
- Filling medium not covering the solid food

13. Spur:

- Formed because of lack of proper functioning of seaming rolls to tuck the cover hook completely under the body hook.

14. Breather:

- Some cans will have minute leakage. So, vacuum is lost. Such cans appear normal. These cans may or may not be spoiled. Such cans are called 'breather'.

VI. Causes of spoilage in Fish Cans and its Prevention

a. Physical:

- Overfilling
- Under-exhausting
- Falty retort operation
- Rough handling

b. Chemical:

- Hydrogen swell in acid packs
- Rusting

c. Microbiological:

- Pre-process spoilage
- Under processing
- Inadequate cooling
- Post-process spoilage due to leakage

d. Miscellaneous:

- Corrosion – label paste
- Fluctuation in storage temperature and humidity
- Excessive chlorination of cooling water

Prevention of spoilage:

» Physical damages are generally prevented by use of tin cans which are strong enough to get free from any damage. However, precaution must be taken at every step.

» Chemical spoilage can be prevented when inside of the can is made anaerobic/vacuum. This will prevent oxidation of fat causing rancidity.

» Microbiological spoilage are prevented by creating anaerobic condition inside the can, thermal processing and sudden cooling after heat processing.

1. CANNING OF SEER FISH IN OIL / TOMATO SAUCE

» Seer fish (*Scomberomorus* species) are highly suitable for canning because of their texture, taste and fat content.

» Individual fish in commercial catches may vary in size from half kg to 4 kg or more. Smaller fish are suitable for canning as steaks and the larger ones as fillets.

» Canners should be careful to select fish having no off odour.

Forms of Pack: Seer fish may be canned in three forms, namely

1. Steaks, which are transverse cuts of dressed fish with skin and bone,
2. Fillets with skin, which are quarter fillets from which all bones have been removed but skin is left over, and
3. Skinless fillets, which do not contain either skin or bone.

Lye-peeling:

» The skin of the big sized fish like tuna, seer can be removed more conveniently by the following two methods, which give higher yield and better appearance than the manual method-using knife. They are:

a) Lye Peeling:

» The fish steaks are immersed in boiling sodium hydroxide solution (2%) for 2 minutes followed by immediate washing in running water for about 15 minutes. The fish may later be dipped in 0.2% citric acid for 2 minutes to bring the pH back to the original value of fish muscle.

b) Cook and peel method:

» Here the fish are first cut into steaks and dipped in boiling water for the required period of time (1½ -2 minutes). The products are cooled sufficiently so as to remove the skin by hand operation.

Filling Media:

The fish may be canned in

» Brine (2% table salt solution), or

» Double refined vegetable oil (such as groundnut oil or cottonseed oil), or

» Tomato sauce (either readymade canned tomato sauce or prepared in the cannery).

Containers:

» Round, sanitary containers, such as No.1 Tall can (301 x 409) or 8.0z can (301 x 206), may be used for steaks as well as fillets.

» Rectangular (Quarter Dingley) cans are suitable only for fillets.

» Plain tin cans have a tendency for sulphide discolouration, and therefore cans with internal coating of sulphur resistant lacquer (S-R lacquer) have to be used, as in the case of all sea foods.

Procedure: The canning procedure consists of the following steps:

1) Raw material:

» Select good quality fish with no stale odour, wash, weigh and keep in crushed ice.

2) Dressing:

» Dress the fish by cutting off head, removing gills, guts, fins and tail

i. Cut the dressed fish transversely to obtain steaks of required length for packing.

ii. If fillets with skin are required, first make half fillets by slitting each fish longitudinally, starting from the tail end. Again cut along the lateral line to obtain quarter fillets. Remove all bones and dark meat at the edges. These may be cut transversely to required length for packing.

iii. For making skinless fillets, the dressed fish may be first cut to shorter steaks. These are dipped in boiling hot, 2% caustic soda solution for 2 min. for skin removal. Immediately the pieces are washed in running water for 15 min. and the given a dip in 0.2% solution of citric acid at room temperature. Now fillets of suitable length are made from these pieces.

3) Size cutting:

Suitable lengths of pieces for packing in the three can sizes are as follows:

» No.1 Tall can : 10.5cm

» 8-Oz. can : 5.4 cm

» Quarter Dingley can : 9.5cm

4) Brining:

» The cut pieces are washed in running water to remove blood etc., and then dipped in saturated common salt solution at room temperature, brining times are 30' for steaks 20' for fillets with skin and 15' for skinless fillets. After brining, the pieces may be given a quick surface wash in fresh water.

5) Packing:

» Pack the fish pieces vertically in round cans and horizontally in Quarter Dingley can, in such a way that skin portion is in contact with the can.

» Attention to good appearance should be given in packing.

» Pack weights recommended are;

- No. 1 tall cans - 420 g
- 8-Oz. can - 210g
- Quarter Dingley can - 106 g.

6) Precooking and draining:

» The filled cans are to be precooked in steam at 100^{0}C for period ranging from 30 to 45min. depending on can size.

» After the cook, drain the cook juice / exudates by inverting the cans, taking care not to allow the fish to drop out.

7) Filling with liquid medium:

» Fill the cans with hot (80-900C) brine (table salt) or refined vegetable oil or tomato sauce, so as to get correct net weight of contents.

» Net weights suitable for the three can sizes are:

- No. 1 Tall can : 450g
- 8-Oz. can : 210 g
- Quarter Dingley can : 106g

8) Exhausting:

» Exhaust the cans in a steam exhaust box for 15 min. at 100^{0}C.

9) Can closing:

» Immediately after heat exhausting, close the cans with code marked lids by means of a double seaming machine. Do not allow the can and contents to cool below 80^{0}C before seaming.

10) Processing:

» The closed cans are to be washed clean and then loaded into the retort for processing. After closing the retort, give sufficient time to drive out all the air and raise the pressure and temperature to processing level.

» Then give the following processes.

- No.1 Tall can with steaks: 100 min. at 12 psi (118^{0}C)
    - ◊ for fillets, 90 min. at 12 psi (118^{0}C)
- 8-Oz. can with steaks: 90 min at 10 psi (115^{0}C)

◊ for fillets, 80 min. at 10 psi (115°C)

- Quarter Dingley cans with fillets: 50 min. at 10 psi (115°C).

11) Can cooling:

» After the process time is over, cool the cans in clean water to room temperature.

12) Storing:

» Store the cans only after they are dry on the outside. Label them while dispatching.

Product Code:

» SEF 3: Seer fish fillets in oil

» SEF T: Seer fish fillets in tomato sauce

- Std. Solid Weight: Not less than 65% of net weight

Flow diagram:

Fish	round or semi dressed
Thawing	if it is frozen
Weighing	
Dressing	Filleting with or without skin
Brining	Saturated brine: 10-15 min
Packing	vertically, 210grams
Pre-cooking	100°C for 30 min
Draining	10 min
Oil adding	Hot (80°C) to net wt. + 2g table salt
Can closing	by vacuum seaming or after heat exhausting
Can washing	
Retorting	113°C for 75 min or 115°C for 70 min
Cooling	
Drying	
Storage	

2. CANNING OF MACKEREL IN NATURAL STYLE (SALMON STYLE)

» In the 'Natural' style, fresh fish with skin and bones are packed with minimum pre-treatment.

» Only very fresh fish are canned in this style.

Product Code: MKN

Cans Used :

a) 8-Oz. Can, SR- lacquered

b) No.1 Tall can, SR- lacquered

Std. net weight :

a) 425 grams for No.1 Tall can.

b) 210 grams from 8-Oz. can

Std. solid weight : Minimum 65% of net weight (BIS)

Procedure:

Raw material	Fresh Fish (whole)
Washing	To remove dirt, slime and other unwanted material
Weighing	To determine yield
Dressing	It is an operation involving removal of head, splitting the belly, removal of viscera, cutting of tail and fins, scaling the fish and removal of slime.
Washing	To remove blood and clean the fish
Cutting	The dressed fish are size cut to required length, discarding tail ends (10.8 cms for No.1 tall can; 5.5 cms for 8-Oz. Can).
Washing	To clean the fish pieces
Weighing	to determine dressing yield
Brining	It is done by immersing the size cut fish pieces in saturated brine solution for 8-12 minutes depending on the size of pieces; to remove blood stains, improve texture and appearance, prevent turbidity in the filling media and prevent 'curd' formation on the product.

Dipping	It involves giving a quick dip in fresh water to remove excess of salt from the surface of fish.
Packing	The product is packed axially to net weight + 3 or 2 grams extra.
Salt adding	Add 0.5-1.0 gram dry, powdered table salt for taste.
Seaming	It is done by vacuum or after proper exhausting for 10-12 minutes at 100ºC, using a double seaming machine set to can size.
Can washing	With soap water and then fresh water
Processing	It is done in retorts / autoclaves at 115ºC (10 psi) for 90 minutes in case of No.1 Tall cans and 70 minutes for 8-Oz.. cans.
Cooling	The hot cans are to be rapidly cooled to 37-38ºC, either in running water or ice cold water, which usually takes 45 minutes.
Drying	After cooling, wipe the can surface. The heat remaining within the meat helps to dry the outer surface of the can, thus preventing rust formation.
Labeling	Indicate the batch, date of production, product code, net weight, etc. on the label stuck to the surface of the can.
Storage	In cool and dry place

Precautions:

1. Minimize the handling of fish.
2. Avoid delay at all stages
3. Keep the fish with crushed ice, if the processing is delayed.
4. Clean the dressing table and floor area frequently
5. Keep the fish cutting knives always sharpened.
6. The limit of freshness of raw mackerel was found to be 20mg% VBN. If the mackerel is held at 20ºC, it should be packed and processed within 15 hours.
7. Mackerel meat autolysis more rapidly than does other fish therefore the use of good preservation technique like ice chilling, and freezing or sometime salts is used with ice.
8. The chilling temperature is considered to be best at -6 ºC or below.

Problems encountered

1) Curds Formation:

» Even with fresh raw material, the formation of "curd" is often found in the canned product. Such curd formation is due to the coagulation of soluble protein in mackerel meat. This is more common where mackerel is generally canned without precooking. The meat protein coagulated by heat, adheres to the inner side of the can ends and presents a poor appearance on opening the can. The lacquer may get peeled off while removing the curd from the can ends. Curd is a salt soluble protein that exudes and coagulates during heating.

» Use of raw fish, which is not very fresh and inadequate bring and pre-cooking are some of the reasons responsible for formation of curd.

Prevention:

» Curd formation can be prevented if the raw fish is soaked in 10-15% NaCl solution for 20-30 minutes followed by thorough washing before filling.

2) Adhesion:

» Adhesion means that the meat adheres to the inner side of the cover of the can. The cause of such adhesion is considered to be due to the raw meat coagulated by heating coming in contact with the inner side of the cover.

Prevention:

» To prevent the development adhesion, the inner side of the can is moistened with distilled water or with 0.1% silicon resin solution before filling.

3. CANNING OF INDIAN MACKEREL IN BRINE / OIL

» Canning of Indian mackerel in brine / oil involves addition of a filling medium like brine or oil.

Product code:

MKL-Mackerel in brine

MK3-Mackerel in oil.

Cans used:

a) No. 1 Tall cans, S-R lacquered

b) 8-Oz. cans, S-R lacquered

Std. net weight :

a) 425 grams for No. 1 Tall cans.

b) 210 grams for 8-Oz. cans

Std. solid weight : Min. 65% of Net weight

Procedure:

Raw Material	Fresh fish (whole)
Washing	
Weighing	
Dressing	
Washing	
Size Cutting	Dressed fish are size cut to 10.8cm.for No.1 tall can and 5.5cm for 8-Oz. Can
Washing	
Weighing	
Brining	By immersing in saturated brine for 8-12 minutes
Dipping	Dip the product in fresh water and remove quickly
Packing	Pack the product into the cans to net weight. Add 2 to 3 grams extra.
Precooking	The packed cans are precooked in steam at 100^{0}C (0 psi) for 30 minutes in case of No.1 Tall cans and 20 minutes for 8-Oz. Cans
Draining	Drain the cooked cans by inverting the cans over a perforated wooden tray for 10 minutes
Filling	Add hot (80^{0}C) brine, salt (3%) and oil to net weight +2 to 3 grams extra.
Seaming	By vacuum seaming or after proper heat exhausting
Can washing	
Retorting	At 115^{0}C (10 psi) for 90- 95 minutes in case of No. 1 Tall cans and 70 minutes for 8-Oz. cans (oil filled cans generally require slightly more time than brine packs)
Cooling	

Drying	
Labeling	
Storage	

4. CANNING OF MACKEREL FILLETS IN TOMATO SAUCE

» Canning of Indian mackerel (skinless and boneless fillets) is one of the best products prepared in quarter Dingley cans using tomato sauce as the filling medium.

Product code : MKFT-Mackerel fillets in tomato sauce.

Cans used : Quarter Dingley can (plain)

Std. net weight : 106 grams

Std. solid weight : 75 grams (Min)

Procedure:

Raw material	Whole fresh fish (or good quality frozen fish)
Thawing	If frozen material is used
Weighing	
Lye peeling	fish are deskinned in boiling (hot) sodium hydroxide (1%) solution for 1 – 1½ minute.
Washing	Wash the lye peeled fish thoroughly in running water for 20 minutes to remove all the traces of alkali from fish. Otherwise, the alkali adversely affects taste and flavour due to pH change and also leads to can corrosion.
Precooking	By cooking the fish in steam at 1000C for 20 minutes.
Cooling	The fish are cooled overnight at room temperature to facilitate easy filleting, due to firming of meat.
Filleting and Cleaning	fish are placed laterally on the surface of planks and fillets are obtained by cutting the fish just above the backbone so as to obtain two longitudinal pieces. The fillets are later cleaned to remove dark meat and any traces of skin, blood clots etc.
Trimming	The fillets are trimmed to can size (9.5 cms. for Quarter Dingley cans)
Packing	80-85 grams of fillets are packed, either longitudinally or by packing them across the can in two layers with their flat inner surface facing the bottom and top of the container. Add 2.5 grams of table salt per can.
Filling	Add hot (85-950C) tomato sauce to net weight.

Seaming	Vacuum seaming or after proper exhausting
Can washing	
Retorting	At 115^{0}C (10 psi) for 50 minutes
Cooling	
Drying	
Labeling	
Storage	

5. CANNING OF OIL SARDINE IN OIL / BRINE

» Indian oil sardine (*Sardinella longiceps*) is one of the most important pelagic fisheries of India, in addition to Indian mackerel.

» It is known that, only very little amount is used for preparation of canned products, using brine/oil/curry/tomato sauce as the packing media for supply to the military personnel and the people of North Eastern states of India.

Product code :

SA3*: Sardine in oil

SA : Sardine in Brine

* (codes 0, 1, 2 or 3 are given to different oils used in cotton seed, soya bean, groundnut, olive oil etc.

Can used :

a) 8-Oz. can, S.R. lacquered

b) 1 lb. Jam can, S.R. lacquered

Std. net weight :

a) 210 grams for 8-Oz. can

b) 340 grams for 1 lb. Jam can

Std. solid weight: Minimum 70% of net weight

Procedure:

Raw material	Fresh whole sardines (or good quality frozen fish)
Thawing	If frozen
Washing	
Weighing	
Descaling	Removal of scales.
Knobbing	Removal of head and gut by partially cutting the head on dorsal side and pulling the entrails without slitting open the belly.
Size cutting	The knobbed fish is cut to required length, discarding tail ends (5.6 cms for 8-Oz. can & 8.5 cms. for 1 lb. Jam can)
Washing	
Weighing	
Brining	Immerse in saturated salt solution for 8-10 minutes depending on size of pieces; to remove blood stains, improve texture and appearance
Quick washing	Give a quick dip in fresh water to remove excess salt from the surface of fish.
Packing	Pack 210 grams in 8-Oz. cans and 340 grams in 1 lb. Jam cans.
Precooking	Cook the products in steam at 100°C (0 psi) for 20 minutes (8-Oz.. can and 30 minutes (1 lb. Jam can).
Filling	Add hot (80°C) oil or brine (3%) to net weight, + 2 to 3 grams extra.
Can seaming	Use vacuum seamer or after proper heat exhausting for 10-12 minutes at 100°C
Retorting	At 115°C (10 psi) for 70 min. for 8-Oz. can & 80 min. for 1.lb jam can) 5' more for both oil and brine packs.
Cooling	
Drying	
Labelling	
Storage	

6. CANNING OF FRIED SARDINES IN TOMATO SAUCE AND OIL

Product code:

SAFT – Fried sardine in tomato sauce

SAF3 – Fried Sardine in oil.

Can used:

a) Quarter Dingley cans, S-R lacquered

b) 8-Oz. cans S-R lacquered

Std. net weight:

a) 106 grams for Quarter Dingley can

b) 210 grams for 8-Oz. can

Std. solid weight: 70% of Net Weight

Procedure:

Raw material	Fresh sardines or good quality frozen sardines
Thawing	
Washing	
Weighing	
Dressing	Descaling and Knobbing
Washing	
Weighing	
Size Cutting	Dressed fish are size cut to 9.5 cm size for Quarter Dingley cans or 5.5 cm for 8-Oz. can.
Washing	
Weighing	
Brining	By immersing in saturated brine for 10 minutes. Dipping : Dip the product in fresh water and remove quickly.
Drying	Dry in open air for 10 minutes to remove surface moisture.
Frying	Fry in oil for 3-4 minutes at 120^0C
Cooling	
Packing	Pack 85 grams into Quarter Dingley and 147 grams into 8-Oz. cans

Filling	Divide the packed cans into 2 batches. - To the first batch add hot tomato sauce (85-900C) to net weight. - To the second batch add hot oil (800C) to net weight, with a gram of table salt.
Seaming	Vacuum seaming or after proper exhausting
Can Washing	
Retorting	At 115^{0}C (10 psi) for 50 minutes
Cooling	
Drying	
Labeling	
Storage	

Method of frying:

» After drying the fish portions in open air, place them in a wire basket and immerse in the hot oil (120^{0}C) present in a frying pan.

» Frying is completed when backbone could easily be removed and shows no redness.

Note:

» Precooking step is avoided in the fried canned fishery products, since frying in oil before packing into cans not only removes excess moisture present in the tissue but also prevents dilution of the filling medium and improves the texture.

7. CANNING OF INDIAN OIL SARDINE IN CURRY

» Canned sardine in curry is ready to eat food, suitable for direct consumption without any further preparation.

» The method differs with the canned sardine in oil / brine from the packing media and packed weight point of view.

Product Code : SA 7

Can used : 8-Oz. Can, S-R. Lacquered

Std. Net weight : 210 grams

Std. Solid weight : Fish to curry Ratio; 60: 40

Procedure:

Raw Material	Fresh whole fish
Thawing	If frozen
Weighing	
Dressing	Removal of scales, head and gut by knobbing process
Size Cutting	5.2 cms
Washing	
Weighing	
Brining	In saturated brine for 10 minutes
Washing	
Packing	Pack 140-150 grams into each can
Precooking	Cook in steam at 100^{0}C for 15 minutes
Draining	Invert the cooked cans over a wooden tray for 10 minutes
Weighing	
Filling	Add hot curry to net weight +2 grams extra.
Can Seaming	Using vacuum Seamer
Can Washing	
Retorting	At 117^{0}C for 90 minutes
Cooling	
Drying	
Labeling	
Storage	

Curry Requirements:

a) For Paste

Ingredients	% in paste		Ingredients	% in paste
Refined oil for frying	4.0		Turmeric powder	1.6
Long dried chillies	3.2		Tamarind	8.0
Short dried chillies	4.8		Onion	14.0
Coriander	2.0		Garlic	1.6

Mustard	1.6		Table salt	8.8
Cumin	0.8		Water for grinding	48.0
Pepper	1.6			

b) For Curry Dilution (For 25 kg of prepared paste)

Ingredients	Quantity in kg		Ingredients	Quantity in kg
Refined oil	7.0		Tomatoes	9.0
Onions	7.0		Water for ketteling	36.0

Method of preparation of curry:

» The spices and other materials shown in the table above are cleaned and trimmed.

» Fry the spices in refined oil and grind into a paste.

» The paste is later added to fried onion and diluted with water and boiled.

» Finally add sliced tomatoes and continue boiling for some time.

» The curry is used for filling into cans containing pre-cooked fish.

8. CANNING OF SHRIMP (DRY PACK)

» Canning of shrimp in the form of 'dry packs', is not a common method of processing.

» In this mode of canning, the blanched shrimp are wrapped in parchment paper and canned without any filling medium.

Product Code: SRD

Can used :

a) 4½ -Oz. Shrimp can, S-R lacquered

b) 5-Oz. Can, S-R lacquered

Std. Net weight :

a) 128 grams for 4 ½-Oz. Shrimp can

b) 141 grams for 5-Oz. Can

Procedure:

Raw Material	Fresh raw shrimp
Washing	
Weighing	
Beheading	
Peeling	Removal of shell (exoskeleton) on the abdomen
Deveining	Removal of gut by deveining blade
Washing	
Weighing	
Blanching	By boiling in 8- 10% salt solution containing 0.3% citric acid for 8-10 minutes
Drying	By fan for around 3-5 minutes
Grading	According to size.
Weighing	
Packing	Pack 128 grams of the blanched shrimp in a parchment paper and place in the can (Net weight + 2 grams extra)
Seaming	By vacuum seaming or after proper exhausting
Can washing	
Retorting	At 121^{0}C (15 psi) for 20-25 minutes.
Cooling	
Drying	
Storage	

Problems encountered with canned shrimps

a) Iron Sulphide blackening:

» Blackening of the can interior and the product due to Iron sulphide is a problem met with in canned shrimp, lobster, crab, tuna, etc.

» Blackening occurs because of the reaction of the sulphur–containing constituents released from the fish during heat processing with any exposed iron of the can resulting in the formation of Iron sulphide that is black in colour.

» Trimethylamine present in marine fish will dissolve tin exposing iron for reaction with the sulphur containing compounds to produce iron sulphide.

Prevention:

- By maintaining proper acidity in the can contents by adding tartaric or citric acid in the can, will retard generation of hydrogen sulphide.
- A parchment lining is commonly used in shrimp cans aiming at preventing the sulphide and ferrous ions from coming into contact.
- Uniform lacquering of cans and avoiding the possibility of any exposed tin / iron or the other remedial measures that helps to prevent the occurrence of this Iron sulphide blackening problem.

b) Softening:

- The cause of softening of canned shrimp is the decomposition of the protein to soluble non-protein substances. That takes place in the raw shrimp when freshness declines.

Prevention:

- Strictly fresh raw material should be employed
- Maintenance of high degree sanitation
- Avoiding delays between different stages in processing.

c) Loss of colour:

- The original pink colour of canned shrimp declines during long storage periods.

Prevention:

- The raw shrimp should be dipped in an acid solution (0.1% citric or tartaric acid) for about 15 min. after the shell is removed.

9. CANNING OF SHRIMP IN BRINE (WET PACK)

- Shrimp is generally canned as 'wet pack" where it is canned in brine.
- Wet pack is the most popular form, many species of Penaeid shrimp are suitable for canning.

Product code:

- SRL - Shrimp in brine.
- SR3 - Shrimp in oil.

Can used:

- 8-Oz. Shrimp can, S-R Lacquered.
- Std. Net Weight. : 210 grams
- Std. Solid Weight : Not less than 64% of water capacity of container (= 128 grams for 4 ½-Oz. can)

Procedure:

Raw Material	The shrimps should be as fresh as possible. For this the shrimps should be taken for processing as soon as possible after the catch
Washing	They should be washed thoroughly with fresh water to make them free from slime, blood, mud etc. and then stored in ice immediately. The handling should be minimum as far as possible.
Weighing	
Beheading	The head is removed. This constitute about 40% weight of total weight of the whole shrimp.
Peeling	Removal of shell/ exoskeleton from the abdomen
Deveining	Removal of the gut using deveining blade
Washing	Wash thoroughly with chilled water
Weighing	
Blanching	The dressed prawns are then blanched in 8-10% brine solution. The brine solution is made first and heated to boil. The shrimps are then dipped in the solution (5-7 min.). The blanching helps to inhibit the enzymic action and cleansing the material about 0.1% citric acid is also added (1000 ml water + 80 to 100gm salt +1g citric acid)
Fan Drying	for 3 minutes cooling is usually done under the fan
Grading	The blanched shrimps are cooled and graded for the different sizes
Weighing	

Packing	After grading the prawns are filled in the S-R lacquered and cleaned cans to the desired weight. In case of 4 ½-Oz. cans weight should pack 128 grams + 3grams extra while filling the cans a headspace of about 1/8" is left. Care should be taken that the cans are not over filled.
Filing	2% brine solution is made with fresh potable water. The salt used in this solution should be pure (over 98% NaCl and free from calcium and magnesium) 1000ml water + 20gms salt + 2gm citric acid. The brine solution should be hot (about 970C while filling).
Seaming	Vacuum seaming or after proper exhausting
Can washing	
Retorting	At 115^{0}C (10 psi) for 20-25 minutes. The purpose of processing is to destroy the bacteria and spores which would cause spoilage of the product and which might cause food poisoning to the consumer.
Cooling	Cans are cooled by potable water in a tank or by water spray till the temperature of the cans reaches about 40^{0}C. A small amount of chlorine (1 ppm) may be added to cooling water for additional precautions.
Drying	
Labeling	The cans are labelled according to the requirement. The label should cover the entire sides of the can and separate label may be put on the top. An attractive label will have a greater consumer appeal.
Storage	The finished cans should be stored in a cool, dry place.

10. CANNING OF SHRIMP IN MASALA

Product code : SRM

Can used :

a) 4½-Oz. Shrimp can, S-R Lacquered

b) 8-Oz. can, S-R lacquered

Std. net weight :

a) 190 grams for 4½-Oz. Shrimp can

b) 210 grams for 8-Oz. can Std.

Solid weight : Fish to Masala :: 60 : 40

Procedure:

Raw Material	Fresh raw shrimp
Washing	
Beheading	
Peeling	
Deveining	
Washing	
Weighing	
Blanching	By boiling in 8% table salt solution, containing 0.2% citric acid for 6 minutes
Fan Drying	
Grading	
Packing	120 grams for 8-Oz. can and 110 grams for 4½-Oz. Can
Filling	Add hot curry (800C) to net weight + 3grams extra
Seaming	By vacuum seaming or after proper heat exhausting
Can Washing	
Retorting	At 115^{0}C (10 psi) for 60 minutes
Cooling	
Drying	
Labeling	
Storage	

Masala curry preparation (for 1Kg of shrimp):

Garam Masala : 30g
Chilli Powder : 30g
Salt : 2%
Refined oil : 150g

Preparation:
Fry the peeled and deveined shrimp in the oil for a short time, mix with other ingredients, mix and dry for a short time and take out for packing.

10 Fish Paste Products

» Fish paste products are prepared from deboned fish meat. The fish paste is obtained by mixing the deboned fish meat with spices and other seasoning materials.

» **Minced meat**: Any meat which has a uniform particle size without bone is called mince meat.

» Fish sausage, a paste product was invented by Japanese scientist Wataru simidu in 1937.

» Trash low cost fishes such as croakers, pink perch, lizard fish etc are mainly used for fish paste products.

I. DEBONED FISH MEAT / PICKED MEAT

» This is obtained manually or mechanically.

» Mechanical deboning is done using a machine called fish deboner or meat picking machine.

» This is of two types:

- Drum type
- Stamp type

a. Drum type meat picking machine:

» A stainless steel perforated drum is rotated by a motor. Another rubber belt is allowed to touch it tightly and it rotates. Fish is pressed between the drum and the belt. A scraper touching the drum scrapes down the skin and bone to be collected by a waste box. The fish meat entering into the drum moves forward and is collected in a meat collecting box.

b. Stamp type meat picking machine:

» In a stamp type machine, a rubber pad is attached to an iron plate which moves up and down to perform a reciprocating motion. Another horizontal circular disc with perforation remains below the pad. When fish remains in between the iron plate and rubber pad, the stamp during its downward movement pressed the fish on to the perforated disk. The meat comes out below the disk and is collected. The skin and bone is removed by a scrapper touching the upper surface of the disk.

II. WATER WASHING OR WATER BLEACHING

» Water washing is done to remove water soluble proteins, fat, blood pigments, broken skin, mucus, enzymes, bacteria and all other micro-oorganisms etc.

» Water soluble protein (WSP) hinders the formation of gel network during gelation process. Gelatin is important properties of all minced products.

» The deboned fish meat is subjected to water washing using chilled water at 4^0C with 3-4 times the weight of the meat (Fish meat to water ratio = 1:5 to 7). Then it is allowed to settle for 5 to 10 mins to prevent denaturation of protein. Thereafter, it is to be pressed or centrifuged until moisture content of 81% is to be maintained. This type washing is to be repeated for 3 to 4 times.

» The whole operation is to be conducted at a temperature of less than 15^0C.

» **Strainer**: The excess water during water bleaching is removed by strainer.

III. SURIMI

» Surimi is a wet frozen concentrate of myofibrillar protein of fish muscle (Lanier, 1986). It is deboned, washed and stabilized fish mince. Like many other fishery products it originated in Japan.

» Deboned fish meat after water bleaching which is the concentrate of myofibrillar protein is known as **surimi.**

» The Myosin group of protein (MGP) gives elasticity to the product whereas, water soluble proteins (WSP) gives taste of the product.

» MGP are long chain proteins linked with peptide bond (-NH-CO-).

» Surimi contains about 80% moisture.

» For frozen storage, surimi is mixed with cryoprotectants.

» The cryoprotectants prevent protein denaturation during freezing

» The cryoprotectants are

- Sucrose – 4%
- Sorbitol - 4%
- Sodium tri-polyphosphate – 0.02%

» Mixing is done using a silent cutter, a mixer.

» After mixing, it is put into trays and frozen in horizontal refrigerator and stored in a cold store maintaining -18^{0}C temperature.

Procedure of surimi production:

» Fish – Washing – Dressing (Beheading, scaling and gutting) – Deboning – Water washing – Straining- Mixing with cryoprotectants in a silent cutter – Panning – Freezing – Storage in Cold store.

» The surimi, thus produced, is used for different surimi based products commonly known as '**Kamaboko**'.

Kamaboko (meat loaf/ calei):

» It is an elastic, jelly, flexible, heat pasteurised white coloured fish paste product.

» In Kamaboko, fat addition is optional where as in Surimi, fat addition is compulsory.

» Kamaboko is of various types which are all Japanese products. These are:

- Kamaboko: A steamed product
- Chikuwa: Boiled product of tubular shape

- Hampen: Boiled product
- Satsumage: Fried in oil. Disc/ Oval/ Rectangular shape

» Distinctive eating characteristics of Kamaboko is its **elasticity** which depends on:

- Species used
- Freshness of raw material
- Type of ingredient
- Processing method employed

» Colour of Kamaboko is white.

» Elasticity is increased if sulfhydric group in kamaboko is reduced.

Preparation of Kamaboko:

Fish paste is prepared by mixing surimi with salt at 2% and grinding it in a silent cutter at low temperature. The meat paste is put on wooden planks made from pine or cedar. It sticks to the plank due to the natural adhesiveness of the meat. Then it is steamed for 80 to 90 mins for large size and 20 – 30 mins for smaller size. The steamed product is called kamaboko. As surimi is a wet concentrate of myofibrillar protein, when 2% salt is added the proteins get solubilised in the salt and forms a sol and on heating it becomes to gel.

Preparation of Chikuwa:

It is obtained by wrapping the ground fish meat paste around a brass tube and broiling. Broiling is done by gas or electricity. The m eat is placed near the flame of the gas or electricity so that it gets heated up on the surface. The brass tube is extracted. Broiled chikuwa of good quality is white inside and golden brown in the outside surface.

Preparation of Hampen:

It is a white coloured spongy boiled product. The sponginess is obtained by mixing air into the meat paste during the grinding process. Usually 6% to 10% ground yam is also used to promote the intake of air. It is then formed into8 cm^2 and boiled at 80^0C to 90^0C. It floats in water.

Preparation of Satsuma-age:

Fish paste is prepared by mixing surimi with salt at 2% and grinding it in a silent cutter at low temperature. The meat paste is then made to various shapes like a disc, oval or rectangular. They are the fried in oil to get satsuma-age.

IV. FISH SAUSAGE

» Fish sausage is ground fish meat with additives like salt, sugar, starch, chemicals, spices, colour etc, stuffed into the casing, sealed and boiled product. In fish sausage, protein acts as best emulsifier. The cross section of fish sausage gives a smooth appearance.

» The fish sausage has good taste, flavour and smooth and attractive appearance apart from providing valuable fish protein at comparatively cheap cost. Due to proper blending of fish and additives, the undesirable smell of fish is suitably counteracted and made into a wholesome and acceptable product.

Fish Ham:

» Fish ham is ground fish meat as in sausage with pieces of animal meat at a ratio of 1:1 (Fish paste : animal meat cubes = 1:1), stuffed into casing, sealed and boiled product. In fish ham, gelatin is used as emulsifier. The cross section of fish ham gives a mosaic appearance.

a. Preparation of sausage:

Raw material	**Fresh or frozen**
Thawing	If Frozen
Dressing	De-scaling, beheading, gutting
Washing	Cleaning the fish
Meat picking	Either manually or mechanically by using meat picker
Water bleaching	Washing the meat with chilled water (Fish meat: Water = 1: 5) two to three times
Draining	Drain out the excess water by pressing or twisting in a cloth
Mincing	Picked meat are minced using a meat mincer

Mixing	Minced meat are added with ingredients in a silent cutter within a period of <12 to 15 mins so that proteins do not get denatured due to heat generated.
Stuffing	Stuffed into casing of 3 cm diameter. Precaution to be taken that air bubble does not get trapped.
Sealing	Sealing the casing using a clipping machine called 'Ringer'.
Heat processed	For synthetic casing in boiling water at 85⁰C to 87⁰C for 60 mins For natural casing in water at 80⁰C for 20 min.
Cooling	Cooled in water at <20⁰C for 20 mins. Here, wrinkles are formed.
Soaking	Soaked in boiling water for 1 min to remove wrinkles
Drying	Fan dried for 15 mins
Storage	Stored at low temperature of 5-8⁰C.

b. Chemical composition of fish sausage:

» Moisture = 67% – 68%

» Protein = 14% - 16%

» Fat = 5% - 6%

» Starch = 8% - 9%

c. Raw material:

» The fish generally have either red coloured meat or white coloured meat

i) Red coloured fish meat:

» Tuna, mackerel, sardine etc have red colour meat

» Good taste because of WSP

» High fat content

» Myosin Group Protein (MGP) is less

» Colour is to be fixed by use of sodium nitrite

» Red colour is mainly due to haemoglobin (O_2 carrier) and myoglobin (O_2 storage).

- $Mb + NO_2 \rightarrow$ Nitroso Mb (Brightred) $\rightarrow$ Heat oxidation $\rightarrow$ Nitroso hemochrome (Pink colour) Stable

» Ideal pH = 7.3 to 7.5. If pH is less than 6.0 => Less elasticity

» Spoilage limit = < 30 mg% TVBN

» If sardine and mackerel are used, it is limited to 10% only. Because –

- It becomes soft soon after catch
- Both have 20% to 30% black meat
- Very high quantity of fat
- pH is 5.6 to 5.8

ii) White coloured fish meat:

» White colour meat

» Shark, rays, croaker, Lizard fish, flat fish

» Less fat content

» More MGP

» Strong elasticity

» Shark and rays contain more urea content.

» Prawn contains more AAN which gives good taste.

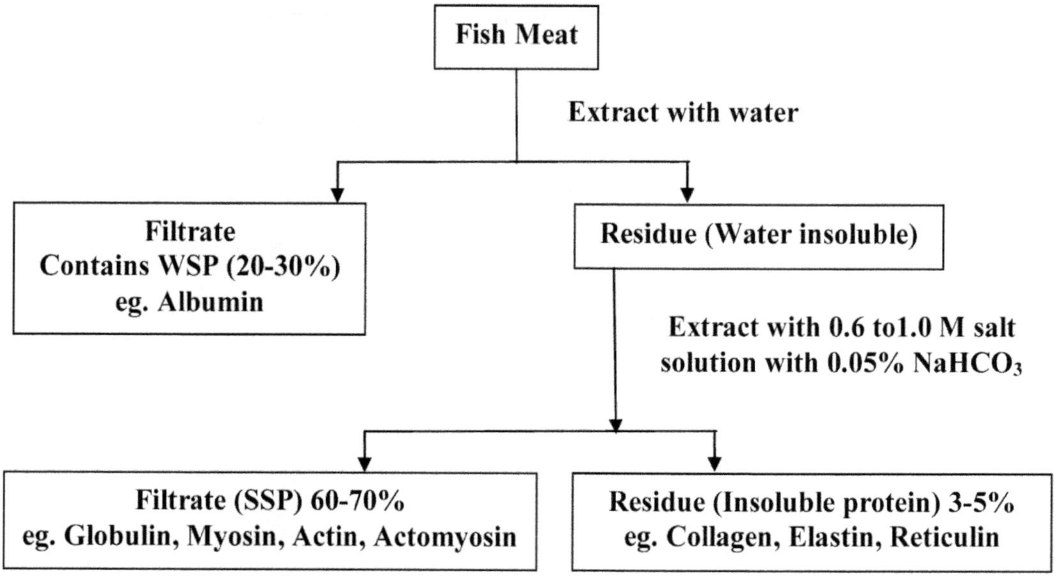

d. Protein:

» For preparation of paste products, highly active protein i.e. quality of protein is to be maintained. Here, proteins should not get denatured. Quality mainly refers to elasticity or jelly strength of the protein.

» For wafers/ spirals/Jams, denatured proteins can be used. In this case protein consistency is disintegrated. Proteins get hydrolysed partially or fully to amino acids. Here proteins are used as nutritional purpose.

» Protein hydrolysis is done by:

- Chemical method –
  - ◊ Using acid (HCl/ H_2SO_4)
    - Concentrated HCl/ H_2SO_4 are not used because amino acid, Tryptophan will be effected producing black substrate called **'Humin'**.
  - ◊ Using alkali (NaOH/KOH)
- Enzymatic method –
  - ◊ Proteolytic enzyme – Ex. Papain

- Pepsin (acid condition)
- Trypsin, Cathepsin (Alkaline condition)

e. Factors of protein denaturation:

» Due to reduction in pH (During rigor mortis, Glycogen breaks down through EM Pathway producing lactic acid and ATP gets released resulting increase in energy followed by increase in body temperature. Further, due to production of lactic acid, pH decreases).

» Extrinsic factor eg. Environmental temperature, Grinding time

» Intrinsic factor eg. Fish body temperature

» Mechanical damage leading to bacterial attack

Tests carried out to find out denaturation of protein:

» Physical method: Physico-chemical test, simple method

» Chemical method: SSP test

» Biochemical method: ATPase test

1. Physical method: (Simple method for commercial purpose)

» 50g minced meat + 1g NaCl = 51g → Mix → Grind → Packed into casing →Boiling for 10 min → Take weight (meat + salt + Exudated water (A) – Drain → Take weight (Meat + salt) (B)

» Find out the Loss % = {(A-B)/A}x100

» If loss is < 10% => Good quality protein with high jelly strength

» If loss is > 10% => Protein is denatured.

2. Flow Bi-refrengence method: By Okada *et al.*(1954)

3. Viscosity:

a. By use of Oswald's viscometer

b. By use of Wade's tilting viscometer

f. Chemistry of fish meat with respect to elasticity formation:

- » Minced meat (Fibres broken) + sol stage → on heating →gel stage
- » Sol stage:
  - a. Linkage broken
  - b. Opening of chain
  - c. Water enters inside
- » Gel stage: (Elasticity formation)
  - a. Relinking of chain
  - b. Net work formation
  - c. Water is bound inside

g. Maximum extraction of myosin group of protein:

a. 7 – 8% salt (1 Mole) – But becomes too salty and brittle in character. Therefore, use of salt up to 2.5% is preferred.

b. Network formation is maximum at pH 7.3 to 7.5. At pH 5 to 6, extraction of MGP is minimum because this is the iso-electric point where all electric charges are neutralised. On either side of this pH, MGP extraction goes on increasing. Fish meat is having a pH of 6.5. To increase pH, alkaline phosphate is to be added (pH = 8.0).

c. Fishes in Rigor mortis stage →Maximum extraction of MGP →Very good Jelly Strength

   Enzymatic stage →MGP extraction is less → Denaturation is More than 50%

   Decomposition stage →MGP extraction is very low →Not suitable for sausage

Myofibrillar proteins:

- » Actin, Myosin, Actomyosin, Connectin, Tropomyosin, Troponin, Desmin
- » Actomyosin can be extracted by 0.6 M KCl phosphate buffer and centrifuging at 4000 rpm at low temperature of +4^{0}C.

Double sterilisation:

a. 45⁰C for 30 min to enhance the formation of network structure

b. 86⁰C for 30 min

h. Jelly strength:

» Good Jelly strength can be obtained by

» Salt = 7-8%

» pH =7.3 – 7.5

» Mixing time = Less than 15 min in silent cutter to prevent protein denaturation

» Fish meat (–) Conc. WSP → Good Jelly strength

» Fish meat + Conc. WSP after denaturation by heat → Good Jelly strength

» Heating effect:

- 45- 50⁰C for 30 min →Protein coagulates, Good elasticity but not bactericidal
- 80 – 90°C for 60 min →Reasonable good elasticity and bactericidal except thermophiles

i. Setting Phenomena:

» Fish paste at RT gives elasticity →on heating → Brittle (No elasticity)

» Fish paste (Viscous) →on heating → Good elasticity

Setting:

» When (ground fish meat +salt) is left at room temperature without heating, fish paste develops jelly having elasticity, called Setting.

Returning:

» When fish paste having acquired elasticity as a result of heating is left at room temperature for a long time, elasticity is lost and meat becomes fragile (B_1→C) called Returning. Also when ground fish meat is heated at a slow temperature, meat looses elasticity (A → C).

Frozen storage of meat:

» For frozen storage, Minced meat + salt + sugar + Polyphosphate are used.

Salted minced meat	Actomyosin turns the gel to a thick solution. Formation of elasticity
+ Sucrose (10%)	Prevent setting during frozen storage
+ Poly phosphate	Chelating agent Buffer action Antioxidant Antibacterial Good binder
Sodium citrate (1-1.5%)	Good chelating agent

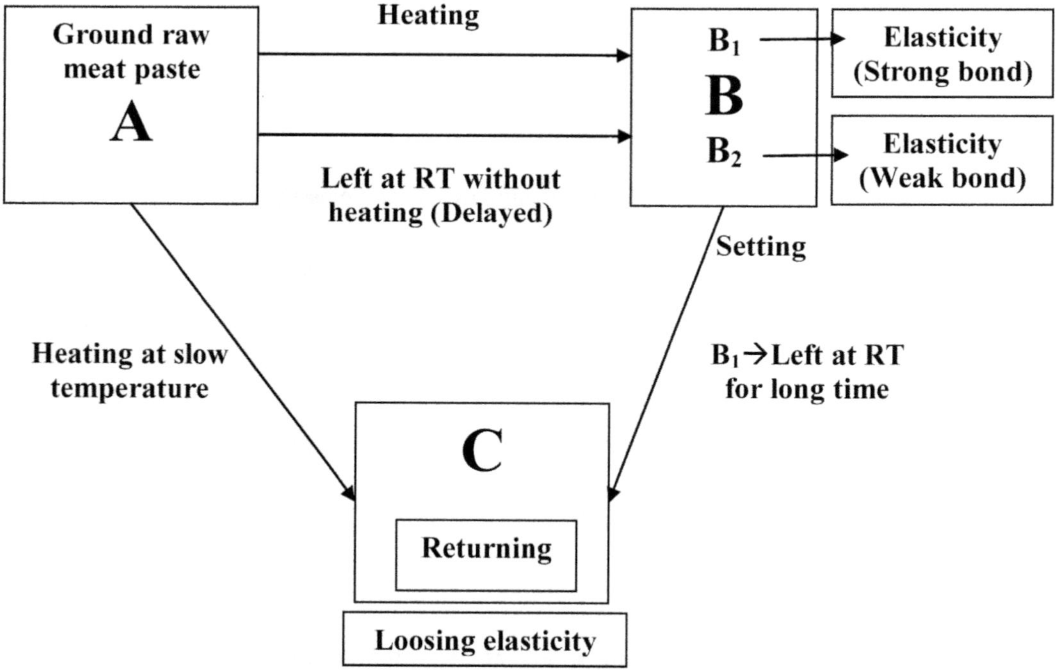

Figure 10.1: Setting phenomena

j. Increase of cross linking is due to:

a. Hydrogen bond

b. Ionic bond

c. Disulphide bond

Strengthening of re-linkage:

» Use of chemicals like $CaCl_2$ and $KBrO_3$.

k. Additives:

i) Seasoning agent:

S. No.	Item	Quantity	Purpose
1	Salt	2-2.5%	Extraction of MGP Giving taste
2	Sugar (Glucose)	1.5%	Gives taste Modify texture of meat
3	Phosphate Sodium tripolyphosphate/ Sodium pyrophosphate	0.15 – 0.2%	Extraction of MGP → Better elasticity Acts as antioxidant Antibacterial Buffer action (When starch added →Decomposed and produce Lactic acid =>pH decreased => But PO_4 Controlled the pH)
4	Fat	4-5%	Gives taste

ii. Taste of the meat:

» The taste of the meat is due to

a. **Extract component**: Amino acid, Nucleotide, Saccharide. Among aminoacids, Proline, Glycine, Alanine, Glutamic acid are responsible for taste of fish.

b. **Fat content**

Binding agent: Starch/ Phosphate/ Gluten/ Soybean flour

Others: Colouring matter/ Chilled water/ Spice extract

Smoke extract:

» It is a light brown coloured powder used to provide flavour and taste to the produce.

- » Limit of usage: For sausage: <1%; For ham: (1.5 to 2.0)%
- » Smoke extract is obtained by:
    - Dry distillation of wood
    - Passing by glass filter
- » Smoke extract powder is obtained by:
    - By addition of water to burning wood at < 400°C
    - By metallic screen filter
- » Advantages:
    - Acts as preservative
- » Smoke extract should be free from 3,4-Benzopyrene, which is carcinogenic in nature. It is to be eliminated during preparation of smoke extract powder.

iii) Taste enhancer:

- » It does not have own taste but enhances the taste when used with spices and other ingredients.

1. MSG – 0.15%
2. Inosininc acid (Nucleotide)
3. Succinic acid
4. Smoke extract – Phenol acts as antioxidant
   Aldehyde acts as preservative

iv) Starch:

- » Desirable content is 10%.
- » Pure carbohydrate is having no taste and no protein.
- » More than 10% starch, taste will go down.

$$\text{Starch + Water (sol stage)} \xrightarrow{\text{Heat at } (70-90)^0\text{C}} \text{Gel structure (Viscous)}$$

- This reaction is known as **Dextrnisation of starch** (α – Starch).
- At > 90^{0}C, elasticity will be lost.
- α – Starch is easy to digest and on standing for a long time changes to β- starch.
- α – Starch to β- starch → Moisture separates out.
- This is known as **Syneresis** or **Ageing of starch**.
- This depends on temperature. Maximum at 0^{0}C. At -20^{0}C or at 40-45^{0}C, it is nil.
- Most of starch flour is heavily loaded with thermophilic spore forming bacteria.
- Most of them are *Bacillus* and *Clostridium* group.

v) Spices:

- Desirable content is maximum 2% (Onion 1% and others 1%).
- Spices include

Ginger (stem part)	Cinnamon (bark)
Cardamom (seed)	Coriander (seed)
Cumin (seed)	Mustard (seed)
Chilly (fruit)	Curry (leaf)
Clove (flower)	Saffron (flower)

- **Function:**
  - Give flavour to food.
  - Reduce undesirable smell of fish.
  - Increase appetite.
  - Have limited preservative effect.
  - Have some medicinal effect.

» Flavour of the spice is due to aromatic component.

» All spices are heavily loaded with bacteria. So, needs sterilisation.

» Sterilisation of spices are made by

- Dry method: 60^0C for 15 min
- Wet method: Steaming for 15 min and then dried.
- Gas method

» To prevent spoilage, spice extract is used.

» Spice extract are obtained by

- Distillation: Essential oils of spices
- Extraction with alcohol: **Oleoresins**

vi) Antioxidant:

» Desirable antioxidant is Sodium ascorbate (0.01 – 0.02%)

» Advantage:

- Helps in colour change
- Prevents rancidity of fat

vii) Preservative:

» Desirable preservatives used are

- Sorbic acid or its salt: 200 ppm
- Benzoic acid or its salts: 0.02%
- Glucono delta lactone
- Fat coated sorbic acid.
- AF_2 (Furyl Furamide): 20 ppm (banned since carcinogenic in nature).

» Acids are generally not used as it lowers pH. So, its salts are used.

» The best preservative is Potassium sorbate (200ppm).

» Advantages of Potassium sorbate are:

  - pH = 6.7
  - Highly soluble in water
  - Stable against heat
- Characteristics of a good preservative:
  - Must control the growth of microorganisms for a longer period
  - Stable against processing temperature
  - Should be harmless for human consumption
  - Should not decrease the nutritive value of food
  - Soluble in water
  - Easy to use

viii) Colouring material:

- Colours are used to improve the appearance of the product.
- The colour of paste products is measured by using 'Hunter's scale'.
- Desirable colouring materials are coal tar dyes.
- The colouring materials at 5% colour solution are used at 2g/kg of the product.
  - Ponceau 4R
  - Carmosine
  - Red 6B
  - Fast red
  - Amaranth
- Characteristics:
  - Solubility to water
  - Purity of colour
  - pH
  - Heat resistance

1. Casing:

- » Casings are used for sausage.
- » **Desirable characteristics of a casing:**
  - Should be harmless for human consumption
  - Should have proper heat resistance
  - Minimum vapour permeability
  - Impermeable to gases like oxygen and other gases
  - Oil proof
  - Should have proper shrinkage characteristics
  - Easy sealable
  - Should have minimum tensileforce, elasticity.
- » There are two types of casing.
  - » **Natural casing:**

    Advantages:

    - » Since it is permeable, it gets spoiled quickly.
    - » High shrinkage coefficient
    - » It can be smoked
    - » Easy to prepare

    Example: This is the alimentary canal of sheep and goat.

    Disadvantage:

    - » Bacterial contamination
    - » Shelf life is 3 to 4 days
  - » **Synthetic casing** – Two types

    1. **Cellulosics**: It is used to prepare frankfurters (meat sausage)

- » Permeable for smoke
- » Colourless
- » Printed prior to stuffing
- » Before using the cellulosics, they need to be soaked in water for 30 min to get better stretches and easy stuffing
- » Devoid of bacterial contamination

2. **Plastic films**:

- » Most commonly used.
- » Thickness < 0.0254 cm
- » Saran and Krehalon are used in sausage making
- » It is prepared industrially
- » Bacteria can't pass through
- » Shelf life is 1 year.
- » Characteristics of Saran (PVDC – Polyvinylidine Chloride).
    - Tensile force = 800 – 1000 kg/cm^2
    - Extensional length = 30 – 80%
    - Negligible moisture permeability
    - Withstand temperature upto 95^0C

m. Stuffing:

- » Fish meat paste ground with all sorts of additives is stuffed into the casing by three ways. They are:
    1. Mechanical stuffer: Home made industry
    2. Air stuffer: semi-automatic
    3. CAM - stuffer: Fully automatic

» Precautions to be taken:

- Time gap between mixing and stuffing be minimum
- Paste must be filled tightly. No air bubble being trapped

n. Sealing:

» Sealing of the sausage casing is done by automatic sealing machine, called **'Ringer'**.

V. MEAT CURING

» Production of thermally cured meat pigment by the addition of NO_2 and other curing agents.

» Curing methods:

- Dry curing: Rubbing with agent
- Wet curing: Curing in water for 4-5 hrs.
- By injection: Through needle

(A) Myoglobin (Purple red) + O_2 → Oxymyoglobin (Bright red) → Metmyoglobin (Brown colour) → Heat → Myochrome (Brown colour) → in presence of air àHematin (Greyish brown)

(B) Myoglobin (Purple red) + NO → Nitric oxide Myoglobin (Red colour) → On heating → Nitrosyl hemochrome (Pink colour) (Stable).

(C) In presence of Ascorbic acid (Curing adjunct)

» Myoglobin (1) (Purple red) + O_2 → Oxymyoglobin (Bright red) → In presence of Na/K ascorbate→ Peroxymyoglobin → On reduction → Met myoglobin→ In presenceof acetic acid or Ascorbate → Myoglobin(1).

» So, (Ascorbic acid +$NaNO_2$) is desirable.

» $NaNO_2$ + Lactic acid(from meat) → HNO2 + Sodium lactate

» HNO_2 → On decomposition under acetic condition → NO + NO_2 + H_2O

Curing agents:

- NO_2 (Nirogen dioxide): Maximum 200 ppm.
  - For sausage $NaNO_2$ is used at a level of 0.01 to 0.02% (50ppm) and for Ham, 150 ppm
- NO_3 (Nitrate): 100 ppm
- Ascorbic acid and its salts (D-iso ascorbic acid: 6 to 10% w/w)

Curing Adjunct:

- Ascorbic acid prevents formation of Metmyoglobin and does not fix the colour. So, called Curing Adjunct.
  - Ascorbic acid: pH = 3.7 to 3.9
  - Sodium ascorbate: pH = 7.0 to 8.0

Rate of curing reaction depends on:

- Rate of diffusion of curing mixture
- Size of cut
- Amount of fat in the meat
- Temperature of curing reaction

VI. FISH HAM

- It is described as small cubes of red meat of tuna or dermal fat of pork, mixed with ground fish meat paste (as binding meat) and packed in synthetic casing.
- Ratio of Meat cubes to fish paste = 1 : 1
- Cross section gives a mosaic appearance.
- **Different types of Ham:**
  - Delux ham
  - Hidelux ham

- Ham cheese ham
- Ham burger

» Meat cubes:

- Meat cubes of tuna, whale/ mutton, dermal fat of pork are generally used.
- The colour of meat cubes needs to be colour fixed with NO_2 treatment

Treatment of pork fat:

» Pork fats are classified to 3 types based on melting point:

- Abdominal fat: Melting Point 37°C – Used for sausage.
- Dermal fat: Melting Point 70°C – Used for ham
- Kidney fat: Melting Point 37^{0}C – Used for sausage

» Fat cubes →Cut to 2 cm cubes → Heat treated at 80 to 90^{0}C for 4 to 5 min with frequent agitation to remove melted fat. →Cooled→colour fixed →used in Ham.

Colour fixation (Curing):

» $NaNO_2$ at a level of 0.01 to 0.02%

Other curing agents are:

» Salt: 2 to 2.5%

» Sugar: 2 to 3%

» Phosphate: 0.2 to 0.3%

» Spices: 0.4 to 0.6%

Ground fish meat (Binding meat):

» Fish meat paste is used as a binding agent.

» Fish meat paste is ground adding preservative, vitamins, seasoning agents like sugar, MSG, spices, smoke extract etc in a silent cutter within 15 mins to prevent denaturation of protein and to get high elasticity.

» Mixing of cured meat and binding meat is done in the mixer along with starch at a level of 5 to 10% for absorption of moisture and good binding with meat cubes.

Packing and sealing:

» The fish meat paste along with processed meat cubes of 140g to 150g are packed in ham casing having diameter of 7 cm.

» Sealing the casing and tying a string over the casing to keep the mass in tight condition.

Heat processing:

» 88^0 to 90^0 C for 60 to 80 min (In general 88^0C for 90 min)

» The temperature and time varies with varied packing weight

Packing weight (g)	Temperature (^{0}C)	Time (Min)
100 to 250	88 to 90	60 to 80
300 to 1000	88 to 90	85 to 100
1000 to 1500	88 to 90	100 to 150

Cooling:

» Cooling is done at 15^0C to 20^0C for 30 to 40 min for 100 to 1000g weight pack and 40 to 60 min for 1000 to 1500g weight pack.

Preservation:

» The preservatives used are:

- Glucono delta lactone
- Fat coated sorbic acid

Shelf life: 3 to 4 weeks or 1 month at Room Temperature at 8^0C to 10^0C.

Nutritional value:

» Protein: 13 – 14%

» Fat: 10 – 11%

» Moisture: 63 – 64%

» Ash: 2-3%

VII. SPOILAGE

» Spoilage of fish sausage and ham is mainly

- Due to growth of nicroorganisms
- Due to unhomogenous mixing of preservative

» Spoilage is indicated by TVBN content > 30 mg%

Types of spoilage	Causes and observation	Prevention
Gas formation	Due to production of NH_3 and CO_2 by various species of bacteria. (a) Along with outer surface discoloured to yellowish white - Due to entry of thermophiles through opening of mouth of casing (b) Inner part discoloured - Due to heat resistant bacteria from raw material (meat/ fat/ spices etc)	Taking enough care during production process. Use of raw material after due sterilisation.
Acidic spoilage	Due to Non-gas former, *Bacillus stearothermophillus* Resembling to 'Flat sour' of canned food originating from raw material	Use of raw material after due cleaning and sterilisation
Swell	Decomposition of protein by *Clostridium* sps and *Bacillus* sps.	Use of preservative
Putrefactive spoilage	With/ without gas formation mostly by proteolytic bacteria originating in the raw material	Use of preservatives Use of raw material after due cleaning and sterilisation
Softening	Softening of meat to mud like consistency along with sour taste Due to action of Dextrinised stage produced by decomposition of starch by *Bacillus circulans* which grow anaerobically.	Use of preservative and polyphosphate treatment

Discolouration	Yellowish white discolouration Due to thermopliles originating from raw material	Use of preservative Use of raw material after due sterilisation.
Spot type denaturation	Due to *Bacillus coagulans, Bacillus subtilis*	Use of preservative
Rancidity	Bt fat oxidation	Use of antioxidant

VIII. QUALITY OF PASTE PRODUCTS

» Quality parameter of the surimi may be divided under following three categories,

- Physical quality
- Chemical quality
- Biological quality

1. Physical quality:

» Physical parameter like pH, impurities, gel strength, etc are important parameters to assess the surimi quality.

a. pH:

» For the measurement of pH, add 90 or 190 ml of distilled water to 10g of test sample as needed to disperse. Homogenize the mixture and then measure pH of the suspension with glass electrode pH meter.

b. Impurities:

» The impurities are scale, bones, black membrane, fragments of intestine etc.

» Take 10g of thawed surimi. Put it into a polythene bag and stretch it to 1 mm thickness using a rolling pin.

» Objects over 2 mm diameter are calculated as 1.

Impurities	Grade	Impurities	Grade
0	10	12-15	5
1-2	9	16-19	4
3-4	8	20-25	3

5-7	7	26-30	2
8-11	6	>31	1

c. Gel-strength:

» There are number of methods used to assess the gel-strength and other rheological properties of surimi. The most popularly used method for measurement of gel-strength by **folding test, teeth cutting, expressible water percentage** and using instrument like **Okada- gelometer.**

i. Folding test:

» Round slices of sausage with 3 mm thickness and 10 mm diameter are taken for the test. The disks are folded into half between the thumb and forefinger of one hand. Grades of the products are given according to the observation as detailed as below.

» Observation Grade

Grade	Operation	Remark
AA	No crack after folding twice	Very good
A	No crack showing after folding into half	Good
B	Cracks gradually when folded into half Here crack is shown only on a part of folded edge	Fair
C	Cracks immediately when folded into half Crack runs through folded edge	Poor
D	If it breaks apart completely	Very poor

ii. Teeth cutting test:

» The sample for testing is prepared as for the folding test. The springiness of the sample is felt by biting the slice between upper and lower incisors and the scoring is made on hedonic scale.

iii. Expressible water %:

» Sample piece = 2 cm dia x 0.3 cm thickness; Weight = around 1g;

» Cover the sample piece with pre-weighed filter paper (Initial weight).

- Impart a pressure of $10kg/cm^2$ for 20 secs.
- Take the weight of filter paper after absorbance of the moisture from the sample piece (Final weight).
- The expressible moisture was calculated as: 100 × (Final weight – Initial weight) / Initial weight.

iv. Okada- Gelometer:

- The Okada-Gelometer has become the standard instrumental method used in surimi industries.
- In this test, a piece of surimi of about 25 mm thickness is placed under a plunger with a round top (5 mm). When pressure is applied through the plunger to the test piece, it gets gradually deformed and ultimately broken. The recorded peak force (F) at break and the depth of penetration are used describe the gel properties. Often these two values are multiplied together to give the gel- strength.
- Expressed in g.cm
- When gel strength is >1000 g.cm, it is considered as of good quality

2. Chemical quality:

a. Protein content:

The proteins are usually classified in to three categories on the basis of their solubility characteristics; Sarcoplasmic protein, myofibrillar protein and stroma protein. During the surimi processing removal of sarcoplasmic protein by washing cycle, increases the concentration of myofibrillar protein and thus enhances the gel-forming ability. Ultimately the quality of surimi depends on the protein content. The percentage of protein content of surimi is dependent on the species of fish in which the surimi is prepared.

b. Fat content:

Standard method for the measurement of fat content, precisely weight 5-10g of test sample with approximately the same quantity of anhydrous sodium sulphate and a small amount of refined sea sand, put in a mortar. Mash the material uniformly into dry powder and put it in a cylindrical filter paper. Extract and determine the fat content according to soxhlet method and calculate a value according to following formula:

$$\text{Fat (\%)} = \frac{(W_1 - W_0)}{S} \times 100$$

Where,

S: Quantity of test sample taken (g)
W_0: Weight of receptacle (g)
W_1: Weight of receptacle after has been extracted (g)

c. Moisture content:

The standard method for moisture content determination is by AOAC, Oven Drying method. Utilizing an overnight (18 hrs) drying period at 100°C, take 10gm of surimi sample and put into the oven and after 18 hrs calculate the moisture content by following formula:

$$\text{Moisture (\%)} = \frac{\text{Pre-dry weight (g) - After dry weight (g)}}{\text{Pre-dry weight (g)}} \times 100$$

d. Eber's test:

- » This test is done to know the spoilage or decomposition of the sausage.
- » Eber's solution = (10 ml Conc. HCl + 10 ml Ether + 30 ml Alcohol) Mix all and kept in a container.
- » When a piece of spoiled sausage is brought close to the solution, a dense white flume/ cloud is formed due to formation of ammonium chloride (NH_4Cl).

3. Microbiological quality:

Micro-organism	**Minimum level (per g)**	**Maximum level (per g)**
Aerobic plate count	10^5	10^6
Escherichia coli	11	500
Staphylococcus aureus	10^3	-
Salmonellasps	0	-

IX. ACTIVE FISH PROTEIN POWDER (AFPP)

» It is a spray dried fish meat powder which retains functional properties of protein.

Procedure:

» Fresh fish ($<6\%$ fat, salt free) → Dressing →Washing →Separating the meat →Mincing in a mincer →Minced meat →Water bleaching →washed meat→Colloidal mill (To which saccharide or sugar alcohol and fatty acid ester of Oleic and Palmitic acid is added) → pH is adjusted to (6.6 – 6.8) in the mixing tank by use of Acetic acid / Citric acid (If pH is <5.0 or >7.0, coagulation of protein occurs) →Spray drying for less than 10 sec (Air temperature at the entry point is 170^{0}C, at drying chamber 60^{0}C to 70^{0}C and outlet point 30-40^{0}C) (No denaturation of sugar protein) →AFPP

Addition of carbonic acid during spray drying:

» Addition of carbonic acid during spray drying helps

- Bringing down pH (Nearer to alkaline 6.4 – 6.6)
- Proteins won't be denatured
- Volatile acid gets evaporated

Proximate composition of AFPP:

» Moisture: 5-7%

» Protein: 65%

» Fat: 4%

» Viscosity: 20,000 centipoises

Different conditions:

» AFPP → Rehydrate with 2 to 3 times of water → Minced meat (Similar to that obtained from fresh meat).

» Fresh fish → Minced meat → Water bleaching → Grinding (5% sugar + 0.2% Polyphosphate) → Freeze dried → Product stored at 15^{0}C and RF of 15% with the use of inert gas → Product with Excellent Jelly strength with 3000 g.cm.

Difference between FPC and AFPP:

FPC	AFPP
Prepared by cooking	Prepared by high heat treatment
Proteins get denatured	No denaturation of protein
No Jelly strength	High Jelly strength

X. PARTIALLY HYDROLYSED AND DEODORISED (PHD) MEAT

» Fish→Picked meat → washed in cold water →Boiling or steaming to remove excess fat and other impurities →washing → Hydraulic press to expel water → Hydrolysis (by adding 1N HCl and keep at RT for overnight at a pH of 4.0) → adjust pH to neutral by adding 1N NaOH → Washing with cold and hot water (Repeat for 3-4 times) → Hydraulic press to remove moisture →**PHD Fish meat** → Dried → **FPC** (Fish protein concentrate).

XI. DIFFERENT SURIMI BASED PRODUCTS

1. Fish balls:

» The meat was picked manually from low cost marine fishes and then minced using a mechanical meat mincer after heading, gutting and washing. The fish mince was mixed thoroughly with 1 % salt, 5% starch and a variety of spices. Then balls of 2-3 cm in diameter were prepared and cooked a while in brine. The cooked balls after cooling are battered, breaded and then fried in hot vegetable oil.

2. Fish fingers:

» The picked meat was mixed with sufficient quantity of water containing 0.6% sodium tri-polyphosphate and 1 % common salt to form a pasty mass of hard consistency. They are then spread to a thickness of 1 cm in an aluminum tray and steam cooked for an hour. They are then cut into 8 cm x 2 cm X 1 cm weighing to about 15 grams. They are then battered, breaded and fried in hot vegetable oil.

3. Fish cakes:

» Asian-style fishcakes usually contain fish with salt, water, flour and egg. Fish cakes can be made by combining the ingredients with fish paste. The combined product is then shaped and left to cool. They are then battered and breaded using a machine. Then, they are usually fried with oil at around 180^{0}C and should reach the internal temperature of 75^0C. After the cooking process, they are frozen and packaged and are kept frozen until used.

» Fishcakes in Japan are commonly made from surimi, a paste made primarily from fish meat and starch, egg whites and spices. After formation, they are commonly fried or boiled before packing. Fish Cakes made from Fish Paste should be sterilized for 45 minutes at 80^0C, measured in the centre. After sterilization, the fish paste should also be stored below 10^0C with frozen fish pastes requiring a minimum storage temperature below -15^0C.

4. Shrimp and lobster analog:

» For the preparation of shrimp and lobster style products, the surimi paste is commonly mixed with pre-prepared surimi meat fibers and transferred to a molding machine or cold extruded in a three-dimensional shape. For imparting the color, a color solution is sprayed inside the mold before stuffing. Another way to impart the color is directly using the coloured paste (brushed) on the surface of cooked molded products. In the later method an additional heating is needed to set the colour.

11 Fish Curing

A. CURING OF FISH

- Curing is the traditional, cheapest and oldest method of fish preservation in our country.
- Curing fish is any fish preservation and flavoring processes by the addition of salt, with the aim of drawing moisture out of the food by the process of osmosis. Dehydration was the earliest form of fish curing under controlled condition.
- Methods of curing fish by drying, salting, smoking, and pickling, or by combinations of these processes have been employed since ancient times.
- Fish to be cured are usually first cleaned, scaled, and eviscerated. They are salted by packing them between layers of salt or by immersion in brine.
- The fish most extensively salted are cod, herring, mackerel, and haddock.
- Smoking preserves fish by drying, by deposition of creosote ingredients, and, when the fish are near the source of heat, by heat penetration. Tuna, Herring and haddock are commonly smoked. Kippers are split herring, and bloaters are whole herring, salted and smoked.
- Sardines, pilchards, and anchovies are small fish of the herring family, often salted and smoked and then preserved in oil.
- Fish are dried under controlled conditions of temperature, humidity, and air velocity.
- Since the dried product is relatively unappetizing and rehydration slow, other preservation methods are more common.

Advantages of curing:

» The method is very simple and can be easily adopted by the common man.

» It prevents contamination with harmful bacteria and enhances the storage life of the cured fish considerably.

» The calcium propionate does not affect the colour, smell, or taste of the cured fish in any way.

» It is comparatively a very cheap method. Considering the enhanced shelf life and increased price that can be realised by curing fish by this method, the slight increase in the cost of production can be treated as negligible.

Disadvantages in the present method of curing:

» Generally, poor quality fish is used for cured products.

» Mostly very low quality salt containing a lot of dirt and sand are used. Fish cured using this type of salt, without proper care for the hygienic conditions, are naturally of low quality.

» Good quality water is not available in such fish curing yards.

» The fish landed are just stacked in big cement tanks with alternate layers of salt. The importance of keeping the premises neat is also often not realised. After keeping the fish in salt in this type of tank for two or three days, the fish is taken out and sun dried on the open beach. In this process, it gets contaminated with a lot of sand and this is then heaped on the ground itself without proper packing.

» Fish cured in this way often shows contamination with red halophilic bacteria and these products cannot be stored for more than two or three weeks at the most.

Improved methods developed by CIFT:

» The fresh fish landed is immediately washed in clean sea water to remove slime, adhering dirt, etc.

» These are then taken to the fish curing yard where very strict care is to be taken to maintain hygienic conditions and quality of material. Unlike in the traditional method, all further processing work should be done on carefully cleaned tables to avoid contamination with sand, dirt etc.

- It is advisable to use water chlorinated up to 10 ppm. for all these cleaning operations
- On the processing tables, the fish is dressed, removing the viscera. In the case of fishes like sardines etc. it is advisable to remove the scales also to improve the appearance of the final cured product. The viscera should be immediately removed to the waste baskets kept under the tables. In the case of small fishes, where this is not practicable commercially, fish is salted directly after cleaning it well.
- The dressed fish is then washed in good quality water and the water is allowed to drain completely. This can be easily done in perforated plastic containers.
- After complete draining, the fish is taken to the salting table where good salt is applied to the fish uniformly by hand. Care must be taken to keep the hands of workers clean for this operation. In general, the salt-to-fish ratio can be 1:4 (one part salt to four parts fish).
- After salting, the fish is stacked in very carefully cleaned cement tanks and kept for at least 24 hours in these tanks. After this, the fish is taken out and just rinsed in fresh water to remove excess solid salt adhering to its surface.
- The salted fish is then dried in clean drying platforms. These can be either clean, raised cement platforms or bamboo lattices. If these are not available, drying can be done on clean bamboo mats, but in this case, fish must be dried to a moisture content of 25% or below.
- At every stage, extreme care must be taken to maintain proper standards of hygiene.

Preservatives used:

- Fish dried by this method is then dusted with a mixture of calcium propionate and fine powdered salt.
- This mixture can be made by intimately mixing three parts by weight of calcium propionate with 27 parts by weight of powdered salt.
- Care must be taken to see that the mixture is applied uniformly on all parts of the fish.
- After this, the fish can be packed in suitable weighed lots in sealed polythene bags for retail marketing. For wholesale marketing, the fish can be packed in polythene lined gunny bags. This type of packaging prevents excessive dehydration during storage as well as contamination with harmful bacteria.
- Generally, one kg. of this mixture is required for dusting 10 kg of fish

- When the fish is soaked in water just before cooking to remove excess salt, this preservative mixture is also removed.
- This is thus a very safe, easy and effective method for preserving cured fish for a long time.
- Fish preserved by this method can be kept in very good condition for a minimum period of eight months.

1. Dry curing:

a. Sun drying:

- Generally done for small size lean fish.
- They are first washed in sea water or brine solution and then dried under sun.
- Desirable moisture content is 10%.
- Should be free of sand content. The sand content should never be more than 7%.
- Shelf life is 2 to 3 months

b. Dry salting:

- Preferred for medium and large size fish.
- Belly split open, gills and viscera removed.
- Ratio of Fish to salt content should be (3-10):1
- The salt is to be rubbed over the fish and then stacked layer by layer for 24 hours.
- Then washed in potable water and dried under sun.
- Shelf life is 2 to 3 months.

2. Mona curing:

- Mona curing is mainly adopted for medium to small size fishes.
- Before salting, the intestine and entrails are removed by pulling out through the gill region without split opening the fish.

- The flesh is not exposed during salting thereby causing less contamination and the product has a shelf stability of about two months.
- The yield obtained by this method is about 70%.
- Therefore, contamination is less and gives a product with better appearance and visual appeal.

3. Wet curing:

- Fish is dry salted and allowed to remain under self brine (brine formed when fish is dry salted).
- They are marketed in that condition without drying.
- It is not suitable for fatty fishes like oil sardine which get rancid when exposed to air.
- Once the salted fish is taken out, the shelf life becomes very short i.e. 8 to 10 days.

4. Pit curing:

- It is a crude method of curing which is slowly getting into obscurity due to the poor quality of the product.
- It is still in practice in Tamilnadu.
- Larger fishes are eviscerated and de-gilled. Smaller fishes are used as such.
- The processing is done right on the beach.
- Large pits are made on the sandy beach and lined with Palmyra leaf mat.
- The fish are mixed with salt at a ratio of 4:1 (Fish: Salt) and stacked in the prepared pits.
- The pit is covered with another mat and the sand put over it, so that it remains completely buried.
- After 36 to 48 hours, the pits are opened.
- Fish are then packed in a bamboo basket and transported to market without any drying.
- The quality of fish cured by this technique is poor with a shelf stability of upto three weeks only.

5. Colombo curing:

- » It is so named because the main market for this type of cured product is Colombo, Sri Lanka.
- » It is prepared in Southern Indian states and Sri Lanka.
- » It is essentially a pickling process where some acidic materials are used during preparation. Medium size fishes like mackerel, otoliths, sardine, lactarius are used for the purpose.
- » The fish are gutted and gilled.
- » Then they are dry salted.
- » The salted fish are taken out and dried pods of Malabar tamarind (*Garicinia cambogea*) are placed in the body cavity and stacked in wooden barrels.
- » When the barrels are full, they are filled with saturated brine, closed air tight and exported to Sri Lanka.
- » Thus, the fish retain their qualities as long as they are under the brine.
- » The shelf life is more than 6 months.

B. DRYING OF FISH

Introduction:

- » Drying of fish is the removal of water from fish.
- » The term 'Drying' refers to removal of moisture from solid material.
- » The term 'Evaporation' refers to removal of moisture from liquid.
- » The term 'Dehydration' refers to removal of moisture under controlled temperature, RH and air velocity.
- » In fish, water constitutes about 70-80% and since water is essential for the activity of all living organisms, its removal will facilitate retardation of microbial and autolytic activity as well as oxidative changes and hence can be used as a method of preservation.
- » Water is held in food stuff in 3 different layers.

a. Adsorption layer (Bound water)

b. Diffusion layer (Heterogenous – water less tightly bound)

c. Free layer (Free water)

» Water is held by different types of bonds.

a. Chemical bond

b. Physico-chemical bond (Adsorptive bond and Osmotic bond)

c. Mechanical bond Ex. Capillary moisture

» Micro-capillaries (Radius <10^{-5} cm) have the capacity to condense moisture even from unsaturated air whereas macro-capillaries do not have such.

» In any process of drying, the removal of water requires an input of thermal energy. The thermal energy required to drive off the water can be obtained from a variety of sources, e.g., the sun or the controlled burning of oil, gas or wood, electrical heating etc. The thermal energy can also be supplied directly to the fish tissue by microwave electromagnetic radiation or ultrasonic heating.

Importance of Fish Drying:

» When huge amount of fish are caught either from the inland water bodies like reservoirs, lakes etc or marine waters and it becomes not possible to either consume or sell all the fish, then drying is the best suitable way for preservation.

» Preservation through drying is also needed in such place where there is lack of transportation facility.

» Fish is a highly perishable item. So, drying is a must to keep them free from rotting.

» Dried fish has a great demand in the market in the rural areas where poor people can keep them in home for days together for their consumption.

» Dried fish maintains the nutritional value of that fish and sometimes also being used as nutritious protein food for poultry.

» Dried fish stays for long time.

» As the dried fish weighs less, the transportation cost becomes low.

Principle:

When fish gets dried by removal of moisture from its body, there is a fall in its water activity. Water activity is the water available for activity of micro-organisms or enzymes (a_w having maximum value 1.0). In fresh fish a_w = 0.95. By drying, it comes down to 0.5 to 0.6 where most of the microbes can't grow. As a result fish gets preserved.

Advantages of Fish Drying:

Fish drying has many advantages. Among those some important advantages are listed here.

- Low cost
- No experience required for preserving fish in this method of drying.
- No need of using advanced devices.
- Dried fish is also very tasty and nutritious like fresh fish.
- Drying fish can be kept for long time.

Disadvantages of Fish Drying:

- Fish drying fully depends on weather. In adverse weather condition, it takes long time for drying. Sometimes it becomes not possible to dry the fish hygienically.
- The fish may get infected by worms or insects during drying in open place.
- If fish gets spoiled and then dried, dried fish gives off-odour.
- If case hardening occurs during drying, dried fish gets spoiled soon and also housewives face difficulties while cooking the dried fish.
- The dry fish may damage due to incorrect and non scientific way of storage.
- Cost of machineries for drying is high.

1. Drying process:

- Drying process of fish involves (a) Heat transfer and (b) Mass transfer.

a. Heat transfer: (Inward transfer of heat)

- Emission of heat from the source (Sun/ Mechanical device)

- » Emitted heat transferred to surface of material to be dried.
- » Conduction of heat occurs within the material.
- » Conducted heat provides latent heat of evaporation and the partial enthalpy of dilution of the system which is regarded as a solution.

b. Mass transfer: (Outward movement of water)

- » Migration within the material to the surface
- » Removal of moisture from the surface
- » Mixing of water vapour with the atmosphere surrounding the material
- » Removal of moisture from the vicinity of the surface

At normal temperatures, fish muscle can be considered to be a gel. It remains as a gel until a considerable quantity of water has been removed. During drying, considerable shrinkage takes place, as well as other irreversible changes, and dried fish will not reconstitute to their original condition.

Equilibrium Moisture Content (EMC):

- » When a wet material is brought into contact with air stream of constant RH and temperature in such amount that properties of air stream remains constant and if exposure is for sufficiently long time, then the moisture content achieved by the material on reaching equilibrium is called EMC. In that condition, the vapour pressure of the water held by the product is equal to the water vapour pressure of the surrounding air.
- » EMC depends on direction in which equilibrium is achieved (i.e. from dried material and wet material).
- » EMC varies for any given % of RH and temperature. EMC reduces with increase in temperature of air.

Free Moisture Content (FMC): Moisture content in the sample above the equilibrium moisture content.

2. Drying phases:

During air drying, water is removed from the surface of the fish and water moves from the deeper layers to the surface. Drying takes place in two distinct phases. In the first phase, whilst the surface of the fish is wet, the rate of drying depends on the condition (velocity, relative humidity etc.) of the air around the fish. If the surrounding air conditions remain constant, the rate of drying will remain constant; this phase is called the 'constant rate period'. Once all the surface moisture has been carried away, the second phase of drying begins and this depends on the rate at which moisture can be brought to the surface of the fish. As the concentration of moisture in the fish falls, the rate of movement of moisture to the surface is reduced and the drying rate becomes slower; this phase is called the 'falling rate period'.

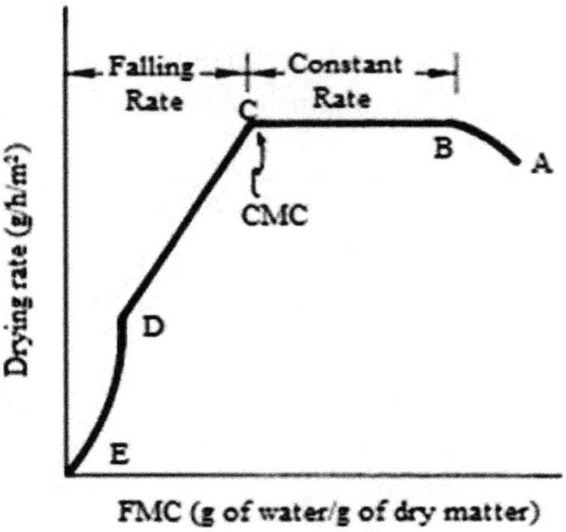

A – B : Initial adjustment period
B – C : Constant rate period
C : Critical Moisture Content
C – D : First falling rate period
D – E : Second falling rate period

Figure 11.1: Drying rate curve

Drying rate 1:

» Drying process of a material can be described as a series of steps in which drying rate plays a key role.

» A is the actual moisture content and section (A – B) is the initial adjustment period of the moisture content of the product before drying.

» Point B represents equilibrium temperature conditions of the product surface.

» Section B to C of the curve, known as the constant rate periods, represents removal of unbound water from the product. The water acts as if the solid is not present. The surface of the product is very wet and water activity equals to one.

- The constant rate period continues as long as the amount of water evaporates is equal to the amount of water supplied to the surface of the material.

Drying rate 2:

- The falling rate period is reached when the drying rate starts to decrease, and the surface water activity falls to less than one.
- The rate of drying is governed by the internal flow of liquid or vapor. This point is represented by C in the figure.
- At this point, there is not enough water on the surface to maintain a water activity value of one.
- The falling rate period can be divided into two steps. A first falling drying rate occurs when wetted spots in the surface continually diminish until the surface is dried (Point D).

Drying rate 3:

- Second falling rate period begins at point D when the surface is completely dry.
- The plane of evaporation recedes from the surface. Heat required for moisture removal is transferred through the solid to the vaporization of moisture in the solid and the vapor moves through the solid into air stream.
- The amount water removed in this period can be relatively small compared to the constant rate and first falling rate period.
- However this period may take much longer than constant rate period because the drying is slow.

Constant rate drying phase:

During this period the rate of drying is dependent on several factors:

1. Air temperature:

At the beginning of drying, the heat energy required for evaporation is balanced by the heat supplied by the surrounding air. Warm air can provide more heat energy and, provided that the air speed and relative humidity will allow a high rate of water movement, the rate of drying will be increased.

2. Relative humidity of the air:

The lower the relative humidity of air surrounding the drying area, the greater will be the ability to absorb water and the faster the rate of drying.

3. Air velocity:

Air velocity has a positive relation with rate of drying. Better the speed of the air over the fish, the greater will be the drying rate. The air around fish consists of an immediate stationary layer above the fish, a slowly moving middle layer and an outer turbulent layer. On saturation of the immediate stationary air layer, the moisture passes into the slowly moving middle layer. The higher the air speed in the outer layer, the thinner the slow moving layer, allowing more rapid movement of water away from the fish.

4. Surface area of the fish:

The larger is the surface area, the faster the rate of drying. By scoring and splitting the fish, the surface area increases relative to the weight/thickness resulting in the rate of drying to be faster.

Falling rate drying phase:

Once the free surface moisture has been removed, the rate of drying depends on the movement of moisture from interior to the surface of the fish.

It is characterised by:

a. Increase in temperature of surface of material

b. Surface is not wet (i.e. surface will be in dry condition)

    » First falling rate period - Here whole surface not completely dried.

    » Second falling rate period – Here surface area is completely dried.

Several factors influence the falling rate of drying:

- » Nature of the fish: a high fat content in the fish retards the rate of drying.
- » Thickness of the fish: the thicker the fish, the further the water in the middle layers has to travel to reach the surface, slowing down the drying rate.

- Temperature of the fish: Diffusion of water from the deeper layers to the surface is greater at higher temperatures.
- Water content: As the water content falls, the rate of movement to the surface layers is reduced.

Critical Moisture Content (CMC): Moisture content at the end of Constant rate period. It depends on thickness of the material.

Optimum conditions for drying:

- Air temperature: 40 to 45°C. If it is more, fish meat gets cooked instead of drying. Thus, meat flakes will break on preservation.
- Relative Humidity: 45 to 55%
- Air velocity: 200 to 300 ft/ min. If it is more or less, the drying rate will slow down.
- Generally 25% moisture is removed during 1 hour, 45% moisture removed during next 3 hours and then 5% moisture removed during 2 to 3 days.

3. Methods of drying:

- There are basically two methods of drying fish.
- The common and traditional method being sun drying, the natural drying method, which is done by utilizing the atmospheric conditions viz., temperature, humidity and airflow.
- In recent times, the controlled artificial dehydration of fish has been developed so that fish drying can be carried out under controlled conditions, regardless of weather conditions by use of mechanical dryer.

i. Natural or sun drying:

- Drying is a method of food preservation that works by removing water from the food, which inhibits the growth of microorganisms.
- In this type solar and wind energies are utilized as the source of energy.
- Open air drying using sun and wind has been practiced since ancient times to preserve food.

- Beach Drying or drying on the ground
- Mat drying
- Platform drying
- Rack Drying

a. Beach drying:

In this method, fish is dried directly on the sandy beach. When catch is in huge quantity and there is no facility for drying so much fish at a time, the fish are dried either on the beach or in an open space. It takes one to 2 days for drying them.

Disadvantage:

» Contamination with sand
» Contamination with microbes
» Poaching by birds and animals
» Bird droppings on dry fish
» During drying air movement at ground level is usually very slow.

b. Mat drying:

Drying fish on mats made of coconut leaf, split bamboo, palm leaf, coir ropes etc. It reduces sand and microbial contamination. However, sand contamination occurs when wind blows and deposit sand on the fish.

c. Platform drying:

Drying fish on cemented platforms. Here, sand contamination is almost negligible.

Disadvantage:

» Drying is slow and case hardening may take place.

d. Rack drying:

» Drying on racks erected on the ground. The frame is either made up of wood, bamboo, PVC pipes or iron.

- The rack is made by making a horizontal surface using plastic nets, garden fencing or stainless steel wire mesh.
- Sometimes coconut leaves are also used to make the rack.
- It is the best method of natural drying.
- Racks should be sited away from forests or high buildings which will reduce the air movement or cast a shadow over the racks.

Advantage:

- Contamination by sand or dust is very less.
- Drying is very fast as moisture is removed from lower to upper side.
- Moreover, water drops down to the ground which evaporates naturally.
- Raised racks provide some protection from animals.
- Fish dried on racks can be more easily protected from rain by covering them with plastic sheets. Fish on the ground can be covered for protection against rain but not against water on the ground.
- During rain, sloping racks allow any surplus water on the surface of the fish to drain away.
- During drying of salted fish, the fish will take up moisture from the surrounding air if the relative humidity rises above 75 per cent. It may, therefore, be necessary to remove the fish from the racks at night, or during rain, when the humidity tends to rise. If the fish are piled and covered in plastic overnight, the absorption of water is minimised until the fish can be put out for further drying the next day. If the fish are press-piled at night by placing weights on top of a stack of fish, movement of water to the surface of the fish is encouraged and the subsequent drying rate will be increased.

ii. Artificial / mechanical dryers:

In case of drying fish using mechanical dryer, the temperature, RH and air velocity remained under controlled condition. Several types of mechanically powered driers have been developed and used commercially in different parts of the world. Fish are dried in a fan driven air-stream; the air is usually heated and, in some cases, the air can be recirculated to control the relative humidity.

- Solar drying using Solar tent dryers, Solar cabinet dryers
- Hot air dryers
- Cabinet dryer
- Tunnel dryer: Solar, Multi deck tunnel, Contact Dryers
- Vacuum dryers: Rotary dryers, Drum dryers
- Freeze dryer

Factors affecting drying operation in a dryer:

a. Design of a drier
b. Insulation
c. Volume of air flow
d. Proportion of recirculated air
e. Quantity of wet material fed at a time
f. Nature of material to be dried

a. Solar driers

- In recent years, the solar powered driers have been developed for drying fish.
- Solar powered units are simple, inexpensive and efficient for drying fish.
- In these, the energy of the sun is concentrated to produce higher temperatures resulting increased rate of drying.
- Raising the air temperature increases the amount of water the air can hold, thus the relative humidity will be reduced and the air will be able to absorb additional water vapour.
- The relative humidity becomes too high for rapid natural drying.

There are two basic methods of collecting and concentrating the solar energy:

i. Parabolic reflectors: Sunlight falling on a mirror is focused to a point, where the temperature becomes much higher. This is not used for fish dryers since the normal requirement is to slightly increase the temperature of large volumes of air.

ii. Absorption units: A black surface absorbs heat energy from the sun far more effectively than a light coloured surface. If an insulated box, painted black on the inside and covered with clear glass or plastic, is placed in the sun, the temperature of the enclosed air is increased considerably. If the box has openings at the top and bottom, the air, as it is warmed, will rise and create a flow. Fish placed in the box will, therefore, be exposed to a flow of air that is warmer and of a lower relative humidity than the ambient air.

The black heat collection units can be connected to a drying chamber to supply a flow of warm air and it is not then necessary to expose the fish to the direct rays of the sun which can cause problems with case hardening and cooking of the fish if the temperature is not adequately controlled.

b. Hot air dryers

» The hot-air drying oven is mainly equipped with a fan, heat source, drying chamber and the trays. This is a method of contact dryer where fish remains in contact with trays.

» When drying, the fan blows the air to a heat source which is usually supplied by electricity.

» In the oven, the fish won't be overheated or under heated with an automatic temperature control system.

» After heating, the hot air goes through the trays and dries the fish.

» Furthermore, the trays provide a dispersed space so that fish can be evenly dried in the hot air, resulting in a uniform color of products. After that, the air temperature reduces and it is sent again to the heat source by a fan for next circulation.

» In the chamber most of the air recycled through convection process so that the drying oven becomes cost-effective.

» Finally, the dried fish will be vacuum packed by vacuum packaging machine for storing and marketing.

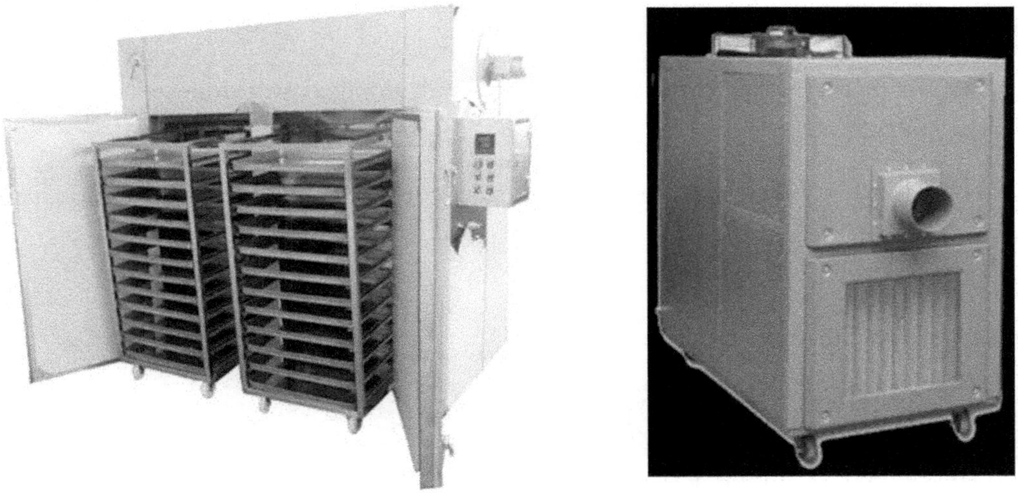

Figure 11.2: Hot air dryers

c. Cabinet dryer:

- Tray-type cabinet dryer was developed to provide small to medium-scale processors which is suitable for drying perishable products, such as fishes.
- The cabinet fish dryer (CFD) consists of a) drying cabinet which serves as the drying chamber; b) set of drying trays where products are placed during drying; c) air moving devices such as fan, chimneys, air inlets and/or outlets for discharging humid air; and d) heating units such as nichrome wire heater, Liquified Petroleum Gas (LPG) or biomass furnace and heat exchange for drying the products.
- The cabinet is constructed with proper insulation to prevent loss of heat.
- The moisture from fish body is removed by application of heat and expulsion of the liberated moisture from the surrounding of fish.
- Air heated by steam or electricity is forced over the fish by a blower carrying away the moisture with it.
- The moisture removing property of the air is directly proportional to its temperature, air velocity and inversely proportional to relative humidity.
- The quality of the dry fish depends on the proper control of fish parameters.

- In a cabinet dryer, fish to be dried as spread in trays and placed in the cabinet in a fixed position.
- In another way, the fish containing trays are arranged on a trolley which is pushed into the drier.
- Hot air is forced over the fish and the moisture laden air is removed by the exhaust.

d. Tunnel dryer:

i) Solar tunnel dryer:

Solar tunnel dryer is used for drying small and medium fish which is constructed using locally available materials and the size of the dryer was 20m×2m with 20m^2 drying area. The solar tunnel dryer consists of a flat plate air-heating collector, a tunnel-drying unit and four small fans to provide the required airflow over the product to be dried. Both the collector and the drying unit are covered with transparent polythene sheet. Black paint is used as heat absorber in the collector. The products to be dried are placed in a thin layer on a bamboo made mat in the tunnel dryer. The whole system is placed horizontally on a raised platform. The air is provided by four DC fans operated by two photovoltaic modules as the air is passed over the products. The black painted collector absorbs heat through transparent polythene from sunlight. The air from the fans blows over the collector and then heated air passes over the products. Solar radiation also passes through the transparent cover of the dryer and heats the products in the dryer. This enhances the drying rate and the temperature in the dryer in the ranges of 35°C to 55°C.

Benefits:

- More consistent quality product can be obtained.
- Protection against contamination (dirt and dust) and also through animals, birds and insects.
- Reduction in drying time upto 60%.
- Drying is much faster than open sun drying and the product retains its nutritive value.

ii. Electric tunnel dryer:

- It is an oblong wooden cabinet which is divided horizontally into two chambers.
- An electric heater of required intensity is installed in upper chamber.
- Air is forced across the heater at high velocity by a bower fitted at one end of the tunnel.
- The hot air is reflected downwards into the lower chamber by an air reflector.
- Fish to be dried is kept in the lower chamber arranged in trays.
- A part of the hot air loaded with moisture is sucked out by an exhaust fan while the rest of it is re-circulated along with the fresh air drawn in the place that is sucked out in order to arrive at the desired relative humidity.
- Small and medium size fishes like mackerel, lactarius, croakers, sardine, anchovies etc can be dried at temperature of 45^{0}C to 50^{0}C, 55 to 65% RH and 100-120 m/ min air velocity in 14 to 16 hours.
- If the RH is high, the fish gets cooked resulting in case hardening. Therefore, air velocity, temperature and RH should be controlled carefully.

iii. Multi-deck tunnel dryer:

- In this case, several tunnels are arranged on one another and air is blown in zigzag manner for three times before it is exhausted.
- Baffle plates are used to regulate the air flow in the desired direction.

e. Fluidised bed dryer:

- It is suitable for drying small fishes.
- Hot air is forced from the bottom of a conveyor belt on which fish to be dried passes.
- Air velocity is sufficiently high to fluidise/ agitate the fish.
- Fish enters at one end and the dried fish comes out from other side of the conveyor belt.

f. Vacuum dryer:

» Vacuum drying is a process in which materials are dried in a reduced pressure environment, which lowers the temperature required for rapid drying.

» Major advantages of vacuum drying are as follows:

- less energy is needed for drying,
- it is highly suitable for heat sensitive food materials,
- faster method than other drying methods,
- it retains integrity of materials etc.

» In general, vacuum drying is performed in combination with other drying techniques.

g. Rotary drum dryer:

» Rotary drum dryer is a double walled cylindrical cell mounted horizontally at a slight angle on bearing so that they can be rotated by an electric motor.

» The dryer essentially consists of aluminium structure bearing hooks and stainless steel ring.

» The aluminium structure is supported by S.S. pipe and bearing. Nut bolts are used to join interconnected parts of the dryer.

» The dryer is operated by 0.75 HP motor having speed regulator.

» The rotating dryer was operated at speed of 90 rotations per minute (RPM).

» The whole system is placed vertically on raised 2" × 2" angel platform.

» Hooks and stainless steel structure are used to hang large size fishes and round shaped plastic net was used as additional platform for drying of small fishes.

» When the system is on, the dryer rotates and ambient air passes over fishes, which enhance the drying process.

» The rotating dryer was constructed using locally available materials.

» About 40 kg of fishes like jewfish, Bombay duck and ribbon fish can be dried at a time in a rotating dryer.

h. Freeze drying:

» Freeze drying or lyophilisation is a dehydration process used to preserve fish and make it into more convenient for transport.

» It is a method of water removal from material by sublimation.

» This drying process is divided into three stages:

(a) pre-freezing of wet material,

(b) primary drying (sublimation of frozen water under vacuum) and

(c) secondary drying stage (desorption of residual bound water from material).

» Freeze drying is initially freezing the material and then reducing the surrounding pressure to allow the frozen water in the material to sublimate directly from the solid phase to the gas phase.

» It is one of the best method of water removal and results in a final product of much higher quality compared to any other drying technique.

» For rapid freeze drying, some heat must be supplied to the fish if evaporation is to proceed at a rapid rate. Moisture must also be removed from the vacuum chamber, otherwise it will become saturated and no further drying will be possible.

» Freeze drying requires a high energy input and is only feasible for very high value products.

» Freeze dried products have the advantage that volume of the freeze dried products will be same to that of fresh one and they can be stored under ambient conditions as long as the packaging is impervious to water.

» Khin et al. (2005) had compared the drying technologies and observed that freeze drying, vacuum drying and osmotic dehydration are considered too costly for large scale production of fried products.

» Types of freeze drying:

- Vacuum freeze drying
- Vacuum contact fat drying

- Accelerated freeze drying
- Vacuum fat drying
- Dehydro-freezing

i. Heat pump drying:

» A heat pump is a device that transports energy from a low temperature source to a higher temperature sink. This transfer requires an input of work which may be supplied mechanically as in a vapour compression cycle.

» The most common type of heat pump operates on the vapour-compression cycle and a basic unit consists of the evaporator, compressor, condenser and the expansion valve. Heat transport is achieved through phase change of the refrigerant. The refrigerant in the evaporator absorbs heat and vaporises at low pressure and temperature. As the vapour condenses at a higher pressure in the condenser, it rejects heat at higher temperature. When used in a drying system, the heat pump dryer cools the process air first to saturation, and then further for condensation of water (dehumidification), thus increasing the drying potential of air.

» In this process, it also recovers low grade heat (sensible and latent) from the air, which is made available at the condenser as sensible heat of higher quality.

» Heat pump dryer consists of a heat pump system and dryer, the performance of the dryer is greatly affected by the performance of heat pump system.

» Heat pump drying is a technology by which materials can be dried at low temperature and in an oxygen-free atmosphere using less energy than common drying methods.

» This drying recorded less drying time than other techniques and it is simple to design.

j. Microwave drying:

» Electromagnetic energy of microwave can directly interact with foods to quickly raise centre temperature since most of the food materials are dielectric materials. It can store electric energy and convert it into heat. It is volumetric heating and quick raise of temperature is possible.

» In microwave drying, microwaves can penetrate materials and resulting heat will help in water removal during drying. Microwave energy at 915 and 2450 MHZ can be absorbed by water containing materials and can be converted into heat. Food materials are dried by the interactions between the electromagnetic energy and polar molecules within the material. Polar molecules rotate in response to the applied oscillating electromagnetic waves. The reorientation of molecules in high frequency electric field occurs frequently and rapidly, resulting in molecular friction that generates heat.

» It is an energy efficient technology and can maintain the quality of food material upon drying.

» Some disadvantages of microwave drying are non-uniformity in drying, limited penetration depth, lack of equipment for large scale production etc.

4. Changes in properties of food materials during drying:

a. Physical changes:

» Increase in solute concentration

» Shrinkage and case hardening – Under certain conditions, where the constant rate drying has been very rapid, the surface of the fish can become 'case hardened' and the movement of moisture from the deeper layers to the surface is prevented.

Effect of shrinkage controls the rate of drying.
If shrinkage is more => Rate of drying reduces
Because:

i. Reduction in surface area exposed to air

ii. Surface hardening occurs

iii. Material gets warped.

To prevent this, moist air needs to be passed through for drying.

» Porosity – It develops by creating steam within the product a case hardened surface through rapid drying

b. Chemical changes:

» Browning reaction: (Amino group + reduced sugar) It proceeds rapidly till moisture is 20% to 15% and it is favoured by temperature and sugar present in the material to be dried.

» Oxidation of fat: It leads to rancidity. It generally occurs in drying of fatty fishes like oil sardine etc.

Process of Small Sized Fish Drying:

There are many small sized fish such as mola, sardine, anchovies etc which are preserved by drying in sunlight. The steps for drying are as follows:

» For small sized fish, there is no need to go for pre-processing viz. removal of gills, intestines, scales etc. from their body.

» But sometimes if needed remove those from the fish by slightly pressuring on their belly.

» Wash the fish properly by using clean and fresh water.

» Then, keep the fish on a bed and make it dry by sunlight.

» It takes about 3-5 days to make the fishes fully dry.

» Keep the fish free from worms, insects and birds while drying. For this, mosquito net can be used.

» For drying the fish fast, turn over those several times a day.

Process of Medium and Big Sized Fish Drying:

Some big sized fish can also preserved by drying through sunlight. The fishes which are preserved by drying are ribbon fish, croakers, lizard fish, mackerel, horse mackerel etc. The steps for drying these types of fish are as follows:

» Pre-processing of fish like removal of scales, fins, gills, intestines etc. from the body of fish.

» Then wash the body with clean water.

» After that cut the fish from their head to tail by knife. Cut in such way so that one part remains attached with another.

» For big size fishes, fillets are made from the whole body and then dried. Sometimes, small pieces are made from large fish.

» Then keep the fish over a bed for drying.

» Use reticular or mosquito net to keep the fish free from insects, worms and birds.

» A large amount of water removed from the fish within three days. Then hang the fish with rope.

» Full drying process takes about 7-8 days depending on the weather, size of fish etc.

» Generally drying fish contain about 10-20% water.

» After drying the fish it is kept in various types of pots.

» Make a thin layer with fish oil inside the pot.

5. Quality evaluation of the dried fish:

» Quality of dried material is judged by estimating its

- Ability to rehydrate
- Speed of rehydration

» Rehydration is the process of replacing water to dehydrated foods.

» Rehydration ratio = W_R/W_D

» Coefficient of rehydration = $(100 - A)\, W_R / (W_D - W_M)$

Where,

W_R = Weight of rehydrated material
W_D = Weight of dehydrated material
A = Original moisture content (%)
W_M = Weight of moisture content in dehydrated material

» % of moisture in rehydration material depends on

- Quantity of dehydrated material
- Temperature
- Time of rehydration

» Organoleptic test which includes colour, texture, flavour, odour and overall acceptance of the product.

C. SALTING OF FISH

Introduction

- Salting of fish is a method of traditional fish processing in most of the countries in the world. It is very often used in combination with drying and smoking.
- In this, fish is exposed to sufficient quantities of common salt i.e. sodium chloride or strong solution of brine which can prevent or reduce bacterial action.
- Thus, the salt from outside solution will penetrate into the fish tissue which has low salt concentration by the process of osmosis.
- When salt concentration in fish tissue becomes around 6 to 10 percent, salt in the tissue will prevent the activity of most spoilage bacteria. By this time, removal of fish tissue water will also reduce the activity of the spoilage bacteria.

1. Salt used in salting of fish:

- Generally commercial salt is used for salting of fish.
- Salt is of two types (a) Pure Common salt (Sodium Chloride) (b) Commercial salt
- Pure salt is not hygroscopic.
- Commercial salt is classified into three types based on its source and method of manufacture.

i. Solar salt: Prepared by evaporation of sea or salt lake water.

ii. Brine evaporated salt: Underground salt deposits are brought to the surface and is evaporated.

iii. Rock/ Mine salt: These are sterile. Natural deposits of salt are ground to varying degrees of fineness without any purification.

2. Chemical composition of salt:

- Besides Sodium chloride, the main chemical impurities are calcium and magnesium chlorides and sulphates, sodium sulphate and carbonate, and traces of copper andiron.
- Solar salts are less pure than mine – evaporated salts.
- Bulk density = 0.9 to 1.6 tons. m^{-3} (Standard is 1.2 tons. m^{-3}).

- Solubility is expressed in terms of
- g per 100g solution (at-21.2^{0}C => 22.4%)
- g in 100g water (at-21.2^{0}C => 28.9%)
- Since rate of formation of solution is much faster than the rate of diffusion, it needs agitation.
- When salt solution is prepared, temperature begins to fall i.e. when 1 kg of salt is dissolved in water, about 20 kcal of heat will be absorbed by dissolution.

Table 11.1: Grades of commercial salts under BIS- 797-1982

S. No.	Characteristics	Requirement for	
		Grade I	Grade II
1	Sodium Chloride (as NaCl) Percent by mass, min.	99.5	98.5
2	Matter insoluble in water percent by mass, max.	0.05	0.2
3	Calcium salts (as Ca) percent by mass, max.	0.03	0.2
4	Magnesium salts (as Mg) percent by mass, max.	0.01	0.10
5	Sulphate (as SO_4) percent by mass, max.	0.2	0.6
6	Iron compound (as Fe) ppm, max.	10.0	20.0

- Usually Grade II salt is preferred for salting small and medium size fish.
- Magnesium chloride is hygroscopic.
- Calcium and magnesium chlorides tend to slow down the penetration of salt into the flesh.
- Fish salted in pure salt may be soft and yellow in colour. Calcium and magnesium salts give a whiter colour but tend to impart slightly bitter taste.
- Trace quantities of copper can cause the surface of salted fish to turn brown colour giving a look of poor quality product.

3. Microbiological purity of salt:

- Solar salts contain large numbers of salt tolerant bacteria called halophiles and counts up to 10^5 per gram.

- » These halophiles cause reddening of wet or partly dried salt fish.
- » Halophilic moulds can grow on fully dried fish and can cause the formation of dark patches called "Dun".
- » Halophilic moulds occur more frequently in rock salt.
- » Addition of preservative agents to salt is very expensive. So not added.

4. Physical properties of salt:

- » Fine grain salt dissolves more rapidly in water and is preferred for making brines.
- » If fine grain salt is used directly on fish, it may cause a rapid removal of water from the surface which becomes hard and prevents penetration of salt to inside of the fish. This is called 'Salt burn'.
- » For dry salting, a mixture of large and small grain sizes is recommended.

5. Principle of salting:

a. When salt enters tissue, it causes high osmotic pressure outside the bacterial cell => Plasmolysis of cell => Bacteriocidal action.

b. Dehydrates the tissue => a_w (Water activity) reduced => Bacteriostatic action.

c. Ionisation of salt givesChloride ion which is harmful to micro-organisms.

d. When salt is dissolved, it reduces solubility of oxygen in solution => reduction in multiplication of micro-organisms.

e. Salt interferes in the actionof proteolytic enzyme.

6. Salt uptake by fish:

- » Salt uptake and replacement with water by the fish depends on various factors.

  a. Higher the fat content, slower the salt uptake.

  b. Thicker the fish, slower the salt penetration to the centre

  c. Higher the freshness of fish, slower the salt uptake.

  d. Higher the temperature, faster the salt uptake.

» During subsequent drying, the presence of salt has following effects.

a. Higher the salt concentration, greater the replacement of water. Thus, less water that remains need to be removed during drying.

b. Higher salt concentration results in slow drying of fish.

c. At RH of more than 75%, when the fish will be dried or during storage of dried fish, fish will not dry further, rather the higher salt content will absorb more moisture from the environment.

» Minimum requirement: 15 percent. But the product will be salty taste. To prevent this during brining, fish needs to be covered with "Cryovac".

» Successful preservation depends on time taken for salt concentration inside the fish, to reach the minimum level to arrest autolysis and growth of micro-organisms. Hence, speed of penetration of salt is important. It depends on specific area (=surface area/ thickness).

» Brining time required (T): $T = \frac{2.303}{KQ^2} \times \log \frac{C_b}{C_b - C}$

Where,

Q = Specific area
C_b = Concentration of salt in brine
C = Concentration of salt in fish
T = Time of brining

» Salting Coefficient (K): 0.11 to 0.20. It depends on (a) Histological structure, (b) Temperature, (c) Viscosity of the medium and (d) Fat content.

» Concentration of salt in fish (C): It can be determined by the formula

$$C = \frac{100 \times S}{S + W}$$

Where,

C = Concentration of salt in fish
S = Salt content in fish
W = Water content in fish

- Salting time: Period during which fish is kept in contact with salt or brine to attain the required salt content.
- It depends on (a) Fat content, (b) Histological structure, (c) Viscosity, (d) Size of fish, (e) Type of processing, (f) Brine concentration, (g) Temperature and (h) Quality of salt used.
- For every rise of 1^0C, rate of salting increases by 2.5% to 3.6%.
- Impurities such as Calcium Chloride ($CaCl_2$) present in salt are responsible for delay in salting process.
- Fick's 1st law: It indicates the quantity of salt diffusing in a period 'T' across 1 cm^2 area.

$$\frac{dm}{dt} = -D \times q \times \frac{dc}{dx}$$

Where,

Concentration gradient
D = Diffusion coefficient
Q = Area/ thickness
m = Quantity of salt diffusion

- Fick's 2nd law: It indicates the increase in concentration of material being diffused in the same period of time 't'.

$$\frac{dc}{dt} = - D \times \frac{d^2c}{dx^2}$$

7. How should fish be prepared for salting?

- The raw material used for salting preservation should be of the best quality to make the best products.
- Stale or spoiled fish should never be used for salting.
- To salt effectively, fish must be prepared so that water removal and salt uptake can take place quickly and easily.

» The following guidelines should therefore be followed:

- the thickness of flesh of the fish must not be more than 25 to 30 mm (about 1 inch);
- small, whole fish need only be gilled and gutted if they are thinner than this;
- medium-sized fish should be split through the back, then gilled and gutted;
- large fish can either be filleted, with the skin left on or split and the flesh scored deeply at 25mm (1 inch) intervals;
- very large fish must be filleted and the flesh cut into 25 mm (1 inch) strips;
- it is best to scale the fish, because the removal of water and penetration of salt will be easier.

8. Types of salting:

» There are different methods of salting fish. They are:

a. Dry salting:

» This is the most widely used method of fish curing.

» Dry salting is advisable for fishes of any size, except fatty fishes.

» The fish is gutted, beheaded or ventrally split open and the viscera removed followed by washing.

» Scoring is also practiced if the flesh portion is thick for facilitating better salt penetration.

» Granular salt is then applied in the ratio 1:3 to 1: 10 (salt to fish) depending upon the size of the fish.

» The fish is then stacked in clean cement tanks or other good containers layered with salt and weight is applied from top for better salt penetration.

» The fish is kept in this condition for 24-48 hours.

» After salting period, the fish is taken out, washed in brine to remove adhering salt and drained.

- It is then hygienically dried to a moisture content of about 25%.
- Yield of the product by this method is about 35-40% with a storage stability of upto three months under ambient conditions.
- *Advantage*: Moisture removal is quick as it produces a strong gradient.
- *Disadvantage*: Texture becomes tough.

b. Wet salting/ Brine salting:

- The initial stages of processing are the same as for dry curing.
- In this, the fish are immersed in a saturated brine solution containing 4 to 5% sodium propionate. The sodium propionate will restrict growth and multiplication of halophiles.
- When fish are placed in saturated brine, the concentration of brine begins to fall as soon as salt begins to penetrate the fish and water is removed.
- Unless plenty of brine is used and the fish are stirred frequently, the rate of salt penetration and water removal may be seriously reduced.
- The presence of impurities may reduce the actual concentration of sodium chloride in solution and, in practice, the brine strength ranges between 80 to 100 per cent, which corresponds to 270 to 360 grams of salt to each litre of water.
- However, the fish kept in tank is allowed to remain in self brine till marketing without further drying.
- For marketing, as per the demand the wet salted fish is drained and packed in palmyrah leaf baskets or coconut leaf baskets.
- This method is particularly suitable for fatty fishes like oil sardine, mackerel etc.
- Wet salted fishes have short shelf stability with a moisture content of 50-55% and a salt content of around 25%.
- *Advantage*: Texture becomes soft.
- *Disadvantage*: Here moisture removal will be slow as it produces a weak gradient.

c. Mixed salting:

» Here fish is dry salted by rubbing salt over the fish and then put immersed in saturated brine solution.

» *Advantages*: Moisture removal is moderate and texture is also moderate.

» The moisture exudates that dilute the brine is compensated by the salt in the vat.

» This is the best method of salting.

d. Pickle salting:

» Pickle curing is a type of wet salting where the fish is layered by granular salt which, dissolves in the surface moisture of the fish forming solution which penetrates into the fish removing moisture from the fish.

» Sufficient salt is then available to go into solution and maintain the pickle at saturation point as salt penetrates the fish and water is removed.

» The water extracted from the fish also contains blood and other compounds that help to reduce the rate at which fat in the fish is oxidised.

» The fish is allowed to remain in this self brine. If the self brine is not sufficient, saturated brine is added to immerse the fish.

e. Kench salting:

» In this method, salt is rubbed on to the surface of the fish and stacked in layers of salt and fish.

» The self-brine formed is allowed to drain away.

» This method cannot be recommended for general use in the tropics as the fish are not covered by the brine or pickle and are therefore more susceptible to spoilage and insect attack.

» Exposure to the air and the presence of salt also encourages the rate of fat oxidation which gives rise to discoloration and the characteristic rancid flavours.

9. Maturation of salted fish:

» Many fish lose their raw odour and flavour after salting. Then flesh becomes tender and easily separates from bones and acquires a pleasant taste and flavour which is referred to as 'Bouquet'.

- » During maturation, protein breakdown takes place due to autolysis by tryptic enzyme from intestine.
- » When fish feeds more, quick maturation due to more enzyme takes place.
- » Active (pelagic fish) muscle are more rapidly subjected to autolysis.
- » Depends on microflora.
- » During autolysis, fats broken down to fatty acids forming formic acid/ acetic acid/ propionic acid/ Butyric acid and results in matured taste and flavour to product.
- » Maturation is accelerated by (a) adding blood pickles, (b) enzymes from certain moulds and (c) adding MSG.
- » An objective method of estimating maturation is by estimating protein breakdown products i.e. NPN and TN.

10. Care to be taken to prevent spoilage:

In order to prevent spoilage, care and attention must be used in 3 areas:

1. Raw material
   a. Fish
      - » It must be as fresh as possible.
      - » Fatty fish is best wet salted, while lean fish is best dry salted.
      - » Care to be taken not to damage fish during handling.
   b. Salt
      - » It must be clean and dry.
      - » For dry salting, mixture of 1/3 small crystals and 2/3 large crystals may be used.
      - » Fine crystals are usually better for wet salting.

c. Water

» Use clean and potable water.

2. Processing methods

a. Hygiene

» Ensure that hands, clothing, cooking utensils and work surfaces are perfectly clean.

b. Time

» Time required for each step of the salting process must be taken care of.

» Quantity of salt or brine required must be ensured from the beginning.

c. Containers

» Containers must be clean, and possess a secure lid.

3. Handling of finished products

a. Finished product with shelf life:

When dried finished product has

15% moisture and salt content of 5 to 20% => Shelf life is around 1 year at RT.
20% moisture and salt content of 10 to 20% => Shelf life is 4 months at RT
25 – 35% moisture and salt content5 to 20% => shelf life max. Upto 2 months
40%moisture and salt content 5 to 15% => Shelf life max upto 2-3 weeks

b. Packaging

» Dry salted fish can be enclosed in clean/dry plastic bags or wrapped and secured inside dry banana leaves.

c. Storage

» The salted fish must be stored in a clean, and if possible, cool place.

» It must be kept away from dust, insects, rodents and direct sunshine.

How to cook salted fish?

The concentration of salt is so high in salted fish products that it is necessary to remove much of the salt before the fish is cooked. This can be done by putting salted fish in clean, fresh water, for a minimum of 12 hours. The water will need to be changed about 3-4 times to encourage the removal of salt. The fish thus, will take up some water. Thereafter, it can be cooked.

D. SMOKE CURING OF FISH

» Smoke curing is a traditional method of fish preservation.

» It is a ready-to-eat product. Generally, 2% of the total fish production goes for smoke curing.

» India produces MASSMIN, the smoked tuna in Lakhadweep.

» Smoked salmon is a highly delicacy product in Europe.

» Smoked fish products are popular in Europe, Africa and North East states of India.

» Smoking is a method of preserving fish which combines three effects:

   a. Preservative value of the smoke: the smoke produced from burning wood contains a large number of compounds, some of which will kill bacteria, e.g., phenols.

   b. Drying: the fire which produces the smoke also generates heat and this will dry the fish.

   c. Cooking: if the fish are smoked at a high temperature, the flesh will be cooked and this will destroy the enzymes and kill bacteria.

» The long storage life of some smoked fish products is mostly due to drying and cooking than to the preservative value of the chemical compounds deposited on the fish from the smoke.

» Smoking is generally performed in specialised chambers called 'kilns'.

» There are two types of kilns – (a) Traditional kiln and (b) Mechanical kiln.

Traditional kiln:

» It is made of a brick structure. Smoke is generated either on the floor of the kiln or in a separate smoke generator and allowed to enter into the kiln. Here, there is no provision of temperature control.

Mechanical kiln:

» Here, there is a system of heaters using which the temperature can be controlled. This will burn the wood at a regulated temperature and the smoke produced will directly pass through the chamber where the fishes will be hung to get smoked.

1. Wood smoke:

» Wood smoke provides pleasant aroma and taste. Also it gives golden yellow colour mainly due to deposition of aromatic compounds (Polycyclic Aromatic Hydrocarbons – PAH).

» Generally hard wood are used for smoking which includes oak, birch, mahogany, sawdust, coconut husk, mango wood etc.

» Soft wood such as pine, poplar etc are not used as it contains turpentine components which are responsible for bitter taste.

» Hard wood must be free from fungal attack and mould growth.

» Wood contains:

- Combustible substance: Cellulose, Hemicellulose, Lignin, Tannin, Protein, Resin, Terpins etc.
- Non-combustible substance: Moisture, Ash and Mineral

» Cellulose:

- Chars at 280^{0}C and at more than this temperature, it hydrolyses to give glucose which gets charred and gives Oxymethyl furfural (unstable).
- This furfural further breaks down to give Formic acid, Levulinic acid, ill-defined humic substances.
- It imparts colour to smoked fish.

» Lignin:

- Dissociates at 350^{0}C. On pyrolysis gives rise to
  i. Tar – Contains phenolic components viz. Guaiacol, Cresol, Catechol, Phenol etc.

    ii. Methyl ester of Pyrogallol

    iii. Dihydric Phenyl ester

  - It has preservative action.

- Hemicellulose:
  - Polymer of pentoses and hexoses
  - Pentosoans are more and these are least heat resistant.
- Resins:
  - It is present in the form of 'balsam' (Solution of solid resin in volatile oil)
- Wood:
  - When moisture content is <25% => Dry wood
  - When moisture content is 25% to 35% => Medium wood
  - When moisture content is >35% => Wet wood
- Combustion of wood at more than 295^{0}C results in release of volatile components (80 – 87%).
- Combustion of wood is affected by:
  - Fuel structure
  - Thickness of wood
  - Depth of ash layer
  - Velocity of air over the burning wood
- When wood is burnt at 280^{0}Cto 380^{0}C, two types of compounds are produced.

  a. Aliphatic compounds (Alcohol, aldehyde, Acid, Ketones etc) – has bacteriocidal effect

  b. Aromatic compounds (PAH) (Catechol, Hydroquinone, Guiacol, Pyrogallol etc.) – has anti-oxidant effect.
- Relative humidity of smoking kiln should be less than 70%.

- Generally fatty fish are preferred for better taste.
- Shelflife of the smoked fish is 3 to 5 days at room temperature and 1 month at 10 to 15^{0}C.

Physical properties of wood smoke:

- Aerosol – The colloidal suspension of the smoke particles in air or gases which gives grey colour of the smoke.
- Wood smoke has 2 phases:

  (a) Particle phase – contains high molecular weight substances.

  (b) Vapour phase – contains low molecular weight substances.
- Particle size in wood smoke = 0.1 to 0.14 micron

Chemical properties of wood smoke:

Wood smoke consists of 4 different types of products viz.

a. Gases – Carbon dioxide, Carbon monoxide, Hydrogen and Hydrocarbon

b. Aqueous distillates – contains

   i. Organic acid –Formic acid, Acetic acid, Butyric acid

   ii. Alcohols –Methyl alcohol, Ethyl alcohol, Propyl alcohol, Isobutyl alcohol

   iii. Aldehydes- Formaldehyde, Acetaldehyde

   iv. Ketones – Acetone, Methyl ethyl ketone

   v. Hydrocarbon – Xylene, Cumene, Cymene, 3,4- Benzopyrene

   vi. Phenol – Catechol and its derivatives

   vii. Organic base – Pyridine, methyl pyridine

c. Tar distillates-

   i. Low gravity Tar oil –Valeric aldehyde, Furanes

   ii. High gravity Tar oil – Phenols and its derivatives

d. Carbon – Black colour

Preserving effect of wood smoke:

- » The preserving effect of smoking is attributed to a combination of surface drying, salting and deposition of phenolic and other anti-microbial constituents of smoke.
- » Smoke imparts a characteristic, attractive colour of golden yellow and the special smoky flavour relished by many people.
- » Phenols (2,6 – Dimethoxy 4-methyl phenol and 2,6 – Dimethoxy 4-ethyl phenol) has Antioxidant property.

Antimicrobial action:

a. Bacteriostatic action by – Organic acid, Phenols and its derivatives

b. Bacteriocidal action by – Combined effect of chemical, heating and drying

2. Smoking procedure:

a. Pre-processing of fish for smoking:

Washing – Gutting – Splitting or filleting and skinning – Brining with saturated salt solution containing permitted dye or dry salting – Hanging the fish for allowing to drip.

b. Technology of smoking:

Smoking either traditional kiln with natural convection of air or mechanical kiln with forced convection and temperature control – For cold smoking temperature is below 40°C, For hot smoking temperature is 75 to 80°C, Liquid smoking using natural smoke flavour extract (additives) and Electrostatic smoking using a high voltage current and infra red heating.

c. Technology of drying:

Drying of the smoked fish goes in line with smoking to achieve a characteristic texture of the product and enhance the shelf life of the product.

d. Storage life:

Depends on freshness of fish, degree of smoking, salting and drying and conditions of storage after smoking. Shelf life is about 3 to 10 days.

3. Smoked product:

- » Optimum moisture content in the product giving good and attractive yellow colour is 6 to 10%.
- » Colour of the smoked fish product is due to deposition of Tar (Phenolic contents)
- » Development of odour and flavour is due to phenolic compounds viz. Guaiocol, 4-Methyl Guaiocol and 2,6-Dimethoxy phenol.
- » Guaiocol gives smoky taste to the product.
- » Syringol gives smoky odour to the product.
- » Nutritive value of the smoked product-
  - i. Low available lysine content
  - ii. Higher biological value of protein due to protein hydrolysis during storage.
  - iii. Vitamin loss is more.

Disadvantage:

- » Aromatic compounds are volatile. So, if not packed properly, they will absorb moisture and facilitate (a) mould growth and also cause (b) rancidity.

Carcinogenic effect:

a. 3,4-Benzopyrene:

- » The aromatic compound of tar fraction of wood smoke contains a typical chemical i.e. 3, 4 - Benzopyrene which is carcinogenic in nature. So, it is desired that 3, 4 - Benzopyrene content in the final smoked product should never be more than 0.1microgram per kg of smoked fish.
- » When tar fraction increases, 3,4-Benzopyrene content will also be more.
- » When wooden logs are burnt at a high temperature, more quantity of 3,4-Benzopyrene will be released.
- » The 3,4-Benzopyrene can be removed from the product by use of metallic filters and also burning wet logs at low temperature.

b. Nitrosoamines:

» It is produced due to Nitric oxide and Nitrates in vapour phase.

» It can be eliminated by

i. Addition of ascorbic acid and

ii. By decreasing temperature.

4. Methods of smoking:

» There are four methods of smoking viz. (a) Cold smoking, (b) Hot smoking, (c) Liquid smoking, and (d) Electrostatic smoking.

a. **Cold smoking:** This is the conventional type of smoking using traditional chimney kilns. Here the temperature is never raised above 40^{0}C. The fish to be dried is hung on the top or kept in mesh trays. The wood is burnt at the bottom of the kiln. The smoke travels upward and gets deposited, being solubilised on the fish body. The fish does not get cooked. Duration of smoking is 36 to 72 hours. The cold smoked products are more stable and have a pleasant odour.

b. **Hot smoking:** It is done in a mechanical kiln which is of a tunnel type. The fish to be smoked is kept inside on the trolleys and heat is supplied either directly or indirectly. Electric heaters are placed at different intervals of the smoke path to maintain a uniform temperature throughout the chamber. The fish to be smoked is dried at a temperature of 75 to 80^{0}C in order to cook the flesh. The fish is then smoked ata temperature around 100^{0}C depending upon the type of the product. The flesh of hot smoked fish is delicate, succulent and has a longer shelf life.

c. **Liquid smoking:** Use of smoking liquid or concentrates are used in developed countries. The liquids are prepared by dry distillation of wood. The fish is salted and given a dip treatment in a smoke concentrate and then dried in an ordinary kiln drier. The liquid thickens and coats the fish, thus imparting the particular flavour to the fish. Here, smoking alone can't preserve fish rather; smoke is deposited to give flavour to the fish.

d. **Electrostatic smoking:** It is mainly used for smoking small fishes. Here, the process of drying, smoking and packing of fishes is carried out within 12 to 18 minutes by means of using a high voltage current and infra red heating. Here, the salted fish is passed through a drying chamber heated by infra red lamps and then passed into a smoking oven by conveyor belts. Smoke is generated from the bottom and gets deposited on the flesh of the fish. Then the fish passes through a baking oven where the fish is dried to required moisture level.

5. Smoked products:

a. Massmin:

» Massmin, the smoked tuna is prepared from tuna in Lakshadweep and Minioy Island.

» A similar product of Japan is 'Katsuobushi'.

» The massmin is produced and exported to Sri Lanka, Malaysia and Hong Kong where it is priced delicacy. The method of preparation is simple.

» Fresh tuna is beheaded, tails and fins are removed.

» Ventral wall splits and viscera removed completely. It is then cleaned and filleted. Each fillet is split longitudinally to get pieces of 12 – 14 cm length and 4.5 cm thickness.

» Each fish is then wrapped in green coconut leaf free from splinters.

» This is meant to prevent any possible disintegration during subsequent cooking.

» The wrapped pieces are cooked in 3% brine or sea water for 1 hour and left overnight for cooling. They are spread on frames and exposed to smoking.

» The smoke is generated by smouldering of coconut husk and saw dust for 3 to 4 hours.

» Coconut leaves are then removed and the meat dried in the sun for 1 day.

» Alternate drying and smoking are repeated 3 -4 times when a dark brown hard product is obtained.

» Since the moisture content is around 10% (preferably less than 10%), its shelf life will be for more than one year.

» A type of fungus is sometimes allowed to grow on it which imparts an ashy appearance and longer shelf life. The product remains for 3 –4 years if packaged properly.

» Thin shavings by scrapping are made from the product by a sharp knife which are soaked in water and used in chutney preparation and added to other caulinary products.

b. Kippers:

- Very fresh, or good quality frozen, fatty herring will make first class kippers.
- The whole herring is washed to remove loose scales. Then they are prepared by splitting the fish down the back and removing guts and roes. The split fish are washed again before brining.
- Brining time depends on size of the fish and fat content. For example, split herring of medium size containing 20%fat require about 15 mins in 80^0 brine. A lean winter herring of similar size requires a shorter brining time and larger fish will require longer. A permitted dye, Brown FK, is usually added to the brine to colour the fish.
- Thereafter, during smoking in traditional kiln fish will impart uniform colour in appearance. The proportion of salt increases slightly in the flesh during smoking. The salt content in the flesh should not be below 1.5% to give good taste.
- The brined fish are hung on tentersticks for kippering, left for dripping for an hour before smoking.
- Smoking is done in a kiln at 30^0C for about 4 hours during which it may lose 15% of its dripped weight. Un-dyed kippers take 6 hrs to attain 20% moisture level. During last half hour, the temperature may be raised to 35^0C in order to develop a deeper colour and to bring fat to the surface.

c. Bloaters:

- Bloaters are made from whole, ungutted, slightly salted herring. The characteristic flavour of the product is described as 'gamy', due to the presence of gut.
- During salting for overnight fish lose roughly 6% of their weight. Then, they are washed to remove surface salt.
- Fish are then threaded on speats, either through the gills and mouth. They are then exposed to drip and then put into the kiln in the wet condition.
- Smoking and drying of bloaters take about 4 hrs at a temperature of 30 to 32^0C.
- During first 3hours, fish gets dried and last hour only lightly smoked. The colour of the product becomes bright silvery appearance with a faint flavour of smoke.

d. Finnan Haddock:

- These are prepared from split haddock of medium size.
- Brining is done by shoaking in an 80^0 brine for 7 to 15 minutes. Dye is not added.
- Smoking is done in a kiln where the temperature of the kiln should not exceed 27^{0}C and time 4 to 6 hours.
- A pale, straw coloured finnan darkens to a good colour within a few hours of cooling.

e. Buckling:

- This is hot smoked herring, a German product, very popular in Europe.

E. SPOILAGE IN CURED FISH

a. **Reddening:** The fish takes on a red colouring with putrid odour. It is caused by halophilic bacteria. These organisms are usually found in solar salt. These do not grow in brine or in fully dried fish. They are aerobic and proteolytic in nature. These grow best at 36°C by decomposing protein and giving out an ammoniacal odour. Spoilage appears on the surface as slimy pink patches. However these bacteria are not harmful in nature. Usage of good quality salt mixture containing salt, sodium benzoate of 0.25%, sodium acid phosphate of 1.5%, potassium sorbate of 0.5% and BHA (Butylated Hydroxy Anisole) of 0.2% is recommended to avoid this condition. This spoilage is mostly found in heavily salted fish and absent in unsalted fish. It can be prevented by increasing rate of salt penetration.

b. **Dun:** In salted fish, brownish black or yellow brown spots are seen on the fleshy parts, referred to as "dun". This is mainly caused by growth of halophilic mould called *Sporendonema epizoum* and by the use of impure salt and unsanitary practices during preparation. This gives the fish a very bad appearance. Moulds usually grow at relative humidity above 75%. The optimum temperature for growth is 30-35 °C. During the initial stages of appearance of moulds on the fish, it is possible to remove them manually. In advanced stages it penetrates into the flesh. To avoid the mould growth it is necessary that the fish be dried, packed and stored properly to avoid uptake of moisture. Chemical method of prevention includes dipping the fish in a 5% solution of calcium propionate in saturated brine for 3-5 minutes depending upon the size of the fish.

c. **Souring:** This is characterised by formation of slime with bad odour. Fish which has soured has a bitter taste bad odour. Souring is mainly by microorganisms caused due to improper salting which results in the uneven distribution of salt throughout the muscles of the fish. It can be prevented by dipping salted fish in 3% acetic acid solution for 10 to 15 mins.

d. **Salt burn or rust:** The fish is extremely dry and cannot be rehydrated. It gives unpleasant taste and rancid smell. This happens when too much fine salt has been used. This salt draws out the surface moisture so rapidly that protein in the fish becomes solid, which stops the fish from taking in water later on. If it occurs on surface, it can be prevented by washing with $NaHCO_3$/$kHCO_3$. If rust penetrates to deep muscle, it becomes inedible.

e. **Slime:** The surface of the fish acquires a slippery coating of slime. This usually occurs in brined fish because of inadequate salting, lack of freshness of fish, and other factors. This is caused by red halophilic bacteria, *Serratia salinaria*. This bacteria is an obligate aerobic and cannot grow at$<10^0$C.

f. **Insect infestation:** It is caused by maggots of *Drossophila casei*. This can be prevented by maintaining proper sanitary condition. Sometimes, just before drying, the product needs to be kept immersed in 5ppm pyrethrum or piperonyl butoxide solution for 30 mins to prevent insect infestation.

F. USE OF PRESERVATIVES IN CURED FISH

» Anti-microbial agents and anti-oxidants are used to prevent spoilage of cured fish.

» A curing mixture is used for salting of fish. It contains the following ingredients.

S. No.	Component	Function	Amount in Kg
1	Common salt	Curing	100.00
2	Sodium benzoate	Prevents reddening due to halophiles and incidence of soft tissue	0.25
3	Sodium acid phosphate	Enhances the activities of sodium benzoate	1.50
4	Potassium sorbate	Prevents mould growth	0.50
5	BHA	Retards brown or yellow discolouration and minimises rancid odour	0.2

» After curing, the cured fish should be stored with proper treatment using Calcium propionate @3% as the preservative.

- Due to Calcium propionate treatment, the shelf life is extended from 2 months to 1 year in case of dry cured products and up to 4 months in wet cured products
- There are 3 methods of using Calcium propionate. They are:
    a. Direct dusting method: Here 3g Calcium propionate powder per 100g cured fish is taken in a mesh cloth bag and dusted. This method is most effective against fungus, red discolouration and off-smell.
    b. Dip treatment: Just after salting, before drying, the cured fish is kept immersed for 30 minutes in saturated brine containing 3% Calcium propionate. The fish is then drained and dried. By this fungal attack is completely prevented.
    c. Calcium propionate mixture treatment: The cured samples are sprinkled with a mixture containing 3% Calcium propionate in refined salt and intimately mixed. The appearance is affected by excessive salt particles on the surface of the fish.

Control of infestation of insects:

- Proper hygiene and sanitation
- Disposal of waste and decaying matter
- Use of physical barriers like screens
- Covers for curing bags
- Use of heat up to 45^0C to physically drive away the insects and kill them.

12 Fish Products

A. FISH MEAL

» It is a highly nutritious concentrated stable feed supplement consisting primarily high grade proteins, vitamins and minerals.

» Raw materials: Trash fish and fish industrial waste.

» Methods of production: there are two methods of production of fish meal.

- Wet rendering/ reduction process –Mainly used for fatty fishes having more than 2.5% fat where, fish meal, fish oil and fish soluble are produced at a time.
- Dry rendering/ reduction process – Used for lean fishes having less than 2.5% fat and other fish waste where, only fish meal is produced.

a. Production method:

» It involves the following major steps.

1. Heating or cooking to coagulate protein and release water and oil.
2. Pressing to separate liquids from solids
3. Drying
4. Grinding to produce a powdered or granular end product

i) Wet rendering or reduction process: It is a continuous process.

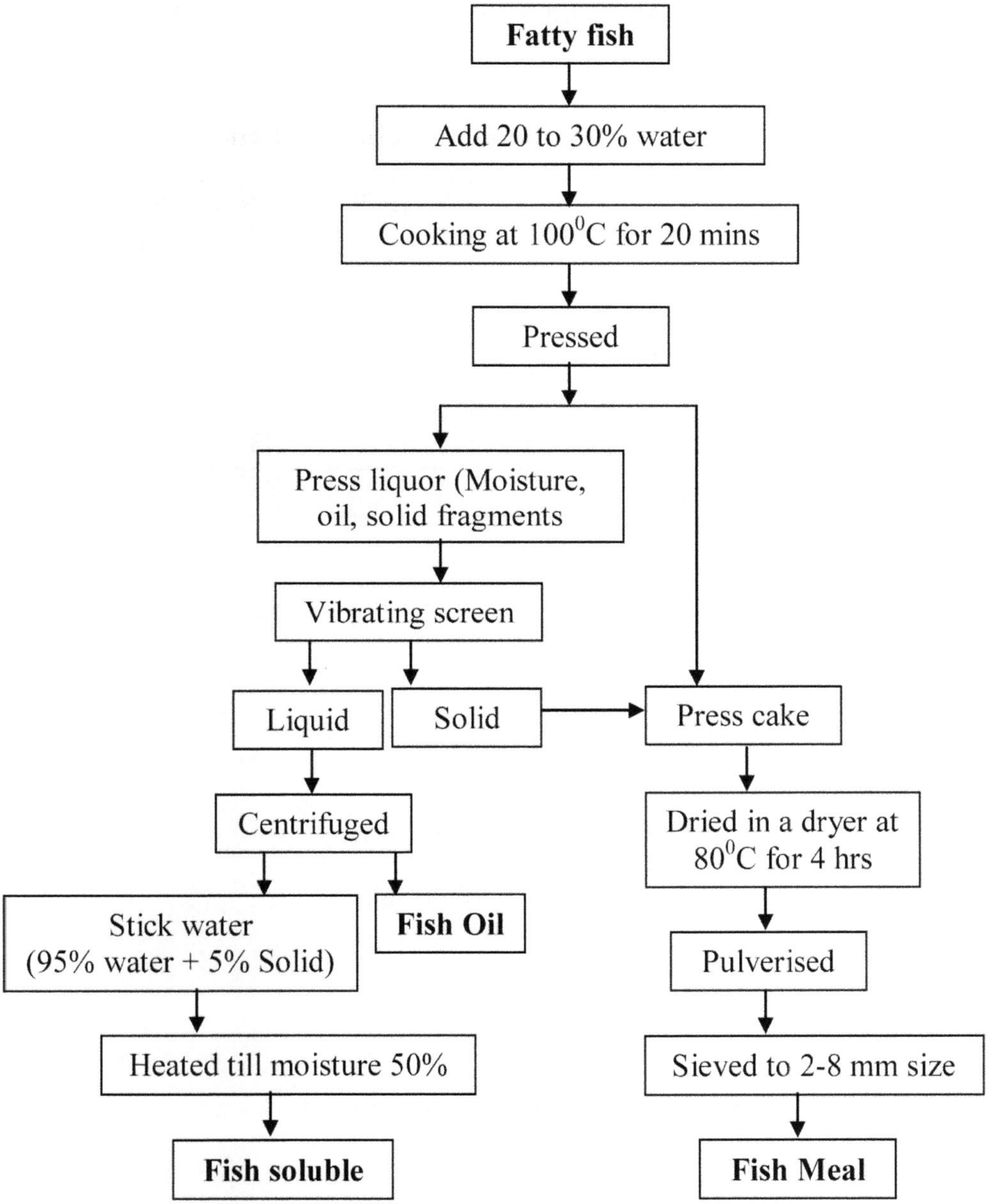

Figure 12.1: Production process of fish meal by wet reduction method

» The fish soluble contain 50% solid and 50% moisture. The solid contains degraded protein, polypeptides, amino acids, vitamins and minerals. The fish soluble can be preserved by adding 2-5% formic acid for 1 week.

» 25% water soluble and 75% fish meal together are called as 'whole fish meal' and this can be used as animal feed.

Table 12.1: Composition at various stages of fish meal manufacture

S. No.	Material	Water %	Solids %	Fat %
1	Raw fish	70	18	12
2	Press cake	53	44	3
3	Press liquor	78	6	16
4	Dilute stick water	94	5	<1
5	Concentrated stick water	65	33	2
6	Fish meal	9	85	6

Advantages:

» Production of good quality fish meal and oil.

» Faster process

» Lower installation and operating cost

» Suitable for processing large quantities of material

» Can yield a valuable by-product, i.e. fish soluble.

Disadvantages:

» Lower yield

» Fish meal is low in water soluble unless concentrated stick water is added back.

ii) Dry rendering or reduction process:

» It is produced in batches. Product is the white meal.

» This process is now-a-days not used commercially.

» The essential steps in the dry rendering process are as follows.

- Fish are coarsely ground in a grinder.

- The ground fish are cooked in a steam jacketed cooker with a stirrer.
- Then dried using a dryer and pulverised to get white meal.

Advantages:

» Production of high quantity of oil even for non-fatty fish.

» Suitable for small batch operation

» Easy to manipulate cooking/ drying times

» Greater flexibility

» Produces whole meal including solubles

Disadvantages:

» Fish Oil is of inferior quality

» High installation and operating cost

» Slow production

b. Proximate composition of fish meal:

Parameter	Quantity	Remark
Moisture	8% preferred (Ranges between 6 to 10%)	More than 12% moisture content leads to mould growth
Protein	65% (Varies between 50% to 75%)	Contain sufficient quantities of all 10essential and 11 non-essential amino acids.
Fat	8% preferred (Ranges between 5 to 10%)	More than 10% fat leads to rancidity during storage.
Ash / Mineral	12% to 33%	Contains Calcium, phosphorus, sodium, magnesium, potassium, iron, copper, zinc, manganese, iodine and selenium
Crude Fibre	Below 1%	Low fibre product

c. Storage:

The fish meal should be packed in airtight, polythene laminated multilayer sacks of 50 kg which will prevent the diffusion of oxygen into the meal. The polythene bags have advantages as follows.

» They prevent rapid movement of oxygen and water.

» Meal is protected from rodents and insect attack and from contamination by moulds and bacteria.

Preservation:

» Fish meal with high oil content creates problems during storage.

» Fish oil will oxidise and the reaction can lead to considerable rises in temperature. This can be prevented by use of antioxidants. Two most common antioxidants are Ethoxyquin and Butylated hydroxytoluene (BHT).

» It is common practice to add 400 to 700 ppm ethoxyquin to fish meal immediately after drying and prior to packaging.

Use of fish meal:

» Fish meal is used as poultry and pig feed at a rate of 5 to 10% of the feed. If more meal is used, it affects the (a) flavour, (b) Loss of appetite and (c) flesh will smell like that of fish meal.

B. FISH OIL

» Oil is the ester of glycerol and fatty acid where as wax is the ester of fatty alcohol and fatty acid.

» Fatty acid: Characterised by C-atom chain of hydrocarbons being terminated by Carboxyl group (-COOH) group.

» **Saturated Fatty acid** (SFA): C-atoms are linked by single bond ($C_nH_{2n}O_2$). Ex.

1. Capric acid ($C_{10}H_{20}O_2$)	5. Stearic acid ($C_{18}H_{36}O_2$)
2. Lauric acid ($C_{12}H_{24}O_2$)	6. Arachidic acid ($C_{20}H_{40}O_2$)
3. Myristic acid ($C_{14}H_{28}O_2$)	7. Behenic acid ($C_{22}H_{44}O_2$)
4. Palmitic acid ($C_{16}H3_2O_2$)	8. Lignoseric acid ($C_{24}H_{48}O_2$)

» **Unsaturated fatty acid** (USFA): C-atoms are linked by double bond at specific position ($C_nH_{2n-2}O_2$). Some of them are

- Oleic acid ($C_{18}H_{34}O_2$) – Double bond position at 9 C-atom (C18:1, ω-9)
- Linoleic acid ($C_{18}H_{34}O_2$) - Double bond position at 9, 12 C-atom (C18: 2, ω-6)
- Linolenic acid ($C_{18}H_{34}O_2$) - Double bond position at 9, 12,15 C-atom (C18: 3, ω-3)

» When degree of unsaturation increases, melting point decreases.

» **Hydrogenation**: Under suitable condition in the presence of catalyst Ni, hydrogen adds on to the fat or fatty acid at the double bond and USFA becomes SFA i.e. oil becomes fat.

» **Fish oil** - It comprises of 25% SFA and 75% USFA

» Fish oil is obtained either from Body or Liver. If it is from body, it is called as 'Body oil' and if it is from liver, it is called as 'Liver oil'.

Factors affecting the quality and quantity of oil in fish:

» Species

» Food

» Sexual maturity of fish

» Temperature of the water

Deteriorative changes taking place in oils and fats:

» Development of FFA (by enzymatic hydrolysis): It is by fat hydrolysis of enzyme Lipase. Thus, FFA is released. When FFA content becomes more than 1mg%, spoilage of oil is observed. The enzyme lipase can be destroyed at a temperature of more than 80°C.

» Development of oxidative rancidity (by chemical oxidative rancidity): It is mainly

- by action of lipoxidase present in fish
- by bacterial lipoxidase activity
- atmospheric condition.

» Micicrobial spoilage: When oil contains >0.3% moisture, growth of micro-organisms occur.

» Typical rancid odour and flavour – due to decomposition products formed by break down of fatty acids.

» Prooxidants viz. Cobalt, Manganese, Lead, Copper ions increase the rate of oxidation.

» **Flavour reversion:** The tendency of various deodourised oils to develop characteristic off-flavour and off-odour is called flavour reversion. It is mainly due to oxidation of fatty acids (Linoleic and linolenic acid). It is prominent in oils having fatty acids with more than 2 double bonds.

Spoilage of fish oil can be prevented by:

» Low temperature

» Low moisture content

» Lipase to be inactivated

» Packing in inert gas like N2

» Use of antioxidants at a dose of 0.02%.

- Natural antioxidants – Tocopherol, Lecithin, NDGA
- Synthetic antioxidants – BHA, BHT, Hydroquinone, Pyrogallate, Parahydroxy benzoic acid (PHBA)
- Synergist – Citric acid and ascorbic acid

a. Fish body oil:

» Oil sardine, mackerel are the important fishes to get fish body oil.

» It should be free from moisture as microorganisms (even at 0.4% moisture) containing enzymes split the triglyceride to release FFA.

» It should be free from sulphur as during processing of hardening of oil, catalyst Nickel is used and presence of sulphur in oil reduces the active life of catalyst acting as a catalytic inhibitor.

- Iodine – Indication of unsaturation of oil
- Oxygen – Cause rancid flavour mainly due to oxidative rancidity.
- Production of fish body oil:
  a. During method of preparation of fish meal
  b. Solvent extraction
  c. Traditional method of settling tank
- Solvent extraction method:
  - Solvents – Trichloroethylene, Ethyl ether, Acetone, Benzene etc.
  - The solvents should be Nontoxic/ Non-corrosion/ Easily purified/ Low Boiling Point
- **Disadvantage of Body oil produced by Solvent extraction method:**
  a. Colour – Dark red (Undesirable)
  b. Elevated temperature to remove solvent will enhance oxidation and loss of Vitamin A
  c. Expensive
  d. FFA remains untouched
- **Settling tank method:**
  - Stick water is settled to get body oil. Here the oil is bright and having low FFA.
- **Assessment of quality of fish oil:**
  a. Unsaponifiable matter (Indication of purity): These are found dissolved in fats and oils. They are soluble in ordinary fat solvents. Ex. Alcohol, sterol, pigments, hydrocarbon etc.
  b. Iodine value: Number of g. of iodine absorbed by 100g oil.
  c. Saponification value: Number of mg. of KOH required to saponify1 g oil.

» **I.S.I. Standard for Sardine oil:**

Parameters	Grade I	Grade II
FFA as Oleic acid%	1.0	3.0
Moisture	0.5	0.5
Iodine value	145-205	145-205
Saponification value	185	185

b. Fish Liver oil:

» Livers of Cod, Tuna, Whale, Elasmobranch (Shark, rays and skates) are used for production of fish liver oil.

» Liver quality: Based on texture and colour

a. Firm and pink both externally and internally → Good quality

b. Firm and grey externally and pink internally → Medium quality

c. Flabby and white both externally and internally →Poor quality

» Preservation of Liver:

- Frozen form (-20^{0}C)
- Cut into 2-3 inch thick pieces and mix with 10% salt
- In 0.5% formalin
- Hopkinson's method – 2% solution of germicide (Alcohol, Resorcinol, phenol)

» Categories of liver:

<table>
<tr><th>S.No.</th><th>Categories of liver</th><th>Fish</th><th>Method of extraction</th></tr>
<tr><td>1</td><td>High oil and Low Vit A</td><td>Cod</td><td>Direct steaming – 80^{0}C for 1 hr
Indirect steaming – Liver: water (2:1) Cooked</td></tr>
<tr><td>2</td><td>Low oil and High Vit A</td><td>Tuna</td><td rowspan="2">Alkali digestion method
Alkali enzymatic method
Acid digestion method
Solvent extraction method</td></tr>
<tr><td>3</td><td>High oil and High Vit A</td><td>Shark</td></tr>
</table>

c. Refining of fish oil:

1. **Winterisation** (Destearination): Stearin is the solid glyceride portion which separates out from the clear cooled oil. Cool the oil and keep for overnight and then filter. This is called 'cold cleaning' of oil.

2. **Degumming:** Removal of phosphatids and other mucilaginous materials from the oil. This is done by treating oil with 2-3% boric acid solution and then washing.

3. **Alkali refining:** By adding 7-18% NaOH at low temperature and then keeping for overnight, the FFA is neutralised.

4. **Bleaching:** Discolouration due to deteriorative changes can be prevented by

   a. **Physical:**

      - Use of activated carbon/ clay/ bentonite/Fuller's earth (1:10) at 100^{0}C and then filter.

   b. **Chemical:**

      i. Oxidising agents – Oxygen, Hydrogen peroxide, $KMnO_4$

      ii. Reducing agent – Zinc and Hydrogen dust

5. **Hydrogenation:** Direct addition of H_2 to double bonds of fatty acids (gaseous hydrogen, Liquid oxygen and Nickel as catalyst).

6. **Deodourisation:** Oxidation of highly USFA of fish oils gives fishy odour. It can be prevented by adding 5% Calcium hydroxide (Slaked lime) + 5% Calcium oxide (Quicklime) to fish oil, agitating and filtering. This de-odourisation of fish oil can also be performed by vacuum steam distillation method.

d. Use of fish oil:

1. Soaps and detergents:

» Soaps are the salts of fatty acids with alkali whereas, detergents are synthetic petroleum by-products.

» Fish oil is hydrogenated to bring down Iodine value to 80 resulting loss of odour and hardening of oil.

- In India, highly hydrogenated oil is preferred as soaps are crystalline in structure which holds more water and must have 40% TFM (Total Fatty Matter).
- Teepol – Non-soap detergent (Soapless soap). They are useful for hard water for dishwashing.

2. Paints:

- Sardine/ pilchard oil is preferred for this product.
- Oil with high unsaturation used for this so that it will absorb moisture from air and dried to form solid film.

3. Varnishes:

- Fish oil used for oil varnishes.
- Factice (Sulphurised oil) are used for this.

4. Linoleum : Floor covering

5. Printing ink: Very quick drying oil are used.

6. Type writer ribbon ink: Whale oil are used for this. Here oil should not dry easily.

7. Rubber goods manufacture: Stearic acid are used for this purpose.

8. Lubricant: Used for watch. Generally Jaw oil of porpoise/ dolphin are used for this.

9. Insecticide

C. NON-FAT COMPONENTS OF FISH OIL

- Non-fat components of fish are the unsaponifiables present in fish oil. These include Vitamin A, Vitamin D, Cholesterol, Lecithin, Pigments, Glycerol ether, Hydrocarbons (Squalene).

1. Vitamin A:

- Precursor Vitamin A is Carotene.
- Phytoplankton contain carotene → consumed by Zooplanktons- zooplankton

convert it to astacene → when fish consumes zooplankton, fish converts astacene to Vitamin A.

» Tocopherol (Natural antioxidant) good forVitamin A in Liver oil.

» Lecithin acts as synergist. NDGA (Nordihydroguaiaretic acid) stabilises Vitamin A in oil.

2. Vitamin D: Found more in Shark liver oil.

3. Cholesterol:

» Present in bile and liver oil.

» Fish lipid, PUFA, unique in contribution of eicosapentaenoic acid and docosahexaenoic acid which help in preventing the coronary heart disease.

4. Lecithin:

» It is the antioxidant. It is of Phosphatid/ phospholipid group

» Use:

- As an emulsifier in production of margarine
- Antioxidant
- As emulsifier in cosmetics
- As dispersing agent in paints
- As a curing promoter in rubber manufacture.
- Improves quality of bakery products

5. Pigments:

» Natural oil soluble pigments

- Carotenoids – Red/ Orange/ Yellow
- Astacin – Red (From Salmon oil/ Liver oil)
- Fucoxanthin – Yellow colour (Pilchard oil)
- Xanthophill - Yellow colour (Pilchard oil)

- Teraxanthine and zeaxanthine – skin of variety of fish

» Colour changes during/ after processing

- Non-enzymatic oxidation – Browning reaction
- Enzymatic oxidation – peroxidises
- Oxidation of pigment
- Oxidation of USFA
- Oxidation of Haemo/ myo-globin

» **due to**

- Spoilage of fish during transportation – Dark colour oil
- Temperature and time of storage – Dark colour oil
- Rancidification during storage – Dark colour oil

6. Hydrocarbon:

- Important is Squalene ($C_{30}H_{50}$), a highly unsaturated hydrocarbon containing 6 unsaturated bonds per molecule.
- Found in shark liver oil.
- Very little nutritive value.
- Presence of squalene in fish renders it free from parasites.
- Imparts a brilliant sheen to natural and artificial silks when incorporated during finishing operations.
- As lubricant for filling thermometers.
- Squalene hexahydrobromide – used in the rubber manufacture and as an emulsifying agent for the textile, rubber and other industries.

D. MARINADING OF FISH

Marinating, or the preparation of marinades, generally involves the preservation of fish by treatment with solution of acid and salt. The process gives the fish a characteristic flavour and texture but the product is often additionally seasoned with sugar and spices. It is very popular in Northern Europe especially in UK. Pelagic fish with high fat content viz. herring, oil sardine is best suitable for marinading. Good quality marinades can be made from high quality fresh raw material.

Principle:

» The acetic acid or vinegar has a preservative action at a pH of 4.5 where, all food poisoning bacteria and most spoilage bacteria are prevented from growing.

» Acetic acid and salt have also an effect on texture of the fish. The acid produces the tenderness characteristic (softening of meat). Salt helps in firming of the flesh by removal of water and coagulates the protein.

» The softness of the flesh due to acid is mainly due to action of certain proteolytic enzymes, cathepsin which cause a partial breakdown/hydrolysis of the proteins with the release of some free amino acids. This hydrolysis depends on temperature. Optimum temperature for marinading is 10-12^{0}C but at 37^{0}C hydrolysis occurs very fast resulting to spoilage. Salt also controls the hydrolytic action and allows it to proceed within desired limits.

» The amino acids thus produced give the products their characteristic taste.

» Some bacteria and enzymes, which are active at pH 4.5, help in ripening the fish contributing characteristic texture and flavour.

» Some of the acetic acid combines chemically with the proteins while the remaining acid controls the salt content, keeps the pH at about 4.5 and allowing the autolytic reactions to take place so that the product has good taste and texture and at the same time becomes safe for human consumption after a reasonable shelf life.

» The fat content of the flesh is also important in giving characteristic flavour.

a. Types of marinades:

» There are three types of marinades.

1. Cold marinade:

» It is a true marinade.

» Raw fish is preserved in a mixture of acetic acid and 10% salt at an optimum pH of 4.5.

» Preparation:

- Fish – Immersed in 8-10% salt concentration for 1 hour for stiffening- They are then headed, eviscerated and deboned.
- Put in Cleaning bath (Exanguination bath – Blood removing bath) containing 3-5% salt + 50g 30% Hydrogen peroxide + 0.5% Acetic acid in 1 litre for 30min – Allowed to drip to remove traces of blood from the flesh.
- Then, treated with Finishing bath (Softening bath) comprising two important solutions (a) Salt : Acetic acid and (b) Fish : Solution.
  - ◊ Both the ratios should be between 1 : 1 to 3 : 2.
  - ◊ Salt gives hardening effect where as acid gives softening effect.
  - ◊ Fish : liquid ratio should be such that final concentration of acetic acid by the end of finishing bath solution will be 2.5%.
  - ◊ For this, 6% acetic acid and10% salt is considered best.
  - ◊ Optimum time for pickling is 3 – 7 days.

» Packing:

- Plastic containers of various sizes/ wooden barrels.
- Cover brine (Aromatic brine): (1-2)% Acetic acid+ (2-4)% salt
- Sweetening agent (Saccharine): It is used to mask too much salty and acid taste.

» Over this Garnish layer (Pickled Onion, carrot, cauliflower etc) over which jelly or jelled brine is poured and on top a parchment paper is given.

» Jelled brine: (4-5)% Gelatin + (1.9-2.0)% Acetic acid + 3% Salt

» Product is not further heated.

2. **Cooked marinade:**

» Very fresh fish can only be preserved by this method.

» Fish → Eviscerated → washed → Added 10% salt solution for 30min to remove blood → Drain → Placed in bleaching bath (1 to 2% acetic acid + 6 to 8% salt) at 85^{0}C → Packed in Jelled brine (Fish : Jelled brine = 1 : 1)(Jelled brine should not be heated at more than 70^{0}C as gelatin will be hydrolysed to give bad taste.

» When Acetic acid is added, pH comes to 4.5 which are most favourable for the enzyme cathepsin present in fish to be active. This will lead to hydrolysis of protein. This hydrolysis also depends on temperature and 10 to 12^{0}C is optimum for marinades. At 37^{0}C, hydrolysis occurs very fast resulting to spoilage.

3. **Fried marinade:**

» Here, fish is first fried in oil in open pan at a temperature of 160 – 180^{0}C for 5 to 12 mins.

» Then packed with cover brine (Fish : Cover brine = 2:1).

» Cover brine contains (2 – 3.5)% Acetic acid and (3 – 5)% Salt

Shelf life: The shelf life is about 1 month at 3^{0}C

b. Spoilage:

1. **Cold marinade:**

- Microorganism, *Betabacterium buchneri* liberate CO_2 by decomposition of glutamic acid and arginine resulting spoilage.
- Red spot formation is accompanied with high pH.
- Preservative: Hexamine (Hexamethylene tetramine) is best preservative for this type of spoilage.

2. **Cooked marinade:**

- Bacteria spoilage (Gm +ve): Through Jelled brine. So, better to use high grade Gelatin.
- Mould growth (Penicillium group): This can be prevented by taking care in filling and sealing.

3. Fried marinade:

- Ropiness and slime formation by *Bacillus subtilis* (Soil bacteria) through wheat floor.

E. FERMENTED FISH PRODUCTS

» Fermented fish products are very popular in South-East Asian countries. Fermentation is the name given to the general class of chemical changes/ decomposition produced in organic

» Substances through the activity of living micro-organisms.

» Mackie *et. al.* (1971) defined fermentation as the transformation of organic substances into simpler compounds by the action of enzymes and micro-organisms. Fermented fish paste and sauce are relished as condiments along with cooked rice in many of South-East Asian countries.

» On daily basis, nobody consumes large quantities of these products but almost every one consumes a little bit every day like pickle.

a. Types of fermented fish:

1. **Fermented fish:** These are the products in which fish retains substantially their original form.
2. **Fermented fish paste:** In this fish are reduced to a paste.
3. **Fermented fish sauce:** Products in which fish are reduced to a liquid.

» Traditionally fermented fish products are basically salt fermented products. Based on the salt content, fermented fish products are classified as:

i. High salt (>20% Salt)

ii. Low salt (4-6% Salt)

iii. No salt (0% Salt) – Alkaline fermentation

b. Method of production:

» Fish is mixed with salt in a ratio of 1:3 to 4 (1 part salt to 3-4 parts of fish). It is allowed to remain in this condition with agitation when required. Fish mass is then filtered. The liquid is called as Fish Sauce. The solid residue is ground to a paste. It is called as Fish paste.

» In modern types, lactic acid bacteria (Lab Cultured) are added to enhance fermentation which requires substrate.

c. Mechanism:

» At high salt concentration, microbial activity is almost arrested. Most of the enzymes get denatured. Some enzymes still remain active in this condition. They hydrolyse the proteins into aminoacids and other volatile substances. Therefore, the NPN value, AAN value and poly peptide nitrogen content get released into the solution.

S. No.	Country	Fish Sauce	Fermented fish	Fermented fish paste
1	Campuchia	Nuoc-Mam	-	Prahoc
2	Korea	Lao – nam pla	Jeotkal	
3	Malaysia	Budu	Pekasam	Belacan
4	Myanmar	Ngapi		Ngapi
5	Phillipines	Patis	Bagoong	Bagoong
6	Thailand	Nam pla	Pedahsiam	Kapi
7	Vietnam	Nuoc-Mam		Mam
8	Hong-Kong	Fish sauce		
9	Sri Lanka		Colombo cured	
10	Manipur, India		Ngari	
11	Bangladesh		Chepa Shutki	

Shelf life:

» Fish fermentation in the Southeast Asian sub-region normally lasts for several months (three to nine months) and the fish flesh may liquefy or turn into a paste.

d. Fermented fish products:

1. Fermented herring:

» Herring (*Clupea harengus*) were formerly used in pickled product.

- The fish were typically dry salted in barrels, proportion of salt being 15% to 36%. The brine was formed and covered the fish in the barrels.
- Fish pickled in this way was kept for more than a year. The fish flesh was then only very moderately softened; ripening continues for more than several months. Such products contain 10-12% salt.
- To encourage fermentation, a proportion of sugar was added to the salt and spices such as peppers, coriander, cinnamon, ginger, cardamom, was added. Benzoic acid was sometimes used as a preservative.

2. Fermented anchovies:

- Cured anchovies are a delicacy popular in Mediterranean area.
- It is prepared from *Engraulis encrasicolus* fish by salting and fermentation.
- The best size of anchovies is that 35 to 40 numbers should weigh 1 kg and should have as high as fat content.
- Fish → beheaded, gutted by pulling off the head → salted in layers using 1 kg salt per 2 kg anchovies and a final salt layer should be put on top, weighted wooden disc to keep the fish well pressed.
- A brine forms, fish sinks down under the brine and additional salt and fish may be added.
- The fish are kept in brine under pressure for a period of 6 to 7 months. During this time, water and fat are pressed out of the fish and form a layer of brine covered with fat. The whole process should be carried out in a sterile condition where no microorganism should be involved.
- The cured anchovies are sold directly from the barrel or they will be packed in small retail cans.

3. Fermented mackerel (Colombo cure):

- Various mackerel species are use for this fermented fish in South Kanara of India known as 'Colombo cure'.
- Absolute fresh fish – gutted, gilled – washed in sea water- rubbed with salt (ratio salt to fish 1:3) – put in cement tanks – about 8 kg of fruit *Garcinia camboges*, which is

similar to tamarind, is added per tonne of fish- fish is kept in brine for 2 -4 months - then exported packed in wooden barrels.

» Shelf life is 1 year or more.

4. Chepa shutki:

» This is found in Bangladesh and literally termed as "pressurized dried fish".

» Puti Fish (*Puntius* sps.) is used to prepare Chepa Shutki. They are sun-dried for 12 to 14 days, kept in a jute sack for one month, after which it is washed and placed in a pressurized earthenware pot for 3 to 4 months.

5. Ngari:

» Ngari is a traditional fermented food of Manipur of India.

» It is prepared by fermenting smaller freshwater fishes with mustard oil and salt. The dried fish are then tightly packed them in a big clay urn which is made airtight. The urn is buried for 30–40 days.

» Ngari is roasted lightly prior to consumption, and then added in many Manipuri dishes, such as *eromba*.

e. Fermented fish paste products:

» Here, the fish and shrimps are sundried and then pounded with a proportion of salt to get a fish paste. The moisture content of the paste varies from 35 to 50% so that almost half the moisture content of the fresh material lost during processing.

» Typical pastes include *ngapi* of Burma, the *pra hoc* and various *mams* of Cambodia, *belachan* of Malayasia and the *trassi* of Indonesia.

1. Bagoong:

» It is a partially or completely fermented fish or shrimps paste product of Philipines.

» Fish bagoong is prepared by mixing salt with fish, and placed it inside large earthen fermentation jars. There it is left to ferment for 30–90 days with occasional stirring to make sure the salt is spread evenly. A food colouring called angkok (Red rice) is added to give the bagoong its characteristic red or pink colour.

» Angkak is made from rice inoculated with a species of red mold (*Monascus purpureus*).

» Some manufacturers grind the fermented product finely and sell the resulting mixture as fish paste. A by-product of the fermentation process is a fish sauce called *patis.*

2. Jeotgal:

» These are salted fermented paste dishes made with seafood such as fish, shellfish, and other marine animals in Korea.

» Depending on the ingredients, *jeotgal* can range from flabby, solid pieces to clear, broth-like liquid.

» Solid *jeotgal* are usually eaten as *banchan* (side dish). Liquid *jeotgal*, called *aekjeot*, is popularly used in *kimchi* seasoning, as well as in various soups and stews.

3. Trassi:

» Fresh fish and small shrimps →sun dried for a day → pounded in a wooden mortar → salted → again dried for a day. It is a village hut product of Indonesia.

4. Belachan:

» It is a product of Malayasia.

» Here, the fresh fish are taken- partially dried under sun – crushed in wooden trays by a treading process – salted – Artificial colour such as Rhodamine B is used to produce the desirable pink colour - Further dried and made the paste – marketed.

» The yield of paste is 40% to 50% of raw shrimp weight.

» Proximate analysis of belachan is as follows:

S. No.	Parameters	Contents
1	pH	7.6 – 7.8
2	Moisture	27% to 40%
3	Ash including salt	20% to 24%
4	Salt	13% to 18%
5	Proteins	30% to 40%

f. Fermented fish sauce:

- » In this process, fermentation offish is carried out for a longer period than in the production of fermented fish and fermented fish paste.
- » The fermented fish sauces are the liquids containing a mixture of amino acids and other protein degradation products. These have high salt content. So, used as condiments for flavouring rice dishes. They may be taken with plain boiled rice and also used in vegetable curry.
- » During the production of typical sauces such as the *Nuoc mam* of Cambodia and the *Nam pla* of Thailand, fish are mixed with salt in tubs or vats and left to stand for a period from 5 to 18 months.
- » It is a clear light or dark brown liquid produced from salted fermented anchovies (*Stolephorus* sp.) or other fish or fishery products such as squid, shrimp and shellfish in Asia. The fermentation period is usually half a year or more, up to one and a half year (the longer the better).
- » The best sauce are made from anchovies (*Stolephorus* spp.) as follows.
- » Fish – washed in sea water – mixed with salt (Fish to salt = 1:5 to 1:1) in tubs or vats and a weight is put over the fish to keep the fish in brine – keep the vat aside for 5 to 18 months – then, clear liquid is skimmed off the top or drained at the bottom of the vat/tub – the liquid is then filtered and bottled and exposed in the sun until it is ripe- the liquid sauce is finally packed in bottles for marketing.
- » First liquid drained off is the high grade and the residual liquid drained off further are of low grade. The first grade product should contain 20-23g per litre of nitrogen and, of this 50% should be formol titratable and not more than 15 to 20% titratable as ammonia. The salt content should be 20 to 25% and the pH should be less than 6.0.

1. Pekasam:

- » These are fermented half/fully fried mild coarse rices made by pickling several varieties of almost all fresh water fish, mainly *Anabas testudineus*, Tinfoil barb, Snakehead, Catfish in Malayasia.
- » For slightly different recipe and process of making Pla ra, the fish is first cleaned, cut into pieces if the size is too big and preserved with salt for several days and then mix with half fried rice bran plus palm sugar or brown sugar.

- The sugar is claimed as starter to the rice bran to become brewed as result the pekasam have nicer sweet sour smells and salty taste.
- Sometimes tamarind juice also added to make the more sour effect. This is then left (traditionally) in a clay crock covered with a cheesecloth lid, to ferment for at least a week and up to a year.

2. Pla- ra:

- These are fermented fish sauce found in Thailand made by pickling several varieties of fish, mainly snakehead murrel.
- The fish is cleaned, cut into pieces and mixed with salt and rice bran. This is then left in a big jar covered with a wooden lid, to ferment for three months to a year.
- Recently a dried powdered version of pla ra has been successfully marketed.

F. FISH AND PRAWN PICKLE

- Pickling of shrimp and fish is a recent concept and different styles have been prepared to satisfy a wide range of consumers spread all over the country.
- To increase the utilization of underutilized low value fishes, which are rich in nutrients, for human consumption value added fishery products can be prepared.
- Fatty fishes such as tuna, seer, mullet etc are preferred over the lean fishes like croaker, lizard fish etc for the preparation of pickle.
- Apart from fin fishes, shell fishes such as small shrimps which are of low cost can be used.
- Fish pickles can be prepared making cubes from the fish fillets, in three forms viz., hot and spicy, hot and sour and sweet and sour using standard methods. However, for prawn pickle, hot and spicy and hot and sour pickles are generally preferred.
- In case of prawn pickle, the prawns should be of medium or small size, headless, peeled, deveined followed by either hot blanching at 8% salt solution for 10 mins or frying in oil at 180^0C for 2 to 3 mins.
- The raw materials such as fish or shrimp must be very fresh and needs to be washed several times before preparation of pickle.

- The time of addition of fish cubes or prawns during preparation of pickle is same.
- The product has a shelf life of 10-12 months at room temperature and can be used as condiments.

Ingredients used for preparation of pickles:

S. No.	Ingredients	Hot and Spicy	Hot and Sour	Sweet and Sour
1	Fish cubes/ shrimp	1.0 kg	1.0 kg	1.0 kg
2	Green chili	100 g	100 g	80 g
3	Garlic whole	100 g	100 g	100 g
4	Ginger paste	150 g	150 g	100 g
5	Garlic paste	150 g	150 g	100 g
6	Fenugreek powder	10 g	10 g	10 g
7	Mustard seed	30 g	30 g	30 g
8	Garam masala Powder	10 g	10 g	10 g
9	Sugar	25 g	25 g	100 g
10	Red chilli powder	60 g	60 g	30 g
11	Cumin seed powder	20 g	20 g	20 g
12	Asafoetida	20 g	20 g	20 g
13	Turmeric powder	10 g	10 g	10 g
14	Bay leaves	10 g	10 g	10 g
15	Vinegar	500 ml	500 ml	500 ml
16	Mustard oil	500 ml	500 ml	500 ml
17	Black pepper powder	10 g	10 g	10 g
18	Monosodium glutamate	25 g	25 g	25 g
19	Salt	100 g	80 g	80 g
20	Citric acid	10 g	20 g	20 g
21	Tamarind paste	200 g	200 g	200 g
22	Sodium benzoate	40 mg	40 mg	40 mg

Flow Diagram for Preparation of Pickle:

» Raw material should be very fresh.

» Washing

» Fillets are prepared from whole fish. For prawn pickle, headless peeled and deveined shrimps should be made ready.

» Fish Fillets are cut into small cubes (3 cm size)

» Fried fish meat cubes/ shrimps in mustard oil and kept them aside

» Add Mustard oil (300ml) - Heat

» Add Sugar (Half of the desired quantity)

» Add Asafoetida (Half of the desired quantity)

» Add Mustard seed followed by Garlic whole, Green chilli, Bay leaves and stir well.

» Add Ginger + Garlic paste.

» Fry till brownish colour

» Add Turmeric powder + fenugreek powder + Garam masala powder + Red chilli powder +.Cumin powder + Black pepper powder + MSG and continued to stir.

» Fry till brownish colour for good flavour and odour

» Add Rest Sugar + Asafoetida

» Add Citric acid

» Add Tamarind paste – Fry for 2 to 3 min

» Add Vinegar

» Add Salt and allowed boiling till characteristic smell emerges.

» Add Fish cubes/ prawns to the gravy

» Cool and then add Sodium Benzoate when the temperature of the content is little higher than the room temperature and mix thoroughly.

- Heat the rest oil and put over it. After cooling completely, it is kept overnight for maturation.
- Pickles are packed in sterilized glass bottles or laminated LD pouches (400 microns).
- Bottles are sealed air tight so as to ensure that the layer of oil floats on the surface of the pickle.
- Labelled the bottles having statutory requirement and printed on it.

G. FISH PROTEIN HYDROLYSATES

Hydrolysis:

- Breaking down the chemical bond by means of water molecules is called Hydrolysis.
  Protein hydrolysis:
- Protein hydrolysis refers to breaking the peptide bond and adding a water molecule with the help of acid/alkali under the supplied heat or breaking the peptides bond using the biological catalysts called enzymes. In this reaction one hydrogen ion added to the amino group involved in the broken peptide bond and hydroxyl ions added to the carboxylic group.

Protein hydrolysates:

- A mixture of amino acids and peptides obtained after hydrolysis of a protein.

Fish protein hydrolysates:

- Protein hydrolysates prepared using fish protein.

a. Methods of protein hydrolysis:

- The protein hydrolysis can be done by two methods such as (a) Chemical method, and (b) Biochemical method.
- Chemical method:
  - Acid hydrolysis
  - Alkali hydrolysis

» Biochemical method:
  - Enzymatic hydrolysis with added enzymes
  - Autolysis assisted hydrolysis

1. Acid hydrolysis:

» Fish proteins are completely hydrolyzed with 6N hydrochloric acid or in some cases Sulphuric acid at high temperature at 118°C for 18hrs and often high pressure.

» The hydrolysates are then neutralized to pH 6.0 or 7.0 using alkali and concentrated to either a paste or further dried to make powder.

Advantage:

» More preferred than alkali hydrolysis

» Underutilized fish are converted into fertilizer due to low production cost and extensive hydrolysis

» It is a highly water soluble product.

Disadvantage:

» Process is difficult and hard to control.

» Neutralization results in high salt (NaCl) which interferes palatability.

» The essential amino acid tryptophan will be lost during the process.

2. Alkali hydrolysis:

» This process involves hydrolyzing the protein at alkaline pH (12.5) and high temperature (95°C).

» Alkali hydrolysis primarily use the fish protein concentrate as the substrate.

» Sodium hydroxide is commonly used as an alkaline reactant.

Advantages:

» Used to recover and solubulize a broad range of protein

- Aid in modifying properties of insoluble FPC
- Solubilize collagen

Disadavantages:

- Resultant product is poor in functionality and process affect the nutritive value adversely.
- Recovery is low.
- Racemization of L-amino acids produces D-amino acids which are not absorbed by human.
- Disulphide bonds are split with loss of Cysteine, serine and threonine through β-elimination reactions and formation of lysinoalanine, ornithinoalanine, lanthionone, and β-amino alanine occurs.
- The products have an inhibiting effect on proteolytic enzymes.

H. FISH SILAGE

- It is a liquefied fish product.
- It is prepared mostly in Poland, Denmark.
- The final product will be having pH around 4.0.

Production of silage:

1. Acid ensilage:

- Fresh Fish → Comminuted by mincing, cutting or chopping to a dia. of 10 mm → To it added 3.5% formic acid and constantly stirred to get required pH of 3.0 → the container is left covered for the fish mix to liquefy. This can take 3-4 days but the rate depends on species of fish, degree of comminution as well as temperature at which mixture is kept. It should be stirred daily.
- If Sulphuric acid (H_2SO_4) is added, because of very low pH, it needs to be neutralised by $Ca(OH)_2$.
- After liquefaction, oil removal may be necessary where fish with a high fat content are used. For this the temperature is to be maintained at about 65-70^{0}C when

coarse suspended solids can be removed by decantation. This is followed by centrifugation to remove oil.

» The liquid fish silage are packed in bottles and transported to the market.

» In the finished silage upto 2% oil is acceptable.

2. Fermented silage:

» Fermented fish silages rely on the biological production of lactic acid by bacteria to lower the pH. In general, lactic acid bacteria such as *Lactobacillus plantarum* ferment sugars to organic acids (primarily lactic acid), thus lowering pH of the mixture.

» Since fish contains very negligible amount of carbohydrate, it is usually necessary to add suitable carbohydrates for the bacteria to convert to acid.

» Addition of mixture of malt and cereal meal, molasses and cereal meal, malt and tapioca meal, and molasses and tapioca meal have all proved successful.

» The fermentation process for conversion of carbohydrates to lactic acid is anaerobic and can be divided into three stages.

- The starch of the carbohydrate source is hydrolysed to maltose by α and β amylase.
- Maltose is broken down to glucose maltose.
- Glucose is converted to lactic acid by bacteria. Small amounts of acetic acid and alcohol are also formed.

» Lactic acid bacteria can be divided into two types,

a. Homofermentative, which converts one molecule of glucose to two molecules of lactic acid and

b. Heterofermentative, which convert one molecule of glucose to one molecule of lactic acid, ethyl alcohol and water.

Hence, Homofermentative bacteria are preferred to use for silage production.

» Since fish do not contain sufficient lactic acid bacteria, it is essential to add a starter culture usually of lactobacilli for successful fermentation.

» The fermentation should be carried out in full air tight containers so that conditions of anaerobic and successful fermentation is indicated by a rapid drop in pH, as the lactic acid is formed, and the production of gas.

Conditions for fish silage production:

» Where high cost is involved for sending to fish meal plant

» If nearby fish meal plant is busy

» If local market is there

» Where small amount fish is available

Use: Used as animal feed especially for pigs.

I. FISH PROTEIN CONCENTRATE (FPC)

Fish protein concentrate (FPC) is any stable fish preparation intended for human consumption in which protein is more concentrated than the original fish.
Principle:

» Removal of most of water and some or all fat content helps in production of FPC in which protein is highly concentrated.

Types:

» There are three types of FPC. They are: Type A, Type B and Type C.

» Type A FPC is suitable for human consumption where as Type C is closely resemble to fish meal.

Table 12.2: Difference among three types of FPC

S. No.	Parameters	Type A	Type B	Type C
1	Moisture	Max. 10%	Max. 10%	Max. 10%
2	Protein	Min. 67.5%	Min 65%	Min 60%
3	Fat	Less than or equal to 0.75%	0.75 to 3.0%	More than3.0%
4	Pepsin digestibility	Min.92%	Min.92%	Min.92%
5	Average Lysine	Min. 6.5% of protein	Min. 6.5% of protein	Min. 6.5% of protein

Production method:

- **Chemical method:**
  - **By use of solvent:**
    - ◊ Product is of tasteless, colourless and odourless.
    - ◊ Chemical solvents are used to remove the water, fat and fishy tasting components, either from raw fish or from fish meal. The solvents most successfully used to make this type of FPC are the alcohols, eg. Ethanol, isopropanol. Normally a high proportion of the solvent is recovered and used again as it is highly expensive.
    - ◊ The solvent extracted product, often called as 'fish flour' has none of the traditional attributes of a foodstuff, namely smell, flavour, texture and colour. It is off-white, odourless, tasteless and can be used only by incorporation in other foods such as bread, biscuit, soup and stews at a level that does not affect their normal properties.
  - **Viobin process** developed by CIFT, Kochi, India.
    - ◊ Fresh fish – Washing in potable water – Descaling, beheading, gutted and washed again- Meat picking by a meat picking machine – Mincing of picked meat using a meat mincer – Minced meat is suspended in an equal amount of water +0.5% acetic acid and kept at 70-80^{0}C for 30 mins under constant stirring - The slurry is then filtered and pressed in a screw press so that a good amount of liquid is removed. The press cake will contain nearly 60-62% moisture. It is then dried in a tunnel dryer at a temperature range of 75-80°C. to various levels of moisture. The dry matter (moisture content below 10%) is coarsely ground in an ordinary milling machine and used for extraction of fat.
    - ◊ Extraction: The material is charged in to the extraction drum having arrangements for stirring and refluxing.
    - ◊ The solvent, azeotropic mixture of hexane and ethyl alcohol with a boiling point of 58.68°C is added in two parts by volume to one part by weight of material and kept at the boiling temperature of the solvent for 30 minutes, under constant stirring.
    - ◊ The solvent is drained off and fresh solvent added in 1.5 parts by volume to

one part by weight of original material and refluxed under constant stirring for 30 minutes.

- ◊ After draining off the solvent the process is repeated once again.
- ◊ After final draining the adhering solvent is distilled off and the last traces removed by steam stripping under reduced pressure.
- ◊ The dried matter is pulverized to fine mesh size using a milling machine.

» **Biological method:**

- ◊ Fermentation process or enzymatic hydrolysis.

» **Physical process:**

- ◊ By use of high voltage current

» **Yield**: Approximately 18% of the total weight of fish

J. IRRADIATION

» Food irradiation is the process for the treatment of food products to enhance their shelf life and to improve microbial safety.

» Electromagnetic radiations, namely gamma, and X-rays having short wavelength (<300 μm) and higher energy than visible light can significantly penetrate the material including foods causing ionization of atoms and molecules by removing electrons from their outer shell.

» The inactivation of living cells (microbial cells) by irradiation is essentially due to scission of single or double strands of DNA, which is essentially caused by the OH radical formed by radiolysis of water.

» In addition to DNA damage, ionizing radiation has also been shown to cause damage to the membrane and other structures causing sub-lethal injury.

» There are two types of radiation considered for use.

a. Particle or electron beam type: Cathode ray and β-ray (limited energy and poor penetrative power)

b. Electromagnetic wave type: X-ray and γ-ray (deeper penetrative power)

- The high frequency electromagnetic waves, high energy electrons and beams of the heavier atomic particles, being capable of inducing nuclear transformations in the atoms of the target food, are not acceptable.
- Irradiation by Ultraviolet of wavelength 2000 – 2950 A0 is generally permitted. It has very poor penetrative power. So, it is limited to surface treatment. It is only bactericidal. In fisheries, its main use is in purifying water.

a. Levels of treatment: Three levels are considered in processing foods.

- ***Radappertisation*** (1-5 Mrads): this level of treatment is the most severe and will destroy all spoilage and pathogenic micro-organisms. However, it will stimulate off-flavours and odours and possibly cause textural damage. Such changes can be reduced either by blanching or by including antioxidants.
- ***Radurisation*** (0.5 Mrad): This will eliminate most non-spore forming bacteria and give a significant reduction in the number of spoilage micro-organisms, thus extending shelf life. But enzymes are not denatured by this.
- ***Low dose*** (50 Krad): this inhibits sprouting in vegetables and cereals, and kills tapeworms and insects.

Irradiation of fish and seafood:

- In general at doses higher than 0.3 to 0.5 Mrads, discolouration, production of off-flavour would make irradiation unacceptable.
- Limiting doses of various pathogens:
- 0.1 to 0.25 Mrads: *Shigella* sps, *E. Coli*, *Proteus vulgaris*
- 0.3 to 0.5 Mrads: *Streptococcus* spp., *Staphylococcus aureus*, *Salmonela* spp.

b. Chemical changes during irradiation:

- With the radurisation treatment of 0.3 – 0.5 Mrads, there is some destruction of vitamin and amino acid Cysteine, and oxidative changes are observed.
- Treatment of mackerel by doses of 50Krad to 4.5 Mrad caused no changes in the Biological Value (BV) or Net Protein Utilisation (NPU) of the fish protein.
- Thiamine (Vitamin) is known to be destroyed by doses in the range of 0.6 to 1.0 Mrad.

» TMAO is degraded during irradiation by doses of 0.6 Mrad upwards.

c. Changes during storage:

» In Bangladesh, fresh water carp irradiated at a level of 200 – 250 Krad were found to have an extended shelf life of 8 -10 days as compared to untreated fish (1 day).

» Shrimp irradiated at 200 Krad had less than 3 ppm carbonyls immediately after irradiation and this value remained constant even after 28 days of storage. But, non-irradiated sample containing 3ppm carbonyls increased to 5 ppm within 21 days showing spoilage characteristics.

» Generally, Irradiated samples show no increase of TMA or TVB during storage, which might be due to destruction of spoilage micro-organisms.

K. COATED FISH PRODUCTS

The present day consumers from urban sector are showing much interest in ready-to-eat and ready-to-cook products. These food items are called as convenient food products and the global demand for them is increasing rapidly. This has led to the development of several fish based products with varied taste, texture and appearance. Battered and breaded products popularly known as coated products are one such group having high consumer appeal. This battering and breading technique has contributed significantly to value addition of fish and fishery products.

A coated food product is one which is coated with another foodstuff. Coating by battering and breading enhances the characteristic of the food such as appearance, flavour and texture. Coating acts as a moisture barrier, minimises the moisture loss during freezing and frozen storage and microwave re-heating and retains the natural juices of the foods. This ensures the final product tender and juicy on the inside and at the same time crispy on the outside. The first commercially successful coated product was fish finger.

There are several ingredients used in the formulation of coatings. Each ingredient performs its functions to contribute to the unique characteristics and functionality of coatings. The commonly used ingredients are grouped under five categories. These are polysaccharides, proteins, fats, seasonings and water. Besides, small quantities of leavening agents, spices, colour etc are added to provide specific effects. A batter can be defined as a liquid mixture composed of water, flour, starch and seasonings into which food products are dipped prior to cooking. The breading is normally a bread-based crumb.

1. Ingredients:

a. Water:

» The addition of water in minced meat is required to maintain acceptable texture and minimize the cost of raw material. Water serves initially to disperse the myofibrillar protein molecules, allowing the development of more expanded net work by protein-protein bonds formed during heating. Addition of water causes a great reduction of shear stress (gel hardness), while it has slight impact on shear strain (gel cohesiveness). The quality of water affects the colour, flavour and texture characteristics of the finished products. Further, the mineral content in the water such as Ca^{2+} and Mg^{2+} helps in textural change of the product.

b. Starch:

» High amylase starches obtained from corn, wheat and potato, form a somewhat brittle gel, while high amyl pectin starches of tapioca and maize form adhesive and cohesive gel. In addition, the colour of high amylase starch gel is opaque, while high amyl pectin starches make the gel translucent. In the minced meat combined starches are added up to 4~12%. Starch granules continue to absorb water and expand until gel matrix limits them. Gel strength decreases if too much of starch is used in the product formulation.

» Flour refers to finely ground starchy material obtained from several sources like corn, rice, soy or barley. Functionality of flour in batter and breading systems depends mainly on the major constituents of flour viz, starch and protein. The pick- up of coating on any product can be increased either by adjusting viscosity of batter or by repeating the process of battering and breading. The fish part is expected to be minimum 50% in any coated product.

c. Salt:

» The addition of salt to raw fish paste is not only important for seasoning but also for solubilising myosin in order to form gel. If 3% salt is added to grounded fish paste containing starch, the 'ashi' (gel strength) will be the strongest. While grinding the fish meat, salt is used in the beginning to extract and solubilise myofibril proteins which are essential for formation of meat gel texture. A concentration of sodium chloride between1.2% to 1.5% gives maximum extraction of myosin as well as the best resilient texture to the final product.

d. Vegetable oil:

» Vegetable oil is used to modify the product texture as well as to make the product whiter when dark colour ingredient is used. Soya bean oil or sunflower oil is commonly used in minced meat as a texture modifier, colour enhancer and as a processing aid.

e. Chemical seasoning:

» The best known and widely used flavour enhancer is monosodium glutamate (MSG). Generally 0.1% to 0.5% MSG and 0.3% of ribotide (Sodium 5- ribonucleotide) are also added to meat during fish ball preparation.

f. Colorant:

» The colour of the product is an important element in making the product appealing to the consumer. Colours used can be artificial or natural. Caramine is a vivid purple red powder. The powder is traditionally sold with 50% to 53% ceramic acid content and is stable to light, heat, oxidation and during storage.

g. Sugar:

» Sucrose is used as cryoprotectant and its role is to prevent protein denaturation, particularly of actomyosin, during frozen storage. The cryoprotective effect of sucrose is considerably increased when small levels of polyphosphate are added to the mince along with sugar. Sugar causes increase in the surface tension of water, thereby, increases the amount of bound water. It also prevents driploss (withdrawl of water molecules from protein) by stabilizing proteins.

2. Preparation Stages:

» The production of battered and breaded fish products involves several stages. The method varies with the type of product and pick up used. In most cases, it involves 7 steps - portioning / forming, pre-dusting, battering, breading, pre-frying, freezing, packaging and cold storage.

i. Portioning / Forming:

» It is an important stage in the production of coated fishery products. A recent innovation for the catering sector is formation of skinless and boneless fish fillets into predetermined shape and size using specially designed forming machines. The shapes vary from conventional fillet shapes to several other imaginative ones.

ii. Pre-dusting:

» Pre-dusting is done to prepare the surface, so that batter can adhere uniformly. It consists of a very fine raw flour type material. A more sophisticated and expensive pre-dust may contain spices and seasonings for functional and flavouring purposes.

iii. Battering:

» Conventional batters are of low to medium viscosity and hence can be applied with total submersion or overflow batter applicators. Low viscosity batters are normally applied in an overflow configuration. However, medium viscosity may require a total submersion system depending on the product requirements.

» Egg contains albumin, a heat coagulable protein, that is useful in binding both breading and batter to the product and to itself. The yolk protein contains lecithin, which being an emulsifier, also contributes to batter stability.

» Salt and sugar are added to many batter/breading formulae as a flavouring agent. They also complete with flour proteins for water and tend to slow the rate of protein hydration.

» Pepper in combination with small quantities of other spices are frequently included at a 3-5% level of batter/breading formulae. However, the spices are not known to interfere with the functionality of the batter/ breading ingredients.

iv. Breadings:

» The secondary coating is referred to as 'breading' although it may not always be derived from bread. The original crumbs are ground dry bread which is used as the major secondary coating. However, a wide variety of breading materials are available now in different sizes and colours. They can be used separately or in combination with flours, starches and flavouring materials.

» The functional characteristics of breading depend on the specific physical and chemical attributes built into the particular breading. The important considerations are the particle size, area to volume relationship, browning rate, moisture absorption, colour and texture and oil absorption.

» There are many types of breading applicators available and the appropriate machine depends on the ingredients used. The speed of the breading machine is adjusted so as to closely match the belt speed of the batter applicator.

» For soft products, the crumb depth should be maintained as thin as possible to avoid product damage when it leaves the breading machine. Frozen, or hard products should however, have a deep bed of crumb.

» Air blowers are used to remove excess crumb from the fish after enrobing. Excess crumb carried into the fryer can cause unsightly black specks on the product. Filters are used to remove small particles from the oil to prevent this.

v. Flash-frying:

» Flash-frying is another very important step in the battering process. In this process, a wet coated product is entered in to the hot oil at a temperature of 180-190⁰C for about 30 seconds.

vi. Freezing:

» The flash-fried fish products are generally frozen is two steps. These are initially individually quick frozen using liquid nitrogen or carbon dioxide. After the initial quick freezing, the products are frozen in a slower manner using a mechanical freezer or even liquid nitrogen or carbon dioxide. Freezing is continued until the internal temperature attained is around -12°C to -15°C.

vii. Packaging and storage:

» Most fish fingers, cutlets, balls, sticks or portions are individually packed into small boxes on a weight basis or specified member of pieces. They are also packaged in bulk form. When individually packed, layers of fish products are separated by waxed paper to prevent further product damage. The box is labelled and over-wrapped a polythene film to prevent moisture loss and freezer burn during storage. The packed boxes are stored at around -20°C.

3. Various Coated Fish based Products:

i. Coated fish fillets:

» Fried coated fish fillet is a prominent food item in the European markets. Sea fish fillet of table size and having minimum pin bones can be used for this purpose. Various stages in the production of coated fish fillet are:

a. Filleting

b. Cold blanching

c. Pre-dusting

d. Coating with batter

e. Breading

f. Pre-frying

g. Freezing

h. Packaging

i. Storage

a. Filleting:

» A fish fillet is a skinless, boneless fish loin cut along the central bone frame and trimmed free of loose or hanging meat. This can be prepared manually as well as using filleting machine. Generally manual filleting gives better yield (around 40 to 50%) than machine filleting. Wash the fillets in chilled water and drain.

b. Cold blanching:

» Dip the fillets in 5% brine solution containing 0.1% citric acid for 3 to 5 mins depending upon the size grade and then drain off.

c. Pre-dusting:

» The fillets are then pre-dusted with a suitable pre-dust or dry batter mix itself. The excess pre-dust adhered to the meat is then removed either by shaking or using an air blower.

d. Battering:

» The pre-dusted fillets are then coated with batter uniformly.

e. Breading:

» The batter coated fillets are further coated with bread crumbs. Generally medium size porous crumbs are preferred. The bread crumbs are uniformly applied on the product and the excess crumbs are then removed using an air blower. The coating picks up depends on the viscous of the batter and the type of crumbs. Breading and battering comes to about 30 to 35% of the product.

f. Pre-frying:

» After breading the fillets are flash fried in hot oil at 180 to 200^{0}C for 20 to 30 seconds depending on the size grade of the fillets.

g. Freezing:

» Immediately after flash frying, the fillets are cooled using a fan and then frozen in a IQF freezer

h. Packaging:

» The frozen coated fillets are then packed in thermoformed containers or pouches made of 12 micron plain polyester laminated with 118 micron LDPE. A specific number of such consumer packs are then packed in master carton.

i. Storage:

» The packed cartons are frozen stored in a cold store maintained at -20^{0}C.

ii. Coated fish fingers:

» Fish fingers are regular sized portions cut from regular frozen blocks of fish fillet or fish mince. The fish fingers when fried in vegetable oil develop a golden brown colour with attractive appearance and odour. It has been observed that the sensory quality of fish finger developed from the frozen block of fish fillets is superior to that developed from the fish mince block.

a. Fish finger from fillets:

» The fish are filleted first and then cut into small pieces of about 10 cm in size and blanch in 3% salt solution for 10 to 15 mins. Drain off and pre-dust with batter powder and coat with batter and bread crumbs and fry.

b. Fish finger from fish mince:

» Fresh fish is dressed, filleted and then minced in a meat mincer. Add 0.1% tri-sodium polyphosphate, 0.6% salt and mix in a silent cutter. Then spread the mince in a tray uniformly and freeze. Cut into small pieces of about 10 cm in size in the frozen condition itself.

» Pre-dust the finger with batter powder and coat with batter and bread crumbs. They are then flash fried in vegetable oil at 180 to 200°C for 30 seconds. After cooling, the fingers are frozen preferably in an IQF freezer and packed in thermoformed trays or pouches and stored at cold store at -20°C.

iii. Coated Fish cutlets:

» Fish cutlet is a popular snack at celebrations, household functions etc. The basic raw material required for cutlet preparation is cooked fish meat generally from low cost fish either from fresh water or marine source without any bone.

» Ingredients used for preparation of fish cutlet:

S. No.	Ingredients	Percentage (%)
A. Paste		
1.	Minced fish meat	56.0
2.	Potato	16.0
3.	Wheat flour	3.0
4.	Spice mixture *	3.2
5.	Salt	1.5
6.	Mono sodium glutamate (MSG)	0.2
7.	Onion	10.0
8.	Ginger- garlic paste	2.0
9.	Baking powder	0.1
10.	Sunflower oil	8.0
B. Battering and Breading		
1.	Egg white	
2.	Bread crumbs	

» Composition of Spice Mixture for preparation of Fish Cutlet

S. No.	Ingredients	Percentage (%)
1.	Chilli Powder	0.8
2.	Pepper	0.3
3.	Clove	0.2
4.	Garam Masala	1.5
5.	Coriander leaves	0.2

Preparation of Fish Cutlet:

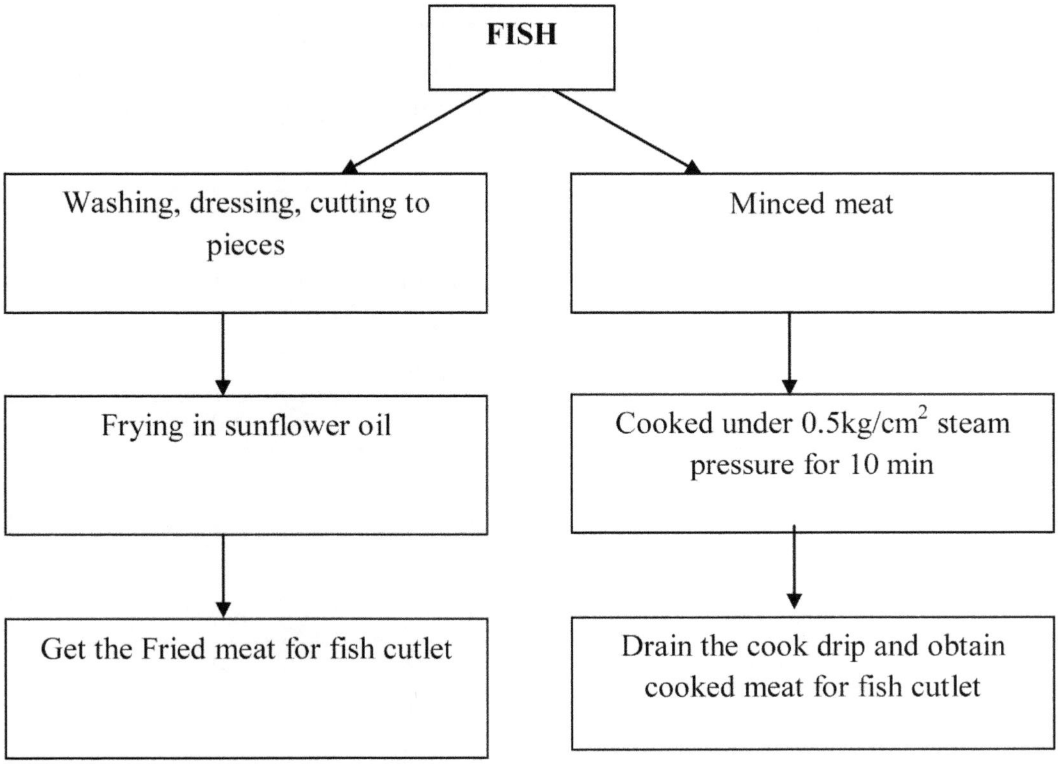

→ Boiled and peeled potatoes were made into fine paste and mixed with cooked meat along with salt, baking soda and spice mixture.

→ Fried onion and ginger - garlic paste were mixed with paste.

→ All ingredients were mixed thoroughly.

→ 25 g of material was shaped into a ball and flattened to 1 cm thickness of any desired shape.

→ Batter the product by dipping in egg white.

→ Bread the product by rolling on bread crumbs powder.

→ Whenever required, cutlets were flash fried at 180 - 200^{0}C for 30 seconds in edible sunflower oil.

→ Cutlets, thus prepared are packed in thermoformed trays and frozen stored at -20^{0}C

iv. Coated Fish ball:

- Fish ball is a convenience fish based product prepared using fish mince from low cost fish and starch that can be processed as a coated product. The fish ball preparation is very simple and requires only few locally available ingredients. The eating quality of fish ball is its jelly strength.
- Ingredients used for preparation of Fish ball:

Sl. No.	Ingredients	Weight in Kg	Percentage (%)
1.	Minced fish meat	1.000	-
2.	Starch(Corn flour)	0.100	10.0
3.	Salt	0.020	2.0
4.	MSG	0.002	0.2
5.	Poly phosphate.	0.002	0.2
6.	Chilled water (1-2°C)	0.050	5.0
7.	Vegetable oil	0.060	6.0
8.	Chilli powder	0.030	3.0
9.	Coriander powder.	0.010	1.0
10.	Turmeric powder	A pinch	
11.	Garam masala powder	0.050	5.0
12.	Beaten eggs	As required	
13.	Bread powder.	0.030	3.0

Preparation of fish ball:

Minced fish meat is ground in a silent cutter for 3 minutes.

↓

Add salt and mix thoroughly.

↓

Add starch, preservatives, sugar, vegetable oil, chilli powder, cardamom powder, turmeric powder, Garam masala powder and chilled water and mix thoroughly in silent cutter for about 15 min.

↓

Forming (10-12 g balls of 3 cm diameter to be made manually)

↓

Set at 30-40°C for 30 min.

↓

Battering the fish balls with beaten eggs.

↓

Breading the battered fish balls with bread crumbs.

↓

Fish Balls thus prepared were flash fried in sunflower oil at 180°C for 30 seconds

↓

They were packed in thermoformed trays for frozen storage.

v. Coated shrimp products:

» Delicious and attractive coated products from shrimps can also be prepared which has good market worldwide. Coated shrimps prepared from medium sized shrimps are served as a snack dish or an appetiser. Coated products such as butterfly, fantail round and peeled and deveined can be prepared from shrimps. The production process involves nine steps as shown below.

a. Preparation of raw material:

» This includes the preparation of shrimps in different forms such as butterfly, fantail round and peeled and deveined. These are then washed and drained.

b. Cold blanching:

» Dipped in 5% brine solution containing 0.1% citric acid for 5 mins.

c. Pre-dusting:

» Coated with a thin layer of fine flour or dry batter mix itself.

d. Battering:

» Usually an adhesive type batter is used.

e. Breading:

» Light coloured coarse crumbs are used for Japanese markets and darker coloured crumbs (yellow-orange) for European and US markets.

f. Flash frying:

» The coated shrimp product is flash fried for 30 seconds at 180 to200^{0}C in refined vegetable oil.

g. Packing:

» The products are packed in thermoformed pouches.

h. Freezing:

» The products are frozen in an IQF machine or in a blast freezer at -40^{0}C.

i. Frozen storage:

» The frozen products in thermoformed containers/ pouches are packed in master cartons at -20^{0}C.

Examination of Frozen Raw Breaded Shrimp:

Product description:

» Frozen raw breaded shrimp are whole, cleaned, headless, peeled and deveined shrimp, coated with a suitable batter and/or breading. Whole shrimp consist of five or more segments of unmutilated shrimp flesh. They are prepared and frozen in accordance with good manufacturing practice and are maintained at temperatures necessary for the preservation of the product.

Composition of the product:

» Frozen raw breaded shrimp shall contain not less than 50 percent by weight of shrimp material when the weight of the shrimp material is determined from the end product.

Styles:

- Style I. "Regular Breaded Shrimp" are frozen raw breaded shrimp containing a minimum of 50 percent of shrimp material.
- Style II. "Lightly Breaded Shrimp" are frozen raw breaded shrimp containing a minimum of 65 percent of shrimp material.

Types:

(a) Type I-Breaded fantail shrimp

1. Subtype A. Split (butterfly) shrimp with the tail fin and the shell segment immediately adjacent to the tail fin.
2. Subtype B. Split (butterfly) shrimp with the tail fin but free of all shell segments.
3. Subtype C. Split (butterfly) shrimp without attached tail fin or shell segments.

(b) Type II-Breaded round shrimp

1. Subtype A. Round shrimp with the tail fin and the shell segment immediately adjacent to the tailfin.
2. Subtype B. Round shrimp with the tail fin but free of all shell segments.
3. Subtype C. Round shrimp without attached tail fin or shell segments.

(c) Type III-Breaded split shrimp.

Grades:

- "Grade A" is the quality of frozen raw breaded shrimp that when cooked possesses a good flavour and odour, and under hedonic scale the total score is not less than 85 points.
- "Grade B" is the quality of frozen raw breaded shrimp that when cooked possesses a reasonably good flavour and odour, and under hedonic scale the total score is not less than 70 points.
- "Substandard" is the quality of frozen raw breaded shrimp that fail to meet the requirements of Grade B.

Definitions and Methods of Analysis

(a) "Fantail shrimp":

» This type is prepared by splitting and peeling the shrimp. Here, for Subtype A, the tail fin remains attached and the shell segment immediately adjacent to the tail fin remains attached. Subtype B, the tail fin remains, but the shrimp are free of all shell segments. Subtype C, the shrimp are free of tail fins and all shell segments.

(b) "Round shrimp":

» This type is the round shrimp, not split. The shrimp are peeled. For Subtype A, the tail fin remains attached and the shell segment immediately adjacent to the tail fin remains attached. Subtype B, the tail fin remains, but the shrimp are free of all shell segments. Subtype C, the shrimp are free of all shell segments and tail fins.

(c) Good flavour and odour:

» "Good flavour and odour", essential requirements for a Grade A product, means that the cooked product has flavour and odour characteristics of freshly caught or well-refrigerated shrimp and the breading is free from staleness and off-flavours and off-odours of any kind.

(d) Reasonably good flavour and odour:

» "Reasonably good flavour and odour" is the minimum requirements of Grade B products. It means that the cooked product may be somewhat lacking in the good flavour and odour characteristics of freshly caught or well-refrigerated shrimp but is free from objectionable off-flavours and objectionable off-odours of any kind.

(e) Dehydration:

» It refers to the occurrence of whitish areas on the exposed ends of the shrimp (due to the drying of the affected area) and to a generally desiccated appearance of the meat after the breading is removed.

(f) Deterioration:

» It refers to any detectable change from the normal good quality of freshly caught shrimp. It is evaluated by noting in the thawed product deviations from the normal odour and appearance of freshly caught shrimp.

(g) Extraneous material:

- It consists of non-edible material such as sticks, seaweed, shrimp thorax, or other objects that may be accidentally present in the package.

(h) Slight:

- "Slight" refers to a condition that is scarcely noticeable but does affect the appearance, desirability, and/or eating quality of breaded shrimp.

(i) Moderate:

- "Moderate" refers to a condition that is conspicuously noticeable but that does not seriously affect the appearance, desirability, and/or eating quality of the breaded shrimp.

(j) Marked:

- "Marked" refers to a condition that is conspicuously noticeable and that does seriously affect the appearance, desirability, and/or eating quality of the breaded shrimp.

(k) Excessive:

- "Excessive" refers to a condition that is very noticeable and is seriously objectionable and the product cannot be graded above Grade B; this is a limiting rule.

(l) Halo:

- "Halo" means an easily recognized fringe of excess batter and breading extending beyond the shrimp flesh and adhering around the perimeter or flat edges of a split (butterfly) breaded shrimp.

(m) Balling up:

- "Balling up" means the adherence of lumps of the breading material to the surface of the breaded coating, causing the coating to appear rough, uneven, and lumpy.

(n) Holidays:

- "Holidays" means voids in the breaded coating as evidenced by bare or naked spots.

(o) Damaged frozen raw breaded shrimp:

- Damaged frozen raw breaded shrimp means frozen raw breaded shrimp that have been separated into two or more parts or that have been crushed or otherwise mutilated to the extent that their appearance is materially affected.

(p) Black spot:

» It means any blackened area that is markedly apparent on the flesh of the shrimp.

(q) Sand vein:

» "Sand vein" means any black or dark sand vein that has not been removed, except for that portion under the shell segment adjacent to the tail fin when present.

(r) Extra shell:

» "Extra shell" means any shell segment(s) or portion thereof, contained in the breaded shrimp except the first segment adjacent to the tail fin for Type I, Subtype A, and Type II, Subtype A.

(s) Net weight:

» Net weight is determined by use of a balance taking the weight of shrimp after removing crumbs.

(t) Uniformity:

» "Uniformity" is determined for packs of various sizes by the ratio of the weights of the largest to the smallest breaded shrimp as outlined by the following schedule:

Up to 10 oz.	3 largest/3 smallest
10.1 oz. to 1.5 lb	6 largest/6 smallest
1.51 lb. to 2.5. lb	8 largest/8 smallest
Over 2 1/2 lb	10 largest/10 smallest

(u) Percent shrimp material:

» Percent shrimp material is calculated as below:

% shrimp material = [Weight of de-breaded sample (20 shrimp) x 100/Weight of sample (20 shrimp)] + 2

(v) Percent shrimp material on-line:

» Percent shrimp material on-line is calculated as follows:

% shrimp material=[Weight of de-breaded sample x 100/Weight of sample]

(w) Cooked in a suitable manner:

» "Cooked in a suitable manner" means cooked in accordance with the instructions accompanying the product. If, however, specific instructions are lacking, the product for inspection is cooked as follows:

- Transfer the breaded shrimp, while still frozen, in a wire mesh deep fry basket sufficiently large to hold the shrimp in a single layer without touching one another.
- Lower the basket into a suitable liquid oil or hydrogenated vegetable oil at 175 - 180°C for 3 minutes, or until the shrimp attains a pleasing golden brown colour.
- Remove basket from the oil and allow the shrimp to drain for 15 seconds.
- Place the cooked shrimp on a tissue paper or napkin to absorb the excess oil.

13 Fish By-Products

A. SHARK FINS AND FIN RAYS

» Dried shark fins are exported to China, Hong-Kong, Singapore and other parts of the world.

» The shark fins are mainly used for production of shark fin rays.

Shark fins:

» The process used for shark fin is similar throughout the world.

» Mainly dorsal, pectoral, ventral and caudal fins are processed.

» Fins are classified based on its size (the length of the anterior margin of the fin).

» For Dorsal, Pectoral and Ventral fins: Grade A = <10 cm size

Grade B = 10 – 20 cm

Grade C = 20 – 30 cm

Grade D = ≥ 30 cm

» For Caudal fins: Grade A = <20 cm size

Grade B = 20 – 30 cm

Grade C = 30 – 40 cm

Grade D = ≥ 40 cm

Production method:

» Fins are cut from sharks of about 125 cm or more in length as soon as they are landed avoiding as much flesh as possible → Any adhering flesh is removed as far as possible → Dusting with salt (Salt : Fins = 1: 10) and set aside for 24 hrs → Washed in water → Dried under sun for a month → The moisture content after drying is generally about 7-8% → Packed in sacks under pressure so as to flatten the fins during storage.

ISI standard of shark fins: (a) White fins and (b) Black/ spotted fins

Shark fin rays:

» Shark fin rays are produced from fresh shark fins or dried shark fins.

» The dried shark fin rays are called as Shisai in Japan and fin rays which are dried under sun are called Taishi.

» Colour of shark fin rays: Golden yellow

» The extracted good quality dried fin rays should be light, yellowish-white in colour, odourless, and have no gelatinous substances on the surface of the strands.

Production method:

» Fresh shark fins are boiled to remove the outer skin and thus, naked fin rays can be obtained.

» Dried shark fins are soaked in fresh water to soften for 4-5days.

» After softening, they are put in10% acetic acid at 60°C for 1-3 hrs so as to swell them, hydrolyse the collagen to gelatine and remove the epidermis layer by scraping off the outer calcareous layer.

» The fin rays are then separated individually from one another manually by hand or forceps and putting them in hot water to remove the gelatinous substance.

» The fin rays are then dried in a dryer at 45-50°C for3-4hrs until a moisture content of 5 – 8% and packed in bags for marketing.

Shelf life: One year

Use: The shark fin rays are used as an essential ingredient in some exotic soups.

B. FISH MAWS AND ISINGLASS

Fish Maws:

» Fish maw is the dried form of fish air bladder.

» Air bladder, also called swim bladder, consists of several membranous layers rich in collagen. Air bladder is located in the abdominal cavity below the vertebral column, helps the fish in regulating its specific gravity.

» Cleaned and dried air bladder is called as Fish Maw, which is an excellent raw material for production of Gelatin.

» Hake, Sturgeons and carps are good sources for air bladder.

» Fish maw contains rich proteins and nutrients such as phosphor and calcium. It nourishes kidney and boosts stamina.

» It is effective in healing weak lung and kidney, anaemia, etc.

» India exports dried fish maws, which form the raw material for the production of isinglass and other such products. Process has been developed to produce the finished products from fish maws.

Production:

» The air bladder of deepwater hake is most suitable for production of maws and then isinglass.

» ***Raw material***: In India air bladders of eel, jew fish, sea bass, lizard fish, croaker and catfishes are used for the purpose.

» The air bladders are separated from fish and temporarily preserved in salt during transport.

» On reaching the shore they are split open, washed thoroughly, outer membrane is removed by scraping and then washed further, air dried to 15% moisture level by hanging or placing in trays.

» Cleaned, desalted, air dried and hardened swimming bladders are called fish maws.

Use:

- Fish maws iare used for production of isinglass, a form of gelatin.
- Fish maw contains rich proteins and nutrients such as phosphor and calcium. It nourishes kidney and boosts stamina.
- It is effective in healing weak lung and kidney, anaemia, etc.

Isinglass:

- The word isinglass is derived from the Dutch and German words, which have the meaning sturgeon's air bladder or swimming bladders.
- Isinglass is purely collagen and only after heating, it hydrolyses to form gelatin and after cooling forms tough gel.
- The fish maws are used for production of isinglass.
- In India air bladders of eel and catfishes are used for the production of isinglass.

Production:

- The dried fish maws are softened by immersing and soaking in chilled water for several hours until it becomes soft.
- They are mechanically cut into small pieces and rolled or compressed between hollow iron rollers that are cooled by water and provided with scraper for the removal of any adhering dried material.
- The rolling process converts the isinglass into thin strips or sheets of 1/8" to ¼" thickness.
- There are processes for the production of isinglass in powder form also.
- Isinglass dissolves readily in most dilute acids or alkalis, but is insoluble in alcohol.
- In hot water isinglass (collagen) hydrolyses to form gelatine and after cooling swells uniformly producing opalescent tough gel with fibrous structure.
- It is used as a clarifying agent for beverages like wine, beer, vinegar etc. by enmeshing the suspended impurities in the fibrous structure of the swollen isinglass.

Types of Isinglass:

There are four types of isinglass. They are,

- Long staple
- Short staple
- Leaf
- Book isinglass

Uses:

- Clarifying agent for beer, cider, wine, vinegar etc.
- Can be used as adhesive waste
- Isinglass dissolved in acetic acid forms a strong cement base
- Can be used as a sizing agent in textiles
- It is an ingredient in Indian ink

C. PEARL ESSENCE

- Pearl essence is the suspension of crystalline guanine in water or organic solvent.
- Guanine is an iridescent material found in the epidermal layers and scales of most pelagic species of fish like oil sardine, mackerel, herring etc.
- These guanine particles are deposited on the inside surface of solid beads, paper weight, ash trays, fishing rods etc. which gives an optical effect similar to that of original pearl.
- The pelagic fish scales (eg. Sardine scale) produce crystalline guanine where as demersal fish scales (eg. Milk fish, croaker) are used for obtaining amorphous guanine.

Production method:

- Freshly removed scales are collected and washed to obtain adhering foreign matter.
- Scales can be preserved in 10- 15% brine which is later drained off and are wiped in muslin cloth.

- It can be stored at 0°C for several weeks without getting them dry.
- Pearl essence can be prepared as aqueous and no – aqueous suspension.
- **Aqueous suspension** can be prepared by agitating washed scales with water containing little ammonia.
- The mixture is then passed through a strainer.
- The pearly substance present as a suspension in the liquor is purified by settling in a cool atmosphere.
- Guanine settles and supernatant is decanted.
- The process is repeated several times and 0.3% salicylic acid is used as preservative.
- **Non- aqueous suspension** can be made by suspending it in organic solvents like acetone, amyl acetate, chloroform or acetic acid.
- It is presented in a form of thick paste of crystals suspended in viscous liquor of cellulose of amyl acetate.

Genuine pearl	Pearl essence
» Very costly because of its luster	» Not so costly
» Consists of layers of crystalline $CaCO_3$ with in between horny organic substance called 'Concholin'	» Guanin is entirely organic in composition
» Produced by molluscs	» Scales/ epidermal layer of pelagic fish

Uses:

- In the manufacture of artificial pearls
- It is used to impart iridescence to a large variety of articles such as shoe, pencil, fishing rod and spectacle frame

D. FISH LEATHER

- Skin: Outer coating of small land animals viz. sheep, goat, or fish like salmon, cod, shark etc.

- Hide: External covering of large land animals viz. cattle, horse or marine mammals.
- The skin has two layers.
  - Epidermis – Epithelial tissue
  - Dermis layer – Collagen layer
- Below the skin layer is the Flesh – Connective tissue, fatty tissue, muscle tissue
- Skins of fishes especially of shark, seal, porpoise, dolphin, skates and rays are suitable for conversion to leather particularly for manufacture of small novelties in different countries like Japan, Vietnam Thailand, Korea, Norway, Sweden etc.
- Skins of shark can be processed into fine leather suitable for manufacture of fancy items.
- Shark skin leather is a by product from shark fishery.
- The main component of skin is collagen.
- Collagen constitutes about 90% of all nitrogenous matter of the skin.
- Both demersal as well as pelagic varieties are suitable for the leather production.

Operations for manufacture of leather:

- The production process is essentially the same as that followed for making leather from animal hides, the principal constituent in fish skin also being collagen.

The protective layer of the shark skin is calcareous material called 'Shagreen'.

1. Beam house operations:

a. Washing and soaking in water

b. Liming and application of sulphide: Liming in saturated $Ca(OH)_2$ solution +2% sodium sulphide to accelerate. By this keratin is removed and collagen is preserved.

c. Splitting and bating: Trimming and bating with proteolytic enzymes Trypsin, being acidified with lactic acid which hydrolyses elastin protein of the skin.

2. Tanning:

» The skin then kept in tanning solution for 1 – 4 weeks.

» Vegetable tan: from bark of tree. It gives heavy leather.

» Chemical tan: Chrome tanning using Chromium sulphate and glucosides of polyphenol. It gives soft and pliable leather. Chemical tannage is preferred for shark skin where active ingredients are tannin.

3. Fat liquoring and dyeing:

a. Fat liquoring: To replace the natural oil removed during beam house operations so as to improve strength and mellowness of leather.

b. Dyeing: Aniline dye is used for colouring the skin.

4. Drying: Dried in a tunnel drier at a slow rate.

5. Finishing: Splitting to desired thickness.

6. Measuring and packing: in polythene bags.

Use:

» It can be used as a suitable raw material for manufacture of suitcases, shoes, belts, vanity bags etc.

E. BIOCHEMICAL AND PHARMACEUTICAL PRODUCTS

» Nucleic acid and Nucleosides: Obtained from fish spermatozoa.

» Protamine: Consist of a polypeptide structure mainly from fish sperm.

a. Salmin – from Salmon sperm

b. Cyprinin – from Carp sperm

» Streptogenin: Growth promoting factor from Lactobacilli – obtained from fish flesh. This is also referred to as Protein Utilisation Factor.

» Glutathione: Present in fish body tissue like kidney, spleen, heart, liver of salmon. It plays an important role for detoxification in man.

» Cortisone: it is a steroid. It is used to alleviate the pain caused by collagenous diseases viz Rheumatoid arthritis. It has anti-inflammatory action. It is collected from blood samples of salmon.

» Bile salt; Bile is produced in liver and stored in gall bladder. Fish bile contains cholic acid and deoxycholic acid. The bile salt is termed as Sodium taurocholate.

» Proteolytic enzymes: Found in pyloric caeca.

Four important biochemical and pharmaceutical products:

a. 5-iododeoxyuridine: Derived from salmon milt

b. Protamin: Derived from salmon milt

c. Desitin: Unsaturated fish oil fatty acid used for skin ointment

d. Cholesterol depressant: fish oil fatty acids and their esters

Insulin:

» It is a hormone which maintains blood sugar level of human being.

» Deficiency of insulin causes Diabetes mellitus.

» Insulin is produced by pancreatic islets called 'Islets of Langerhans' (CAP)

» It is a protein.

» It contains all amino acids except methionine, tryptophan and hydroxyproline.

» CAPs are located on the gall bladder.

Preservation of islets:

» 95% alcohol acidified with 0.3% HCl

» Stored at low temperature of 0-4^{0}C

Extraction method:

» By vacuum distillation

» Refining done by alkali.

F. CHITIN AND CHITOSAN

- Chitin is the second most abundant biopolymer on earth next only to cellulose.
- It is mainly poly (ß-(1-4)-2-acetamide- D-glucose), which is structurally identical to cellulose
- It is a white, hard, inelastic nitrogenous polysaccharide extensively used for several purposes.
- In India, the single largest source of chitin is the shrimp shell and head waste.
- Chitin is produced from the exoskeleton of crustaceans by deproteinisation and demineralization.

Chitin production:

- The residual trace of protein of exoskeleton of crustaceans are deproteinised with hot 0.5% caustic soda followed by boiling with 3% caustic soda for few minutes and filtering off the liquor.
- It should be washed free of alkali before demineralisation.
- The demineralization is done by treatment with dilute hydrochloric acid at room temperature.
- Demineralization reduces the volume of the shell considerably and Chitin is produced.

Glucosamine hydrochloride:

- Chitin can be hydrolysed to glucosamine hydrochloride by adding concentrated hydrochloric acid and warming until the solution no longer gives opalescence and diluting with water.
- The excess acid can be distilled off under vacuum.
- The crude glucosamine hydrochloride is diluted with water and clarified with activated charcoal.
- The solution is filtered and evaporated under vacuum.
- The crude glucosamine hydrochloride can be separated by adding alcohol.

Chitosan:

- Chitosan is a natural polysaccharide comprising copolymers of glucosamine and N-acetylglucosamine, and can be obtained by the partial deacetylation of chitin.
- chitosan is a co-polymer consisting of (ß-(1-4)-2-acetamido-Dglucose and (ß-(1-4)-2-acetamide-D-glucose units with the latter usually exceeding 80%.
- Chitosan has several industrial and medicinal uses.

Chitosan production:

- Chitin is dried or centrifuged or pressed to remove water.
- The deacetylation is done by heating at 90-95^{0}C with 40% (w/w) caustic soda for 90-120 min.
- The reaction is followed by testing the solubility of the residue in 1% acetic acid.
- As soon as the dissolution is completed, caustic soda is removed from the reaction mixture.
- The drained caustic soda can be reused for the next batch of deacetylation by fortification if necessary.
- The residue is washed with water free of alkali.
- It is then centrifuged and dried in the sun or an artificial drier at a temperature not exceeding 80°C and pulverized to coarse particles.

Properties of chitosan:

- Chitosan is almost colourless, light in weight and less soluble in dilute organic acids but soluble in water, alkali and organic solvents.
- It gives viscous solution when dissolved in dilute organic acids such as formic acid, acetic acid etc.

Use of chitosan:

- Chitosan finds extensive applications in various areas viz; food industries, pharmaceutical applications, chemical industries, dental and surgical uses as a haemostatic agent, wound healing, biodegradable films as a substitute for artificial skins for removing toxic heavy metals, wine clarification, Industrial effluent treatment, agriculture, photography, cosmetic applications and textiles, and in nano applications.

Collagen- chitosan membrane for plastic surgery and dentistry:

» Development of sutures from fish gut.

» Developing a collagen chitosan membrane from the collagen of fish air bladder and chitosan from prawn shell is well received by the medicos.

» It finds extensive use in dental surgery also as a material for guided tissue regeneration. Dental surgery uses costly imported teflon membrane.

G. FISH GLUE

» Fish gelatin and fish glue are more or less same and can be prepared from fish skin and fish head.

» Glue from fish head is inferior than that of skin. Glue from head needs bleaching.

» Generally lean fish are used for this. In case of fatty fish, the presence of fat reduces the adhesive strength.

» If required, fish skin can be preserved by salting and drying before processing into glue however, fish head should be processed fresh.

Fish Glue Production:

a. Glue from fish skin:

» Skin whether fresh or salted is washed and soaked in fresh water for 1- 18 hrs depending on the condition of the material (fresh or salted).

» Washed skins are immersed in 0.2% caustic soda solution to open the fibre bundles and remove cementing materials.

» It is then neutralised with HCl and washed again in cold running water.

» Swollen skin is then heated with steam with an equal weight of water.

» Small quantities of acetic acid also may be added to the mixture to hasten the hydrolysis of the stock into glue and to act as a catalyst.

- Cooking is continued for about 8 hrs and the glue liquid is collected from the cooker.
- The second run is made in a similar manner which is then concentrated in open heated pans at atmospheric pressure until the solid content reaches to 50- 55 % and cooled.
- Sometimes small quantities of volatile essential oils may be added to preserve the glue and to mask the fishy odour.

b. Glue from fish head:

- It should be processed fresh with addition of some bleaching agents like sulphurous acid.
- During cooking of the head, some glacial acetic acid is added which softens the head bones.

Uses:

- Can be used in furniture
- Box making
- Sizing agent
- Can be used in special cements
- Photo engraving plate manufacture
- Book binding and small repair work etc

H. GELATIN

- Skin of fish constitute nearly 3% of the total weight and is suitable for the extraction of gelatin.
- Gelatin is a protein that lacks in an essential amino acid tryptophan, and hence cannot be considered as a sole source of protein in animal or human nutrition.
- However, it is a relatively high source of lysine and methionine, which are deficient in cereal proteins.
- The gelatin finds extensive use in food as also in the formulation of some industrial products.
- Gelatin can be extracted from the skin and bones of fish.

Uses:

- Gelatin is used in the food industry as a gelling, stabilising, emulsifying, dispersing or thickening agent.

I. FISH ALBUMIN

- Fish albumin is a product similar to egg albumin in physical and chemical properties.
- It can be processed out of proteinaceous residue from fish scrap or fish waste.
- Two grades of fish albumin are produced, (a) the technical grade and (b) the food and pharmaceutical grade.

Uses:

- Fish albumin is widely used in food and pharmaceutical products as whipping, suspending or stabilizing agent.
- Food grade albumin is an additive in ice cream, soup powder, puddings, confectionery, bakery products, mayonnaise, custard powder etc.

J. SQUALENE

- Squalene is a highly unsaturated hydrocarbon (C_3H_{50}) present in the liver oil of certain species of deep sea sharks mainly *Centrophorus* and *Squalidae* spp.
- The liver oil of these species contain high percentage of squalene (90%).
- This can be isolated and purified.
- The squalene can be used as a dietary supplement, in cosmetics and medicine.

K. SEA CUCUMBERS (BECHE-DE-MER)

- Sea cucumbers are the Holothurians, which occur in coral reefs in tropical areas.
- They are also known as Beche-de-mer, or Sea slugs or Trepang.
- These sea cucumbers vary in size and colour from species to species.
- It is consumed by the humans.

Processing method:

» Fresh sea cucumber – removal of viscera and cleaning the gut cavity – Boiling for 1 to 1½ hrs – cleaning the gut cavity for 2nd time by making a longitudinal cut on the top of the sea cucumber - Drying using a mechanical dryer or under sunlight or can be used for smoking.

Market value:

» The market value depends on the species, size of the animal, colour, appearance, odour and moisture content.

» *Microthele nobilis* is most costly. Bigger the size, cost is more.

» The main market for this dried beche-de-mer is China, Hong Kong, Singapore and Far East countries.

L. FISH ROES

» Fish roes are utilised for production of Caviar.

» True Caviar is made from the roe of the female sturgeon fish but an inferior caviar is made from the eggs of a number of fish such as salmon and cod.

» Some roes from mullet, shad and Spanish mackerel are used as a caviar substitute.

Production method:

» True Caviar:

- Roes are removed from freshly killed fish – Rubbed gently through a sieve to remove the membrane – the eggs are mixed with salt (4-10% by weight), stirred and left for 10-15 mins – they are then drained, bottled and stored at chill temperatures or pasteurised.

» Mullet roe:

- The mullet roes as a whole dry salted or brined (10% salt by weight) for 10 to 15 hrs - after draining roes are sun dried for (5 – 10) days – roes may be smoked to dry them.

Shelf life:

- The dried roes can be preserved for 3 – 4 months. The shelf life can be extended further by coating them with 50 : 50 mixture of beeswax or paraffin wax.

M. SEAWEEDS AND ITS PRODUCTS

1. Carrageenan:

- Carrageenan is one of the commercially important water-soluble polysaccharides extracted from the red algae (Chondrus crispus).
- Carrageenan is a sulphonic ester of galactone (agar agar).
- There are several varieties of carrageen like kappa and lambda which are used in cooking and baking.
- Gelatinous extracts of the Chondrus crispus (Irish Moss) seaweed have been used as food additives for hundreds of years.
- Carrageenan is a vegetarian and vegan alternative to gelatin.

Use:

- It is widely used in foods, dairy products and pharmaceuticals as emulsifying, stabilising, viscous giving compound
- It is also used in insecticidal syrup manufacture.
- In the textile industry, it is used for sizing the cloth.

2. Agar Agar:

- Agar is colloidal carbohydrate present in cell wall of red algae, Gracillaria and Gelidiella.
- It is a mixture of two polysaccharides of gel forming substances viz. (a) agarose, and (b) agaropectin.

Preparation:

- Gracillaria and Gelidiella are used as raw materials for preparation of agar.

- Seaweed is blanched, and cleaned by abrasion to hard substance to remove sand and other traces of foreign matter.
- Seaweeds are soaked in 6-8 litres of soft water for 24 h and wet ground in stone mortar into pulp using soft water.
- The pulp is then leached in soft water of about 6-7 liters for a period of 24 h and separated by filtering with organdy cloth.
- They are then extracted at 90°C in 3 litres of soft water for 1-2 h.
- Hot seaweeds are then filtered with 2-3 layer of cloth and sedimentation is carried out for suspended impurities in the solution.
- The agar solution is cooled at room temperature and shredding is carried out for agar gel.
- Sun dried gel strips on plastic netting can be known as agar agar.

Use:

- They are used in food, confectionary and dairy industries as gelling, stabilizing and thickening agents mainly in the manufacture of sweets, jellies, and ice creams.
- Agar is also used as microbiological media.
- Fresh dried and processed seaweeds are utilized for human consumption. Food value of seaweeds depends on the mineral, trace elements, proteins and vitamin present in them.

3. Algin and Sodium Alginate:

- Algins are polysaccharides such as D-Mannuronic acid and D-Guluronic acid occurring in the cell wall of brown algae like Laminaria sps. and Microcystis sps.
- These polysaccharides are not gel forming substances.
- Algin is used for preparation of sodium alginate which has industrial importance.

Use:

- The sodium alginate is used to prepare the textile printing paste in the textile industry.
- Sizing material in textile industry.

- Used as water proof material in ceiling material.
- In food industry as thickening agent.

4. Kelp:

- Kelps are large brown algae seaweeds that make up the order Laminariales. There are about 30 different genera. Despite its appearance, kelp is not a plant; it is a heterokont.
- Kelp is high in antioxidants, including carotenoids and flavonoids, which help fight against disease-causing free radicals.
- Antioxidant minerals, such as manganese and zinc, help combat oxidative stress and may help protect cardiovascular health and prevent cancer.
- Seaweed can be briny and salty. Its texture can range from rubbery to crispy and crunchy. It's naturally high in glutamic acid, which is a flavoring agent most notably found in monosodium glutamate (MSG) and responsible for its umami taste.
- The FDA recommends a dietary intake of 150 micrograms (mcg) of iodine per day. Raw kelp of 500g can contain up to 2,500 mcg of iodine. Thus good at eating kelp in moderation.

N. CORAL

- Corals are mainly used for the decoration purpose.
- There are two types of corals from the point of commerce. They are (a) Ostocoralla (8 branches) and (b) Hexacoralla (6 branches).
- As per colour, two types viz. (a) Red coral and (b) Black coral.

O. SPONGE

- Principal component is Spongin.
- There are other components like Fibrein and Sericin.
- In addition, sponge consists of Iodine, Sulphur and bromine.
- When sponge combines with iodine, Iodospongin is formed.

14 Food Additives in Fish Processing

INTRODUCTION

A substance or a mixture of substances other than a basic food stuff, which is present in a food during any aspect of production, processing storage or packaging. Substances which are of little or no nutritive value, but are used in the processing or storage of foods, especially in the developed countries; includes antioxidants; food preservatives; food coloring agents; flavoring agents and other similarly used substances.

Food additives are substances added to food to preserve flavor or enhance its taste and appearance. With the advent of processed foods in the second half of the 20th century, many more additives have been introduced, of both natural and artificial origin.

Direct additives are intentionally added to foods for a particular purpose. Indirect additives are added to the food during its processing, packaging and storage.

Food Preservatives are the additives that are used to inhibit the growth of bacteria, molds and yeasts in the food. Some of the additives are manufactured from the natural sources such as corn, beet and soybean, while some are artificial, man-made additives. Most people tend to eat the ready-made food available in the market, rather than preparing it at home. Such foods contain some kind of additives and preservatives, so that their quality and flavor is maintained and they are not spoiled by bacteria and yeasts. More than 3000 additives and preservatives are available in the market, which are used as antioxidants and anti-microbial agents. Salt and sugar the most commonly used additives. Some of the commonly used food additives and preservatives are aluminum silicate, amino acid compounds, ammonium carbonates, sodium nitrate, propyl gallate, butylated hydrozyttoluene (BHT), butylated hydroxyanisole (BHA), monosodium glutamate (MSG), white sugar, potassium bromate, potassium sorbate, sodium benzoate, etc.

Some artificial colors are also added to the foods to give them an appealing look. These coloring substances are erythrosine (red), cantaxanthin (orange), amaranth (Azoic red), tartrazine (Azoic yellow) and annatto bixine (yellow orange). When the food is to be stored for a prolonged period, use of additives and preservatives is essential in order to maintain its quality and flavor.

The excess water in the foods can cause the growth of bacteria, fungi and yeasts. Use of additives and preservatives prevents spoiling of the foods due to the growth of bacteria and fungi. Additives and preservatives maintain the quality and consistency of the foods.

They also maintain palatability and wholesomeness of the food, improve or maintain its nutritional value, control appropriate pH, provide leavening and color, and enhance its flavor.

When we need to store any food for a longer time, it should be properly processed. During this processing, some substances and chemicals, known as additives, are added to the food. Additives consistently maintain the high quality of foods.

A. Technical benefits of food additives:

» To extend the storage life

» To improve flavour, colour, texture, taste etc

» To make the food attractive

B. Major groups of additives:

» There are two major groups of additives. They are:

- Intentional additive - Salt, Spices etc
- Incidental additives – Pesticides, fertilisers, chemicals used for ripening of fish

i. Technical advantage of incidental additives:

» Increase the production, quality and availability

» Prevents wastage during storage

» Prevents losses during transport

ii. Technical advantage of intentional additives:

- » Needs of consumer
- » Consumer acceptance for colour, flavour, texture etc.
- » Storage for long periods
- » Production of convenience foods
- » Prevention of spoilage safety of food
- » To retain nutritional value of food during different processing
- » To fortify foods from nutritional point of view

C. Different additive Groups:

1. Preservative: Salt, spice, sugar, chemicals
2. Nutritional supplement: Aminoacid, mineral, vitamin
3. Colouring agent: Synthetic/ Natural
4. Flavouring agent: MSG/ Oleoresins of spices/ Nucleotides
5. Improved functional properties: Stabilising agent, emulsifiers/ curing and bleaching agent
6. Processing aids: Sanitisers, metal binders, antifoam
7. Moisture control: Humactants like glycerine, Calcium or Sodium silicate etc.
8. Alkalinity: (pH) control: Acid/ alkali
9. Physiological activity control: Ripening agent, anti-metabolic agent

- » Any additive before addition should be passed through a number of tests of GRAS (Generally recorded as safe).
  - Preliminary investigation:
    - ◊ Chemical nature of additive
    - ◊ Condition of use

- Final investigation:
  - ◊ Fixing up of tolerance level
  - ◊ Manner of use

1. Minerals:

a. Iodine:

- » Deficiency causes Goitre
- » Added through salt (Iodised salt) (1part iodine in 20,000 part salt) in the form of KI (in Mine salt) or, KIO_3 (in Marine salt)

b. Fluorine:

- » Deficiency causes dental carries.
- » Added through fluoride toothpaste

c. Phosphorus:

- » For strength of bone and its formation
- » Added to wheat flour in the form of Na/K orthophosphate

d. Ferrus:

- » Deficiency causes anaemia
- » Added to wheat or cereal flour

2. Vitamins:

- » Vitamin B-Complex: Fortified to wheat flour
- » Vitamin A: Fortified to cheese, milk
- » Vitamin C: Fortified to soft drinks, processed food stuff
- » Ascorbic acid is used as synergist to antioxidant
- » Yeast powder (Torula/ Candida/ Saccharomyces) is a very good source of Vitamin B.

3. Amino Acid:

» Required for growth and cell metabolism

» FAO/ Reference protein: Resembles very much to meat

- Protein: has a very high biological value
- Meat protein: A Class – Composed of 18 amino acids. Out of which 8 or 9 amino acids are essential amino acids.

» Corn diet: Deficient of tryptophan and lysine

» Wheat diet: Deficient of Lysine

» So, needs to be fortified.

» Essential amino acids:

» Methionine, Arginine, Threonine, Tryptophan, Valine, Isoleucine, Leucine, Phenylealanine, Histidine and Lysine (MATTVILPHLy)

» Lysine is 6.5% of total protein. Here, free $-NH_2$ group in 5th Carbon position.

- Nutritionally available
- Very sensitive to heat energy
- If Lysine is present =>all amino acids are present.

Addition of specific nutrients for:

» Promotional reason: To make good and more attractive

» As a nutrient

» Restoration of lost nutrients during processing

4. Colour:

» Colour is straight forward from a stand point of chemistry and measurement but it is complex from a stand point of psychology and acceptance.

» It can be observed by one sense organ i.e. eye.

» Colour is of two types;

- Natural colour: Annato, Caramel, Carotene, Chlorophyll, Turmeric, Xanthophil
- Synthetic colour: FD & C Orange No. 1, FD & C Red No. 1, FD & C Yellow No. 3 & 4.

5. Flavour:

» Flavour is complex from a stand point of chemistry and measurement and also from a stand point of psychology and acceptance.

» Here, three sensory mechanisms are involved i.e. Taste, Odour and Texture.

Importance of adding colour and flavour:

» Preserving or restoring the natural colour and flavour

» To create new types of foods

Choice of colour / flavour depends on:

» A thorough understanding of consumer attitude

» Psychological reactions

» Condition response

6. Taste:

» There are four basic tastes. They are:

- Bitter: Vinyl taste
- Sweet: Glucose sweet taste
- Sour: Citric acid sourness
- Salt: Sodium chloride taste

7. Antioxidant:

» To prevent fat oxidation

» Antioxidant must be used for better effect to
  - Good raw material
  - Correct process
  - Proper packaging and storage condition

» There are two groups of antioxidants:

» **Natural antioxidant:**
  - Tocopherol, Lecithin (Use has no limit)

» **Synthetic antioxidant:**
  - Ethoxyquin (0.01%)
  - BHA (0.02%)
  - BHT (0.02%)
  - Propyl gallate (0.02%)

a. Maillard reaction:

- It is an important non-enzymatic oxidation reaction.
- Sugar reacts with any nitrogenous compound to give brown coloured compound.
- Antioxidant can't prevent this type of oxidation.
- However, this type of oxidation can be controlled by:
  - ◊ Careful control of processing temperature
  - ◊ (antioxidant + SO_2) combination is found to reduce this oxidation
  - ◊ Can be inhibited by use of Ascorbic acid/ Citric acid and organic acid.

b. Enzymatic oxidation:

- Enzymatic oxidation occurs by peroxidises and catalyst.
- It is accompanied by off-flavour, off-colour and off-odour (Grassy smell)

- Additives are not effective to this type of oxidation.
  - ◊ Enzymes can be inactivated by Heat treatments

c. Oxidation of Natural pigments:

- Natural pigments include Chlorophyll b, carotenoids, xanthophylls, anthocyanins
- Oxidation of natural pigments causes a different type of discolouration.
- Carotenoids undergo free-radical type oxidation.
- Oxidation process is catalysed by
  - ◊ High temperature,
  - ◊ Metal ions,
  - ◊ Light and
  - ◊ Microbial by-products like H_2O_2, SO_2
- This can be inhibited by antioxidant provided the oxidant is miscible with the pigments.

d. Oxidation of protein pigments:

- Haemoglobin and myoglobin are red in colour.
- Haemoglobin/Myoglobin (Red) + O_2 = Oxy-myoglobin → Under oxidation → Met-myoglobin (Brown).
- Off-colour (Brown) is produced by oxidation of protein pigments.
- Can be inhibited by use of antioxidant.
- Mono- and di-glycerides are used as texture modifying agents.
  Ex. of Phosphatid: Lecithin/ Cephalin
- **Tocopherol:**
  - ◊ It is a natural antioxidant.
  - ◊ More in vegetable oil than animal fat, so vegetable oil is more stable.
  - ◊ Tocopherols are readily oxidised to Tocoquinones which has no antioxidant

property.

◊ Tocopherolsare readily degraded by heat and particularly at atemperature of refining the vegetable oil.

◊ Gossipol and Sesamolines are similar structure to tocopherol and have antioxidant property.

e. Oxidation of USFA:

- Cause rancidity to fatty acids.
- Can be prevented by use of antioxidants

8. Fat/ Oil:

- Fats are used as
  - ◊ Texture modifying agent
  - ◊ Source of energy
- It gets oxidised quickly due to presence of USFA
- End products are ketones, aldehydes and sulphonyl compound which gives off-odour.

» **Fat deterioration:**

- There are Four types of Fat deterioration:

a. Hydrolysis:

- Enzymatic hydrolysis by lipase
  - ◊ End product is FFA
- Non-enzymatic hydrolysis by high temperature and presence of acid
  - ◊ Results in Soapy flavour.

b. Rancidity:

- It is a very common phenomenon.

- It gives off-flavour due to oxidation of USFA
- It can be prevented by antioxidants.

c. Flavour reversion:

- Deodourised fat will undergo oxidationto give off-flavour during storage.
- Mainly due to oxidation of linoleic and linolenic acid Ex. FPC during storage gives fishy odour.

d. Polymerisation:

- Reaction between cross-linking between unsaturated fats between two carbon atoms or by oxygen binding between two fatty acid chains.
- It occurs to USFA only.

» **Prevention:** Antioxidants are effective only to rancidity and polymerisation but not to hydrolysis and flavour reversion.

» **Fat Oxidation:**

RH (Fatty chain)→ R (free radical) + H
R + O_2 → ROO (Peroxy radical)
ROO + RH → ROOH (Hydro peroxide) + R
ROOH → in presence of Carbonyl group → RO + OH

a. Initiation/ Induction:

◊ Free radicals in any food react with O_2 to form peroxy radical.

◊ Free radical + O_2 → Peroxy-radical →Hydro peroxides → peroxides.

◊ Peroxides break fatty chain at the double bond.

◊ Light, heat and metal ions are initiators.

b. Propagation:

◊ Peroxides will break fatty chain at α-carbon of the double bond and gives rise to short volatile monomeric compounds with characteristic flavour and odour.

- ◊ Rate of peroxide formation increases as propagation goes on.
- ◊ Fe, Cu, Co, Ni ions accelerate the reaction.

c. **Termination:**

- ◊ Propagation will continue till completion of fatty chain.
- ◊ It gives rise to short volatile monomeric compounds.

Mechanism of oxidation:

- » Hydrogen ion donation by antioxidant
- » Electron donation by antioxidant
- » Addition of lipid to the aromatic ring of antioxidant
- » Formation of complex between lipid and aromatic ring of antioxidant

Selection of antioxidant:

- » They must have adequate potency
- » They should not contribute any off-flavour/ odour to the product
- » They should be economically attractive

Different methods for measurement of oxidative stability:

a. Active oxygen method (AOM):
   - ◊ It is a swift stability method
   - ◊ Periodical evaluation of peroxide content by titration method

b. Thiobarbituric acid (TBA)

c. Free fatty acid (FFA) content

d. Organoleptic test (Subjective test)

e. Oven storage test

f. Oxygen bomb test

g. Barcriff-Warburg method

h. Ekey method

9. Synergist:

» It is a chemical compound which does not have its own antioxidant property but when it is added to any other antioxidant, it will enhance the antioxidative property of the antioxidant.

» Example: Ascorbic acid, Citric acid, Phosphoric acid

10. Acidulants:

» Acidulants are mostly the organic acids.

» Examples: Citric acid, Acetic acid, Lactic acid, Propionic acid, Tartaric acid, Sorbic acid.

» Functions:

- As a flavouring agent
    - ◊ Intensify certain taste
    - ◊ Mask undesirable off-taste
    - ◊ Blend unrelated off-taste
- Buffers
- Preservative (Against growth of microorganism)
- Increase appetite
- Preserve colour
- Viscosity modifier
- Melting modifier
- Meat curing agent

11. Sequestrants:

- A sequestrant is a food additive which improves the quality and stability of foods.
- There are all organic acids.
- A sequestrant form chelate complexes with polyvalent metal ions, especially copper, iron and nickel, which can prevent the oxidation of the fats in the food.
- Sequestrants are a type of preservative.
- These are considered as '**Chemical curiosity**'.
- Sequester acts as Metal curator (Chelating agent):
  - M + L (Ligand or sequester) ←→ M L (Metal ion complex)
    (Ligand = an ion or molecule attached to a metal atom by coordinate bonding.)
  - When bond angle is between 90^0 to 109^0, seqestrants are effective.
  - Example: Glycyl-glycine, Salicylic acid, EDTA
- **Functions:**
  - Help to establish, maintain and enhance the integrity of many foods.
  - Effective for chelating by reacting with metal ions to form complexes.
  - Establish/ stabilise numerous properties to wholesome foods including colour, texture and flavour.
  - Prevent discolouration
    - ◊ Shell fish contain more Fe, Cu, Zn metal ions
    - ◊ Melanogenesis (Black spot formation)
    - ◊ Blue green discolouration
    - ◊ Greyish discolouration
    - ◊ Struvite crystals
  - All such discolourations can be prevented by use of EDTA + Alum
- Naturally occurring sequestrants in foods:

- Polycarboxylic acid: Oxalic acid/ Succinic acid
- Hydroxyl carboxylic acid: Citric acid/Malic acid/ Tartaric acid
- Polyphosphoric acid: ATP/Pyrophosphate/ Hexa metaphosphate
- Amino acid
- Various macromolecules: Porphyrins/Peptides/ Proteins

» Selectivity of sequestrant:

- Whether permitted for use in foods
- Colour, flavour and substrate compatibility
- Chelating efficiency
- Economical point of view it is cheap

» Widely used sequestrant: Citrate, Pyrophosphate, EDTA

12. Chelating Agent:

» It binds to the metal ions thus, prevents its action in enhancing the oxidation.

» Example: Citric acid, EDTA

13. Starch:

» Uses:

- Texture modifying agent
- Thickening agent (by absorbing moisture)
- Dusting agent in bakery
- Bulking agent
- Moulding agent
- Canning industry – in filling media
- Film making industry
- In the native form: A source of energy

14. Preservative:

» A food additive, which prolongs the shelf-life of a food by protecting against deterioration caused by microorganisms. It prevents or inhibits spoilage of food due to fungi, bacteria and other microorganisms. It stops microbes from multiplying and spoiling the food.

Types of preservatives

i. Antimicrobial preservative

ii. Antimicrobial synergist

iii. Antimould and antirope agent

iv. Antimycotic agent

v. Bacteriophage control agent

vi. Fungistatic agent

vii. Preservative

Some preservatives are:

» Benzoic acid and benzoates - are found in soft-drinks, beer, margarine and acidic foods. They are use to extend shelf life and protect food from fungi and bacteria.

» Nitrites and nitrates - are found in processed meats, such as sausages, hot dogs, bacon, ham, and luncheon meats, smoked fish. They are used to extend shelf life and protect food from fungi and bacteria; preserve color in meats and dried fruits.

» Sulfites - are found in dried fruits, shredded coconut, fruit based pie fillings. They are used to extend shelf life and protect food from fungi and bacteria.

Parabens:

» These are a group of preservative ingredients used in food products.

» They are highly effective in preventing the growth of fungi, bacteria and yeast that can cause products to spoil. Thus, they contribute directly to the quality of the products by extending their shelf life.

» Parabens are derived from Parahydroxy benzoic acid (PHBA) that occurs naturally in many fruits and vegetables such as cucumber, cherries, carrots, onion etc. PHBA also is formed naturally in the human body by the breakdown of some amino acids.

» Other parabens in foods include Propylparabens, Heptylparaben and Butylparaben. All these are under GRAS list.

Polyphosphate:

» Polyphosphates are widely used to prevent drip loss.

» Functions of Polyphosphate:

- As buffering agent
- Increasing water hydration and water binding
- Mineral supplementation

» Tolerance level:

- 0.5% level is tolerated without any adverse physiological effect

» Uses of polyphosphate:

- As a source of nitrogen and phosphorus ➔ Ammonium orthophosphate
- As a leavening agent ➔ Anhydrous mono-calcium phosphate
- As a source of calcium and phosphorus ➔ Mono-calcium phosphate monohydrate
- As a drying agent ➔ Tri-calcium phosphate
- As an iron enrichment compound ➔ Ferric ortho phosphate

D. Dangers of food additives and preservatives:

» Although additives and preservatives are essential for food storage, they can give rise to certain health problems. They can cause different allergies and conditions such as hyperactivity and Attention Deficit Disorder in some people who are sensitive to specific chemicals. The foods containing additives can cause asthma and certain reactions such as rashes, vomiting, headache, tight chest, hives and worsening of eczema.

- Some of the known dangers of food additives and preservatives are as follows:
    - Benzoates can trigger skin rashes and asthma
    - Bromates destroy nutrients in foods. It can give rise to nausea and diarrhoea.
    - Butylates responsible for high cholesterol level causing damage to kidneys and heard.
    - Sodium chloride triggers high blood pressure.
    - Caramel can cause vitamin B6 deficiency.
- To minimize the risk of developing health problems due to food additives and preservatives, you should eat the freshly prepared foods as much as possible rather than processed foods.

E. Effects of food additives:

- Avoiding or minimizing toxins in the diet is an important step toward enhancing our health and lowering risk of disease.
- Effects of food additives may be immediate or may be harmful in the long run on constant exposure.
- Long-term effects may increase the risk of cancer, cardiovascular disease and other degenerative conditions.

Bibliography

Aguilar, P.R., Crawford, D.L. and Lampila, L.E. 1989. Procedure for the efficient washing of minced whiting (*Merluccius productus*) flesh for surimi production.*J.Food.Sci.*54 (2): 248-252.

Amerine, M. A., Pangborn, R. M. and Roessler, E. B. 1965. Principles of sensory evaluation of food. New York and London: *Academic Press*, 602 pp.

Anonymous 2001. Tilapia fillets, *Info fish Int.*, 1:14.

APHA 1976. Compedium of methods for the microbiological examinations of foods (Ed. N.L.Speak), *APHA, N.Y.*, p.701.

Ashie, I. N., Smith, J.P. and Simpson, B.K., 1996. Spoilage and shelf life extension of fresh fish and shellfish, *Crit. Rev .Food Sci.Nutr.*,36: 87.

Ayyappan, S., and Krishnan, M. 2007. Changing consumption patterns . In. *The Hindu survey of Indian agriculture* 2007.pp.101-106.

Balachandran, K.K. 2001. Mince and mince based products . In: Post-harvest Technology of Fish and Fishery products. pp.271-287.

Balachandran, K.K. and Vijayan, P.K. 1988. Development of a process for canning fresh water fish rohu, *Fishery Technol.*, 25: 40.

Balchandran, K. K. and Muraleedharan, V. 1975. Fish pickle preparation using organic acid. *Fish. Technol.* 12: 145-149.

Basu, S., Kasim, D.I., Gupta, S.S. and Rao, C.C.P. 1989. Quality of dry fish from markets in Andhra Pradesh. *Fish. Technol.*, 26(2): 114-118.

Basu, S.2001. Value added products form low cost fish and shell fish. *Training Manual on Fish Processing and Development. 2001.CIFE, Mumbai*.pp.8-18.

Bawa, A.S. and Jathilakan, K. 2002. Fresh water fish processing- a review.*Indian Food Ind.*, 21: 34.

Bligh, E. G., 1971. Specific problems in the quality assessment of fresh water fish, **In**: Fish Inspection and Quality Control, Kreuzer, R.(Ed,) *Fishing News, London*, p.1.

Braekkan, O.R. 1956. Function of the red muscle in fish. *Nature (London)* 178, 747–748.

Bremner, H.A., 2000. Towards practical definition of quality for food science, *Crit Rev.Food sci. Nutr.*, 40: 83,.

Burt, J.R., Gibbsons, D.M., Jason, A.C. and Sanders, H.R.,1976 Comparison of methods of freshness assessment of wet fish. II. Instrumental and chemical assessment of boxed experimental fish. *J. Food Technol*, 11: 73-87.

Chen, H.H. 2000. Effect of non-muscle protein on the thermogelation of horse mackerel surimi and the resultant cooking tolerance of kamaboko. *Fisheries science*, 66: 783-788.

Connell, J. J., 1995. Control of Fish Quality. *Fishing News Books, London,* p.240.

Devadasan, K. 2008. Diversification and value addition are the key words. *The Hindu Survey of Indian Agriculture,* 2008. pp 97-101.

Ehrenberg, A. S. C. and Shewan, J. M. 1953. The objective approach to sensory tests of food. *Journal of the Science of Food and Agriculture*, 4, 482 - 490.

FAO, 1971. "Fermented Fish Products," Prepared by I. M. Mackie, R. Hardy and G. Hobbs, *FAO Fish Rep.*, Vol. 100, 1971, p. 54.

Ganesh, A., Dileep, A.O., Shamasundar, B.A., and Singh. U. 2006. Gel-forming ability of common carp fish *(Cyprinus Carpio)* meat: effect of freezing and frozen storage. *Journal of Food Biochemistry*, 30: 342–361.

Gopakumar, K. 2006. Fishery products. In: Text book of fish processing Technology, Publ. by *ICAR, New Delhi,* pp. 271-302.

Gorski, Z. 1999. Shell fish processing: State of the art machinery in the next millennium, *Info fish Int.*, 6: 44.

Govindan, T.K. 1985. Fish processing technology. *Oxford & IBH Publ. Co. ,New Delhi.* p.252.

Hastings, R.J. 1989. Comparison of the properties of gels derived from cod surimi and from unwashed and once washed cod mince. *Int. J. Food.Sc. Technol.*24(1): 93-102.

Hossain, M.I., Itoh, Y., Morioka, K., and Obatake, A. 2001. Contribution of the polymerization of protein by disulfide bonding to increased gel strength of walleye pollack surimi gel with preheating time. *Fisheries science*, 67: 710-717.

Hsu, C.K., and Chiang, B.H. 2001. Optimum blends of different grades of surimi determined by non-linear programming. *Fisheries science*, 67: 500-505.

Hultin, H. O. 2000. Oxidation of lipids in sea foods **In**: Chemistry, Processing Technology and Quality, Shahidi, F. and Botta, J. R., (Eds.) *Technomic* ,p.819.

Jacobs, M.B.1958. The chemical analysis of foods and food products. *Krieger Publ. Co. Inc., N.Y.*, p. 393-394.

Jaczynski, J., and Park, J.W. 2004. Physico- chemical changes in Alaska Pollock surimi and surimi gel as affected by electron beam. *Journal of Food Science*, 69 (1): 54-57.

Kreuzer, R. (Ed.) 1971. Fish inspection and quality control. Published by arrangement with the Food and Agriculture Organization of the United Nations. *West Byfleet, Surrey: Fishing News (Books) Ltd.*

Kuhn, C.R., Prentice-Hernández, C., Vendruscolo, J.L., and Soares, G.J.D. 2004. Surimi of King Weakfish (*Macrodon ancylodon*) Wastes: Texture Gel Evaluation with Protease Inhibitors and Transglutaminase. *Brazilian Archives of Biology and Technology*, 47(6): 895-901.

Lakshmanan, R., Piggot, J.R. and Paterson, A. 2003. Potential applications of high pressure for improvement in salmon quality, *Trends Food Sci. Technol.*, 14:354.

Lee, C.m. 1984. Surimi process technology. *Food Technology*. 38(11): 69-80.

Lin, D. and Morrissey, M.T. 1995. Northern squaw fish (*Ptychocheilus oregonensis*) for surimi production, *J. Food Sci.*, 60: 1245.

Lin, T.M. and Park, J.W. 1996. Extraction of proteins from Pacific whiting mince at various washing conditions. *J. Food. Sci.*. 61: 432-438.

Love, 1970. The connective tissues of fish III. The effect of pH on gaping in cod entering rigor-mortis at different temperatures. *J. Food Technol.*, 5: 249-260.

Love, 1975. Lipids in the fish of cod (*Gadus morhua*) from Faroe bank and Aberdeen bank in early summer. *Autumn. Mem. Fac. Fish., Kagoshima Univ.*, 24: 123-126.

Love, R. M., 1988. The Food Fishes. Their Intrinsic Variations and Practical Application. *Farrand Press, London*, 1988.

Miyauchi, D., Kudo, G. and Patashnik, M. 1973. Surimi (a semi processed wet fish protein). *Mar. Fish. Res. Paper*. 1026, *Mar. Fish. Rev.*, 35(12): 7-9.

Moller, A.B. 2003. Studies on sea food value addition, FAO/ GLOBEFISH special market study, *FAO of the United Nations, Rome*, 2003, p. 93.

Offer, G. and Trinick, J. 1983. On the mechanism of water holding in meat: The swelling and shrinking of myofibrils. *Meat Sc.*, 8: 245-281.

Park, J.W. and Morrissey, M.T. 2000. Manufacturing of surimi from light muscle fish. In: Surimi and surimi seafood (Ed. Park, J.W.), *Marcel Dekker, New York*, pp. 23-58.

Pillai, K., Valson, A.P. and Nair, R. 1956. Studies on the chemical qualities of cured fish products from the West coast of India. *Indian J. Fish.* 3: 43-58.

Ravishankar, C.N., Setty, T.M.R. and Shetty, T.S.1993. Studies on the utilization of Indian oil sardine (Sardinella longiceps) for the preparation of fish sausages:1. Effect of washing in water on meat characteristics and sausage quality. *Fish. Technol.*30: 46-51.

Reppond, K.D., and Babbitt, J.K. 1997. Gel properties of surimi from various fish species as affected by moisture content. *Journal of Food Science*, 62 (1): 33-36.

Roussel, H. and Cheftel, J.C. 1998. Characteristics of surimi and kamaboko from sardines. *Int. J. Food. Sci. Technol.*, 23(6): 607-624.

Sehgal, H.S. and Sehgal, G.K. 2000. Aquacultural and socio-economic aspects of processing carps into some value added products, *Bioresouce Technology*, 82: 291.

Sen, D.P. 2005. Mince fish Technology. In: Advances in fish processing Technology, Publ. by *Allied Publishers Private Ltd. New Delhi*, pp. 575.

Shewan, J. M., Macintosh, Ruth G., Tucker, C. G. and Ehrenberg, A. S. C. 1953. The development of a numerical scoring system for the sensory assessment of the spoilage of wet white fish stored in ice. *Journal of the Science of Food and Agriculture*, 4, 283 - 297.

Siddaiah, D. 2001. Changes in lipids, proteins and kamaboko forming ability of silver carp mince during frozen storage, *Food Res. Int.*, 34:47.

Sing, R.K., and Balange, A.K. 2005. Characteristics of pink perch (*Nemipterus Japonicus*) surimi at frozen temperature. *Journal of Food Processing and Preservation*, 29: 75-83.

Smruti, K. 2003. Tenderized, dehydrated steaks of fresh water fish, Rohu: Preparations and storage characteristics, *Fish. Technol.*, 40:43.

Snir, I.J. 2001. Value added tilapia products: deliberate policy or no other choice?, In: Tilapia production, marketing and technological developments, Proceedings of the Tilapia 2001, (Subasinghe, S. and Singh, T., Eds.) Kuala Lumpur, Malayasia, 2001. (Subasinghe, S. and Singh, T., Eds.) Kuala Lumpur, Malayasia, 2001 International Conference, 28-30 May, 2001, Malayasia, Subasinghe, S. and Singh, T., Eds. *Infofish Int.*, P. 98.

Subasinghe, S. 1996. Handling and marketing of aquacultured products, *Infofish Int.*, 3:45.

Synowiecki, J., and Sikorski, Z.E. (1988). Heat induced changes in thiol groups in squid proteins. *Journal of Food Biochemistry*, 12: 127-135.

Tabilo-Munizaga, G., and Barbosa-C'anovas, G.V. 2004. Color and textural parameters of pressurized and heat-treated surimi gels as affected by potato starch and egg white. *Food Research International*, 37: 767–775.

Thanonkaew, A., Benjakul, S., Visessanguan, W., and Decker, E.A. 2005. The effect of metal ions on lipid oxidation, colour and physicochemical properties of cuttlefish (Sepia pharaonis) subjected to multiple freeze–thaw cycles. *Food Chemistry*, 95: 591–599.

Thawornchinsombut, S. and Park, J.W. 2006. Frozen stability of fish protein isolate under various storage conditions. *Journal of Food Science*, 71 (3), 227-232.

Thomas, F. and Balachandran, K.K.1987. Fish curing in India. In: K.K.Balachandran (Eds.) Diversification of post harvest technology for low cost fish. Proceedings of a symposium held on 10-11 March, 1987 at Cochin, India, pp. 1-15.

Tongreung, J. 2001. Surimi.1st ed.Chulalongkorn publication, Bangkok. p. 55-77.

Vannuccini, S. 2001. Global markets for tilapia, In: Tilapia production, marketing and technological developments, Proceedings of the Tilapia 2001 , (Subasinghe, S. and Singh, T., Eds.) Kuala Lumpur, Malayasia, 2001 International Conference, 28-30 May, 2001, Malayasia, Subasinghe, S. and Singh, T., Eds. *Infofish Int.*, P. 65.

Wheaton, F.W. and Lawson, T.B. 1985. Processing aquatic food products. *John willey & sons*. N.Y.p. 518.

Wood, C.D. 1984. Report on a visit to the Central Institute of Fisheries Technology (CIFT), India to investigate losses in traditionally cured fish and identify counter measures. *Tropical Development and Research Institute, England*, Report No. 120 2A, 40 pp.

Xiong, G.Q. Ye, L.X. and Zhang, H. 1999. Factors influencing the quality of fresh water fish sausage, *Food Ind.*, 1: 39.

Yu, S.Y., 1990. Acceptability of fish cake prepared from *Aristichthys nobilis* and *Nemipterus* sps, *The 2nd Asian Fisheries Forum, Tokyo, Japan, 17-22 April, 1989, Hirano, R. and Hanya, I., Eds,* 1990. p. 889.

Index

For Product Safety Concerns and Information please contact our EU
representative GPSR@taylorandfrancis.com
Taylor & Francis Verlag GmbH, Kaufingerstraße 24, 80331 München, Germany

www.ingramcontent.com/pod-product-compliance
Ingram Content Group UK Ltd.
Pitfield, Milton Keynes, MK11 3LW, UK
UKHW052224190926
475702UK00007B/1187
* 9 7 8 1 0 3 2 2 0 4 7 2 7 *